Advancing Silicon Carbide Electronics Technology II
Core Technologies of Silicon Carbide Device Processing

Edited by
Konstantinos Zekentes[1] and Konstantin Vasilevskiy[2]

[1]Foundation for Research and Technology, Heraklion, Greece

[2]School of Engineering, Newcastle University, Newcastle upon Tyne, United Kingdom

Cover illustration:

Dense array of vertical 4H-SiC nanowires formed by successive nanoimprint lithography, reactive ion etching and oxidation. See Chapter 5 in this volume for the in-depth review of silicon carbide nanotechnologies.

The figure is taken (with minor editing) from M. Cottat, A. Stavrinidis, C. Gourgon, C. Petit-Etienne, M. Androulidaki, E. Bano, G. Konstantinidis, J. Boussey, K. Zekentes, 4H-SiC Nanowire arrays formation by nanoimprint lithography, plasma etching and sacrificial oxidation, Proc. WOCSDICE 2019, Cabourg France. Reprinted with permission.

Copyright © 2020 by the authors

Published by **Materials Research Forum LLC**
Millersville, PA 17551, USA

Published as part of the book series
Materials Research Foundations
Volume 69 (2020)
ISSN 2471-8890 (Print)
ISSN 2471-8904 (Online)

Print ISBN 978-1-945291-66-6
ePDF ISBN 978-1-945291-67-3

Distributed worldwide by

Materials Research Forum LLC
105 Springdale Lane
Millersville, PA 17551
USA
http://www.mrforum.com

Manufactured in the United States of America
10 9 8 7 6 5 4 3 2 1

Table of Contents

Preface

Chapter 1 **Historical Introduction to Silicon Carbide Discovery,
Properties and Technology**
K. Vasilevskiy, N.G. Wright   1

List of used symbols and abbreviations   3
1.   Introduction   5
2.   Discovery of silicon carbide   5
    2.1   Acheson process   6
    2.2   Silicon carbide in the nature   7
3.   Silicon carbide material properties   8
    3.1   Chemical bonds and crystal structure of silicon carbide   8
    3.2   Crystal structure and notation of SiC polytypes   10
    3.3   Stability, transformation and abundance of SiC polytypes 14
    3.4   Chemical and physical properties of silicon carbide   15
    3.5   Polytypism and electrical properties of silicon carbide 17
    3.6   Silicon carbide as a material for high temperature electronics   18
    3.7   Silicon carbide as a material for high power electronics 19
4.   Silicon carbide in early radio technology   22
5.   Electroluminescence of silicon carbide   24
6.   Silicon carbide varistors   26
7.   Lely platelets   27
8.   Silicon carbide bulk crystal growth   29
9.   Silicon carbide epitaxial growth   30
10.   Emergence of industrial silicon carbide electronics   32
    10.1   Foundation of Cree Research, Inc. and the first commercial blue LEDs   32
    10.2   Industrial SiC wafer growth   33
    10.3   Preconditions and demands for SiC power electronics   34
    10.4 $4H$-SiC polytype as a material for power electronics   36
    10.5 $4H$-SiC unipolar power devices   36
    10.5.1 $4H$-SiC power Schottky diodes   37
    10.5.2 $4H$-SiC power JFETs   38
    10.5.3 $4H$-SiC power MOSFETs   40
    10.6   Development of $4H$-SiC power bipolar devices   42
    10.6.1 Minority carrier lifetime enhancement   43
    10.6.2 Suppressing stacking faults   44

	10.6.3 Advanced 4*H*-SiC power bipolar devices	46
	10.7 Emergence of automobile SiC power electronics	47
Conclusion		48
References		48

Chapter 2   **Dielectrics in Silicon Carbide Devices: Technology and Application**
Anthony O'Neill, Oliver Vavasour, Stephen Russell, Faiz Arith, Jesus Urresti, Peter Gammon   **63**

List of used symbols and abbreviations		65
1.	Introduction	68
	1.1 Interface-trapped charge effects and requirements	68
	1.2 Near-interface trap effects	69
	1.3 SiC MOS interface requirements	70
2.	Dielectrics in SiC device processing	71
	2.1 Silicon dioxide in SiC devices	71
	2.2 Silicon nitride in SiC devices	73
	2.3 High-κ dielectrics in SiC devices	74
3.	Methods of dielectric deposition used in SiC device processing	75
	3.1 Plasma enhanced chemical vapor deposition of dielectric on SiC	76
	3.2 Deposition of silicon oxide films using TEOS	77
	3.3 Atomic layer deposition of gate dielectrics in SiC devices	78
	3.4 Densification of dielectrics deposited on SiC	80
	3.5 Deposition methods conclusion	80
4.	Thermal oxidation of SiC	81
	4.1 SiC oxidation rates and a modified Deal-Grove model	81
	4.2 Interface traps introduced during thermal oxidation of silicon carbide	83
	4.3 High temperature oxidation	85
	4.4 Low temperature oxidation	86
	4.5 Post oxidation annealing	88
	4.6 Thermal oxidation conclusion	89
5.	Other methods to improve channel mobility	89
	5.1 Sodium enhanced oxidation	89
	5.2 Counter doped channel regions	90
	5.3 Alternative SiC crystal faces	90
6.	Surface passivation by dielectrics	91
7.	Summary	93
Acknowledgements		93
References		94

Chapter 3   **Silicon Carbide Doping by Ion Implantation**
           Philippe Godignon, Frank Torregrosa, Konstantinos Zekentes   107

List of used symbols and abbreviations   109
1.   Introduction   111
2.   Ion implantation technique   112
      2.1   Basics of ion implantation physics   112
      2.2   Basics of ion implantation technology   115
3.   Specificities of ion implantation in SiC   119
      3.1   General considerations   119
      3.2   SiC ion-implanted dopants   120
      3.3   Implantation damage   120
      3.4   Hot implantation   121
      3.5   Post-implantation annealing. Activation and diffusion   121
      3.6   SiC devices requirements   123
      3.7   Other SiC implantation reviews   123
4.   *n-T*ype doping   124
      4.1   *n*-Dopant atoms   124
      4.2   Heating during implantation of *n*-dopants   126
5.   *p*-Type doping   127
      5.1   *p*-Type dopants   127
      5.2   *P*-dopant atoms diffusion   127
      5.3   Aluminum doping   128
      5.4   Heating during implantation   129
6.   Post-implantation annealing   132
      6.1   Fast thermal annealing   132
      6.2   Very high temperature conventional (CA) and microwave
             annealing (MWA)   133
      6.3   Laser annealing   134
      6.4   Others techniques   135
      6.5   Optimization of post-implantation annealing in the case of
             Al-implantation   135
      6.6   Surface roughness   137
      6.8   Capping layer   138
      6.9   Electrical activation   139
7.   Crystal quality and electrically active defects   141
8.   Channeling and straggling   143
      8.1   Channeling in SiC crystal   144
      8.2   Lateral/transverse straggling   148

	8.3	Box profile	151
9.		Plasma implantation	152
10.		Ion implantation simulation	153
11.		Diagnostic techniques of implanted layers	155
	11.1	Secondary Ion Mass Spectroscopy (SIMS)	156
	11.2	Electrical measurements	156
	11.3	Rutherford backscattering spectrometry (RBS)	156
	11.4	Transmission Electron Microscope (TEM)	157
	11.5	Raman spectroscopy	158
	11.6	X-ray diffraction (XRD)	158
	11.7	Cross-section imaging techniques	159
12.		Implant services suppliers	161
13.		Implant tools for SiC	162
		Conclusions and challenges	163
		Acknowledgements	165
		References	165

Chapter 4 **Plasma Etching of Silicon Carbide**
K. Zekentes, J. Pezoldt, V. Veliadis

			175
		List of used abbreviations	177
1.		Introduction	177
2.		Gas chemistry – etching mechanisms	178
	2.1	SiC etching gas chemistry	178
	2.2	Surface carbon rich layer	180
	2.3	Cl-based chemistry	180
	2.4	Results relative to the use of different fluorinated gases	181
	2.5	Role of additives (N_2, H_2, O_2, Ar, He) in the gas mixture	183
3.		Etch rate	186
	3.1	Role of pressure	186
	3.2	Role of substrate platen RF power / DC self-bias	188
	3.3	Role of ICP RF power (source/coil power)	190
	3.4	Role of gas flow	191
	3.5	Role of crystal face	191
	3.6	Role of doping type	191
	3.7	Role of chamber/substrate electrode geometry	192
	3.8	Role of substrate temperature	192
	3.9	Loading effects	194

4.	Morphology of etched surfaces/sidewalls	195
	4.1 Micromasking	195
	4.2 Micromasking after deep (> 10 μm) etching	198
	4.3 Ion-etching induced polishing effect of SiC surfaces	200
	4.4 Microtrenching effect	200
	4.5 Isotropic etching	204
	4.6 Sidewall shape	205
	4.7 Sloped walls from sloped etch mask	207
	4.8 Vertical scratches	208
5.	Mask material (adherence, micromasking, selectivity)	209
6.	Surface conditioning prior-to / following etching	211
7.	Carrier of the SiC sample under etching	212
8.	DRIE (Deep RIE) process in SiC: via-hole formation - MEMS	213
	8.1 Continuous etching process	213
	8.2 Bosch process	214
9.	Nanopillar/nanowire formation	215
10.	Electrical properties after etching	217
11.	Main conclusions	220
	Acknowledgements	222
	References	222

Chapter 5 **Fabrication of Silicon Carbide Nanostructures and Related Devices**
M. Bosi, K. Rogdakis, K. Zekentes 233

1.	Introduction	235
2.	SiC nanoparticles	237
	2.1 Si to SiC conversion based fabrication of SiC nanocrystals	237
	2.2 Chemical vapor based fabrication of SiC nanocrystals	237
	2.3 Electrochemical and chemical etching based methods for SiC nanocrystals formation	238
	2.4 Chemical Synthesis of SiC nanocrystals	239
	2.5 Formation of SiC nanocrystals by laser ablation	239
	2.6 Other methods for formation of SiC nanocrystals	239
	2.7 Other (non-cubic) polytype SiC nanocrystals formation	240
	2.8 Formation of SiC hollow nanospheres, nanocages and core-shell nanospheres	240
	2.9 Luminescence of SiC nanocrystals	240
	2.10 Applications of SiC 0D nanostructures	242

3.   Bottom-up growth of SiC nanowires and nanotubes   242
    3.1   General description of bottom-up NW growth   242
    3.1.1 Vapor-Liquid-Solid process   244
    3.1.2 Vapor-Solid process   244
    3.1.3 Solid-Liquid-Solid process   245
    3.2   SiC nanowire growth without a template   245
    3.2.1 Catalysts   246
    3.2.2 Precursors and processes for 3*C*-SiC nanowire growth   246
    3.3   Template assisted SiC nanowire growth   248
    3.3.1 CNTs conversion into SiC NW and NT   248
    3.3.2 Si NW conversion into Si/SiC core/shell NW, SiC NT and
           SiC NW   249
    3.4   Conclusions on SiC NW bottom-up formation   251
4.   Top-down formation of SiC NWs   253
5.   Processing technology of SiC NW based devices   255
6.   Functionalization of SiC nanostructures   256
7.   Applications of SiC nanowires   256
Conclusions   259
Acknowledgements   260
References   260

Keyword Index   276
Author Index   278
About the Editors   280

Preface

This volume is the second and the last part of the book "Advancing Silicon Carbide Electronics Technology". In this book, we concentrated our efforts to provide a detailed review and in-depth analysis of silicon carbide device processing which is a specific and important part of SiC technology. We deliberately excluded silicon carbide material characterization, bulk growth, epitaxy, device design, circuit design and applications from this book as far as these fields of SiC electronics are very extensive and matured and require a separate and comprehensive consideration.

The previous volume is entitled "Advancing Silicon Carbide Electronics Technology I - Metal Contacts to Silicon Carbide: Physics, Technology, Applications" and is devoted to an important part of SiC device processing which is fabrication and characterization of metal contacts. The volume is opened with Chapter 1 focusing on silicon carbide surface cleaning which is the first and essential step in any device processing. It is followed by Chapter 2 describing fundamental physics, electrical characterization methods and processing of ohmic contacts to silicon carbide. The chapter provides detailed analysis of contact resistivity dependence on material properties, limitations and accuracy of contact resistivity measurements, practical advises on ohmic contact fabrication and test structure design, critical overview of different metallization schemes and processing technologies reported up to now. In Chapter 3, the basic physical principles of Schottky barrier formation are recalled and adapted to the specific case of SiC. Next, the important fundamental topic of Schottky barrier inhomogeneity in SiC materials is introduced. Then, a section of this chapter is devoted to the technology and design of $4H$-SiC Schottky and Junction Barrier Schottky diodes. Si/SiC heterojunction diodes are also briefly discussed in this chapter as a particular case of rectifying contacts. Some common applications of SiC Schottky diodes in power electronics and temperature/light sensors are provided in the last section of this chapter. The volume is concluded with Chapter 4 reviewing high power SiC unipolar and bipolar switching devices. The challenges and prospects of different types of SiC devices including material and technology constraints on the device performance are discussed in this chapter to elucidate the main application area of metal contacts to silicon carbide.

The present volume entitled "Advancing Silicon Carbide Electronics Technology II - Core Technologies of Silicon Carbide Device Processing" contains 5 chapters and starts with the introductory Chapter 1 which provides a historical overview of the first synthesis of artificial silicon carbide, its discovery in the nature and the main pivotal steps in the development of bulk and epitaxial growth of silicon carbide. SiC material properties are briefly described in the scope required for reading and understanding other chapters in this book. Current state in commercial production and availability of SiC wafers and epitaxial structures as well as potential market of SiC power devices are also outlined. Finally, the benefits of adoption of SiC devices in power systems are illustrated by estimation and direct comparison of electrical characteristics of two unipolar devices with the same power rating but fabricated of SiC and Si. This chapter is followed by four chapters describing the core SiC device processing technologies.

Chapter 2 reviews the main dielectrics that are used in silicon carbide devices. The most commonly used dielectrics in electronic devices are SiO_2 and Si_3N_4 and so these are introduced first, followed by high-κ dielectrics (i.e. dielectrics with higher permittivity than Si_3N_4). The methods of dielectric deposition are discussed before focusing on SiC thermal oxidation. Different parameters of the oxidation process and post-oxidation annealing, which have an impact on oxide quality and the formation of residual carbon in the SiO_2/SiC interface, are evaluated. Efforts to improve electron mobility in SiC MOSFETs using a variety of dielectric layer formation techniques are reviewed, indicating where the progress has been made. The issues surrounding SiC surface passivation by dielectrics are also discussed.

The next Chapter 3 aims to give all the necessary information for employing ion implantation in the fabrication of silicon carbide devices. It starts with introduction to the ion implantation technique and its application in SiC device processing technology. Namely, the special attention is devoted to the channeling and straggling effects, which are more pronounced in SiC than in Si crystals. The main characteristics of SiC n- and p-type doping by ion implantation and different annealing techniques for dopant activation are also discussed. The chapter also describes the crystal quality and defects formation issues and presents a novel implantation technique for low defect's surface doping. The implantation simulation and characterization aspects as well as some practical aspects such as implantation facilities and equipment are addressed in the last sections of this chapter.

Chapter 4 of this volume aims to be a comprehensive guide for silicon carbide dry etching. Its first part explains why the fluorine chemistry is mainly used for silicon carbide etching. Effects of other gases addition in the process and possible etching mechanisms are discussed. The second part is dedicated to the control of the etch rate through various plasma parameters. The third part covers the issues related to the morphology of etched surfaces. Hard masking materials and especially their selectivity with SiC are the subject of the forth part. In the following sections of this chapter, SiC surface conditioning prior or after the plasma etching, the choice of appropriate carriers for a SiC wafer under etching, and electrical properties of the etched surfaces are discussed. Silicon carbide deep etching for via-hole and MEMS applications as well as the top-down formation of SiC nanowires is also addressed in this chapter.

The volume is completed with Chapter 5 focusing on silicon carbide nanostructure fabrication, processing and device integration. At first, the chapter describes different methods of fabrication of SiC nanocrystals which have been the subject of intensive research due to their potential applications in optoelectronics structures and especially in nanoscale UV light emitters. These methods include chemical vapor deposition, electrochemical and chemical etching, laser ablation. Then, a substantial part of the chapter is dedicated to SiC nanowire (NW) fabrication technologies. They are grouped in two categories: the top-down and the bottom-up approaches. Vapor-liquid-solid, vapor-solid, solid-liquid-solid SiC NW growth techniques using different precursors and catalysts constitute the bottom-up approach and are discussed in this chapter. Then, the commonly employed in other semiconductors top-down technological approach including e-beam lithography and subsequent dry etching is addressed. The following section is devoted to the processing technologies of SiC NW based devices, mostly focusing on the ohmic contacts formation. The chapter is concluded with the brief review of SiC nanowires application in field effect transistors.

The book should be of high interest for technologists, scientists, engineers and graduate students who are working in the field of silicon carbide and related materials. The book also can be used as a supplementary textbook for graduate courses on related specialization.

In conclusion, we would like to thank all authors for their hard work and very valuable contributions to this book to warrant its scientific quality and actuality. We also wish to express our deep appreciation to Thomas Wohlbier of Materials Research Forum LLC who was very flexible and patient to accommodate all our wishes during the editing process and who made his best for timely publication of this book.

Konstantin Vasilevskiy

Konstantinos Zekentes

Advancing Silicon Carbide Electronics Technology II   Materials Research Forum LLC
Materials Research Foundations **69** (2020) 1-62   https://doi.org/10.21741/9781644900673-1

CHAPTER 1

Historical Introduction to Silicon Carbide Discovery, Properties and Technology

K. Vasilevskiy*, N. G. Wright

School of Engineering, Newcastle University, Newcastle upon Tyne, United Kingdom

* konstantin.vasilevskiy@newcastle.ac.uk

Abstract

This chapter reviews the history of silicon carbide technology from the first developments in the early 1890s to the present day and highlights the major developments that have facilitated the emergence of the world-wide SiC electronics industry. Physical, chemical and electrical properties of silicon carbide are also briefly described and discussed. The advantage of silicon carbide over silicon in use for fabrication of semiconductor power devices is illustrated by the rough estimation of blocking layer parameters in unipolar SiC and Si devices and comparison of their characteristics. Current state in commercial production and availability of SiC wafers and epitaxial structures as well as potential market of SiC power devices are also outlined.

Keywords

Acheson Process, BJT, Carborundum, Cree, Crystal Growth, DIMOSFET, Electroluminescence, Epitaxy, GTO, Hexagonality, History, HTCVD, IGBT, Inverter, LED, Lely Platelets, LETI Method, Lifetime Enhancement Thermal Oxidation, Material Properties, Micropipe, Modified Lely Method, Moissanite, MOSFET, Polytypism, Radio Detector, RAF Growth Process, Schottky Diode, SiC, SiC Devices, Silicon Carbide, Stacking Fault, Step-Controlled Epitaxy, Step-Flow Growth, Sublimation Sandwich Method, Technology, Varistor

Contents

List of used symbols and abbreviations..3
1. Introduction..5
2. Discovery of silicon carbide 5

2.1   Acheson process..6

2.2   Silicon carbide in the nature..7

3.   Silicon carbide material properties...8

3.1   Chemical bonds and crystal structure of silicon carbide 8

3.2   Crystal structure and notation of SiC polytypes 10

3.3   Stability, transformation and abundance of SiC polytypes..........................14

3.4   Chemical and physical properties of silicon carbide 15

3.5   Polytypism and electrical properties of silicon carbide 17

3.6   Silicon carbide as a material for high temperature electronics.....................18

3.7   Silicon carbide as a material for high power electronics 19

4.   Silicon carbide in early radio technology 22

5.   Electroluminescence of silicon carbide 24

6.   Silicon carbide varistors...26

7.   Lely platelets...27

8.   Silicon carbide bulk crystal growth 29

9.   Silicon carbide epitaxial growth 30

10. Emergence of industrial silicon carbide electronics 32

10.1 Foundation of Cree Research, Inc. and the first commercial blue LEDs 32

10.2 Industrial SiC wafer growth...33

10.3 Preconditions and demands for SiC power electronics.................................34

10.4 $4H$-SiC polytype as a material for power electronics 36

10.5 $4H$-SiC unipolar power devices 36

10.5.1 $4H$-SiC power Schottky diodes...37

10.5.2 $4H$-SiC power JFETs 38

10.5.3 $4H$-SiC power MOSFETs 40

10.6 Development of $4H$-SiC power bipolar devices..42

10.6.1 Minority carrier lifetime enhancement 43

10.6.2 Suppressing stacking faults..44

10.6.3 Advanced $4H$-SiC power bipolar devices..46

10.7   Emergence of automobile SiC power electronics..47

Conclusion 48

References..48

List of used symbols and abbreviations

D	polytype hexagonality;
d	thickness of blocking layer;
E_g	energy gap between conduction and valence bands;
e-h	electron-hole;
F_{CR}	critical electric field strength;
m	any positive integer number;
n	number of bilayers in a unit cell;
n_-	number of bilayers with negative pseudo-spins constituting 1/3 of the rhombohedral unit cell;
n_+	number of bilayers with positive pseudo-spins constituting 1/3 of the rhombohedral unit cell;
N_D	donor density;
n_e	electron density;
n_h	number of hexagonal double layers in a unit cell;
n_i	intrinsic charge carrier density;
n_k	number of cubic double layers in a unit cell;
q	elementary charge;
Q_{RR}	reverse recovery charge;
r	any integer;
R_{ON}	on-resistance;
$R_{ON\text{-}SP}$	specific on-resistance;
R_S	parasitic series resistance;
S	device area;

V_B	avalanche breakdown voltage;
V_{BL}	blocking voltage;
V_{GS}	gate-source voltage;
ε	relative permittivity;
ε_0	vacuum permittivity;
μ_e	electron mobility;

AC	alternating current;
BJT	bipolar junction transistor;
BPD	basal plane dislocation;
CVD	chemical vapor deposition;
DC	direct current;
DIMOSFET	double implanted MOSFET;
DMOS	Double Diffused Metal Oxide Semiconductor;
EV	electrical vehicle;
GTO	gate turn-off (thyristor);
HEV	hybrid electrical vehicle;
HTCVD	high-temperature CVD;
IC	integrated circuit;
IGBT	insulated gate bipolar transistor;
$I\text{-}V$	current-voltage;
JBS	junction barrier Schottky;
JFET	junction field-effect transistor;
LED	light emitting diode;
MPD	micropipe densities;
MOSFET	metal-oxide-semiconductor field-effect transistor;
RAF	repeated a-face (growth process);
RF	radio frequency;

SCR   space charge region;

SF   stacking fault;

SI   semi-insulating;

TED   threading edge dislocation;

TI-VJFET   trenched and implanted VJFET;

VJFET   vertical JFET.

1.   Introduction

Silicon carbide (SiC) is a semiconducting material with outstanding physical, chemical and electrical properties which make it very suitable for fabrication of high power, low loss semiconductor devices. Moreover, excellent thermal stability, superior chemical inertness and hardness potentially allow operation of SiC devices in harsh environment at high temperature conditions. On the other hand, fabrication of these devices is rather intricate owing to the same properties of silicon carbide like its chemical inertness and hardness. It took more than hundred years to develop SiC electronics up to its modern state when power SiC devices possessing higher efficiency than their silicon counterparts became commercially available and widely used in numerous applications. This chapter briefly describes the history of SiC discovery and technology development. It summarizes also selected properties of SiC relevant to semiconductor device processing.

2.   Discovery of silicon carbide

Silicon carbide has arguably the longest history of all semiconducting materials in relation to electronics. The discovery of silicon carbide is in itself unusual – in that it was discovered not in the nature but as a man-made material. Probably, the first observation of a chemical compound with Si-C bonds was reported by Jöns Jacob Berzelius (1779-1848) in 1824 [1]. Berzelius, of the famous Karolinska Institute in Stockholm, is widely considered to be one of the founders of modern chemistry. He was renowned for his expert experimental technique and this allowed him to perform experiments under unusual conditions. He discovered (in addition to numerous other achievements) several new chemical elements including silicon. Concerning silicon carbide, Berzelius made a very conservative statement that he discovered an unknown material which produced equal number of silicon and carbon atoms after burning [2].

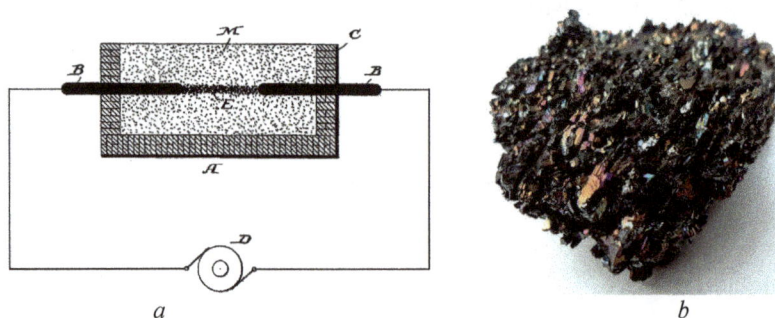

Figure 1. *(a) Schematic sketch of Acheson's furnace showing the use of a conducting graphite core, E, to generate the heat required for carborundum formation [5]. (Public domain image, source: United States Patent and Trademark Office, www.uspto.gov); (b) druse of carborundum grown by the Acheson process (size about 3 × 3 × 2 in).*

2.1 Acheson process

The first verified synthesis of silicon carbide happened accidentally seventy years later [3, 4]. In 1891, American engineer Edward Goodrich Acheson (1856–1931) experimented with a recently invented electrical furnace trying to produce artificial diamonds (highly demanded by the industry as an abrasive material). The furnace was packed with a mixture of clay and coke around a central carbon resistive core which was then heated by electric current (see Fig. 1). Acheson supposed that the carbon, dissolved in melted silicate of alumina, could be crystallized when the mixture was cooled to the temperature of solidification. He said later: "…had I been a chemist, it is probable that such an experiment would not have been thought worthy of consideration, and certainly would not have been attempted." Nevertheless, after the furnace charge was cooled and broken, he found a few small single crystals within the reacted mass. They were bright blue, very hard and very brittle but they were not diamonds. Acheson named this new material "Carborundum" (the combination of words carbon and corundum) as he believed that it was composed of carbon and alumina. Very soon he realized that quantity and quality of synthesized carborundum crystals were determined by silica impurities and replaced the furnace load with a mixture made of coke and glass sand.

These results were so promising that the method was patented in 1893 [5] and the Carborundum Company was organized for manufacturing and marketing carborundum as an abrasive material. The company organized a chemical laboratory and employed Dr

Otto Mulhaeuser, who discovered that the blue-black carborundum samples grown from the carbon mixture with clay were silicon carbide crystals with aluminum impurities while light green samples grown from the mixture of coke and sand were pure silicon carbide crystals. Interestingly, Acheson appears not to be aware of the work of Berzelius and stated "the material that I have designated as carborundum is practically a new compound hitherto unknown to chemistry and in its purity is represented by the formula SiC."

Nowadays, the Acheson process is still in use as the main industrial method for manufacturing silicon carbide. The process works by heating together a source of carbon (typically coke) and a source of silicon dioxide (typically quartz sand) to over 1700 °C in an electrical furnace. The reaction of the carbon and silicon dioxide produces silicon carbide and carbon monoxide, which is liberated from the furnace as a gas. Silicon carbide remains within the reacted materials in the form of small crystals with sizes varying from a fraction of a millimeter to about three millimeters. These crystals are typically colored dark green or black depending on the impurities incorporated from the raw materials (see Fig. 1b). To use this silicon carbide as an industrial material, it is chemically treated to separate from the rest of reacted mass and then crushed and milled into an appropriate grain size.

The world output of synthetic silicon carbide produced by Acheson method is more than 1.5 million tons (about 2,300 M$) per year (2016). It is mainly used as an abrasive material but a large part (about 30%) is also used in the steel industry as a deoxidizing agent and as a refractory material. Silicon carbide is also widely used as a component of ceramic and composite materials owing to its lightweight, extreme hardness, excellent mechanical strength and thermal resistance.

2.2 Silicon carbide in the nature

Naturally occurring silicon carbide was found by Ferdinand Frederick Henri Moissan (1852-1907), famous French chemist, in the Canyon Diablo meteorite from the Barringer meteorite crater in Arizona desert in 1905 [6]. He dissolved a 56 kg piece of the meteorite in strong acids to get a few tiny crystals which were identified as silicon carbide. Subsequently, this mineral form has been found in numerous other meteorite samples and impact craters from meteorite strikes. It was assumed that the heat generated by friction between the earth's atmosphere and the meteorite is the source of the temperature needed to form the silicon carbide from appropriate raw materials already contained in the meteorite. There have also been suggestions that the source of silica (quartz) could be from the material displaced during the moment of impact with the earth's crust in some cases. Curiously, it was also disputed, as it could be just a contamination from the carborundum tools which were already widely used.

Nowadays, the extra-terrestrial origin of this silicon carbide is clearly proven by its carbon isotopic composition. It has the ratio of ^{12}C isotope concentration to that one of ^{13}C about 64. This value is noticeable lower than the $^{12}C/^{13}C$ ratio in materials of terrestrial origin. Today, the mainstream theory states that the interstellar silicon carbide was synthesized in expanding atmospheres of pre-solar carbon stars [7].

The first natural SiC of terrestrial origin was found by B. I. Ozernikova in sediments of the Tyung river (Siberia) in 1956 and by A. P. Bobrievich in diamond-bearing kimberlite pipes in Yakutia, USSR in 1957 [8]. Carbon isotopic composition of terrestrial and synthetic SiC is the same ($^{12}C/^{13}C = 95.14$) and corresponds to the natural abundance of ^{13}C (1.04%). Nevertheless, they were clearly distinguished by markedly different chemistry of trace elements. Details of elemental analysis of micro-inclusions in natural terrestrial SiC and discussion of possible mechanisms of its genesis can be found elsewhere [9].

All naturally occurring silicon carbide got the name "moissanite" to commemorate its discovery by Dr Moissan (he also won the Nobel Prize for isolating fluorine from its compounds). Both circumstellar and terrestrial crystals of moissanite are very rare and small with maximum dimensions not exceeding fractions of millimeter. The name "moissanite" is now also used as the marketing name for synthetic silicon carbide crystals for the gem and jewellery industry.

3.   Silicon carbide material properties

Silicon carbide is the only chemical compound of group IV elements. It has a strictly stoichiometric concentration ratio of silicon (Si) and carbon (C) atoms. It should not be mixed with solid solutions which may be formed by other group IV elements and may have variable component concentration ratios (e.g. Si_xGe_{1-x}).

3.1   Chemical bonds and crystal structure of silicon carbide

Both Si and C atoms in SiC have their four valence electron orbitals sp^3 hybridized which form very strong Si-C bonds with dissociation energy of 3.1 eV. These bonds are mainly covalent due to sharing binding electrons between two atoms. The electron density is higher at carbon atoms because of its higher electronegativity (2.55 eV for C vs. 1.9 eV for Si) resulting in partially ionic character of Si-C bonds. A strict definition and expressions for quantitative description of Si-C bond ionicity can be found elsewhere [10].

Materials Research Forum LLC
https://doi.org/10.21741/9781644900673-1

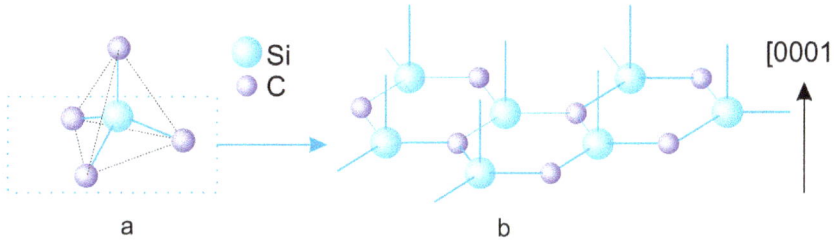

Figure 2.   (a) A silicon atom in SiC crystal lattice with its four first order neighbors which form a regular tetrahedron; (b) 3D sketch of silicon carbide close packed bilayer (Si and C atoms are shown coplanar for better visibility).

Each Si (or C) atom in the SiC crystal lattice is surrounded by 4 carbon (silicon) atoms (first order neighbors) which form a regular tetrahedron shown schematically in Fig. 2a. The distance between Si and C atoms (the bond length) in silicon carbide is 0.189 nm. Four atoms highlighted by the dotted rectangle in Fig. 2a form a silicon carbide close packed bilayer shown in Fig. 2b (note, that Si and C atoms are shown coplanar in this sketch for better visibility). The plane formed by this bilayer is called a basal plane and the crystallographic direction perpendicular to this plane is called stacking or <0001> direction. This direction is polar since only silicon dangling bonds are at one side and only carbon bonds are at the opposite side of the SiC bilayer. When a SiC wafer is cut or grown by the way that its two sides are parallel to the basal plane, this wafer has two surfaces which are physically inequivalent: (0001)Si face and (000$\bar{1}$)C or simply Si and C face. The direction from C to Si face is [0001] and it is called c-axis. These Si and C faces have significantly different properties, for example, the oxidation rate of C face is about eight times higher than that one for Si face.

Fig. 3a schematically depicts the hexagonal unit cell, fundamental translation vectors and main crystal planes in silicon carbide. Fig. 3b shows Si and C atoms' layout in the basal plane of silicon carbide. The length of translation vectors in the basal plane is 0.308 nm. The length of the translation vector along the c-axis is determined by the number of stacked bilayers (n) constituting the unit cell and by the distance between two adjacent bilayers which is 0.252 nm (4/3 of the Si-C bond length). Note that a- and m-planes are nonpolar as the number of Si and C dangling bonds at either side of these planes are the same.

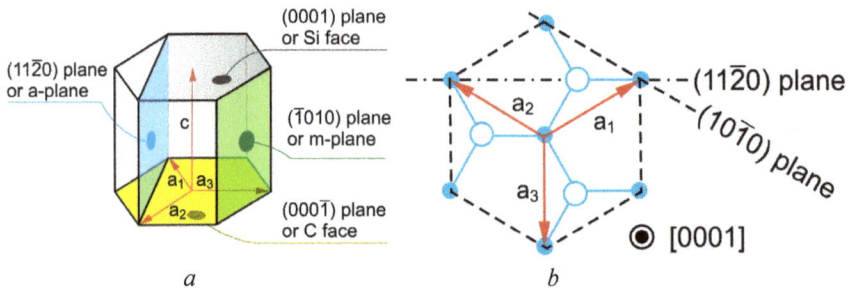

Figure 3.   (a) Hexagonal unit cell and fundamental translation vectors in silicon carbide, (b) Si and C atoms' layout in a basal plane of silicon carbide. Solid and open circles denote C and Si atoms, respectively.

3.2   Crystal structure and notation of SiC polytypes

A SiC bilayer cannot be stacked with itself because of Si bonds in a bottom bilayer have to be aligned with C bonds in a top layer. The top layer has to be either rotated for 180 ° or shifted in the $<1\bar{1}00>$ direction to align the Si and C atoms in two adjacent SiC bilayers as it is schematically shown in Fig. 4. The rotated layers are shown by different colors (blue and orange) in this figure and a positive or negative sign (sometimes called a pseudo-spin) may be attributed to them to denote their orientation. There are many ways to stack shifted and rotated bilayers to form different unit cells of SiC crystal lattice. The most amazing property of silicon carbide is that it occurs in more than 170 crystal structures (called polytypes) with different number of stacked bilayers in a unit cell. The vast majority of SiC polytypes were discovered by x-ray diffraction as micro-inclusions in carborundum crystals but some of them are abundant in the nature and can be grown of reasonable size for industrial use.

Each SiC polytype has unique electrical and optical properties. Contrary to the crystal growth, device processing parameters (like annealing temperatures, oxidation and etching rates, etc.) usually do not depend on the SiC polytype but fabricated devices made of different SiC polytypes have very different characteristics. Consequently, when used in semiconductor electronics, silicon carbide is a common name for a set of materials with different structural and electrical properties. These materials have to be clearly and unambiguously designated. In this section, crystal structure and notation of SiC polytypes are described briefly in a scope sufficient for understanding following chapters in this book. Detailed description of polytypism phenomenon in SiC and other materials is given

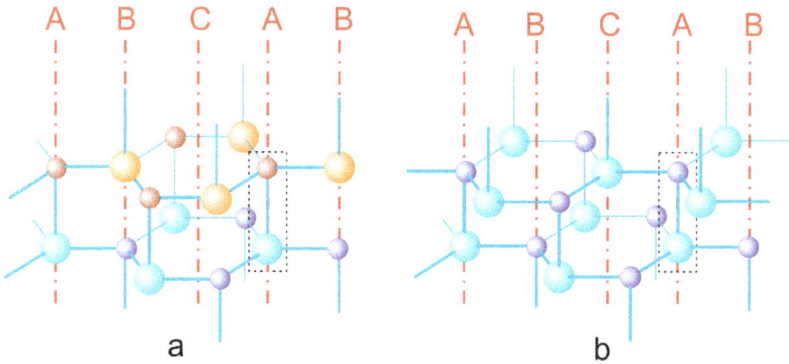

Figure 4.   3D sketch of two successive silicon carbide close packed bilayers which are (a) rotated or (b) shifted to be stacked. Si and C atoms in each bilayer are shown coplanar for better visibility. Large and small balls denote Si and C atoms, respectively. Dash-dotted lines denoted A, B and C lay in a (11$\bar{2}$0) plane which is coincidental with the plane of drawing. Si and C atoms in a single bilayer are shown coplanar for better visibility.

by A. R. Verma and P. Krishna in their classical book [11]. More recent overviews of polytypism and properties of silicon carbide can be found elsewhere [12-15].

Polytypism is usually defined as a one dimensional polymorphism which is in its turn the ability of a substance to crystallize in different crystal structures. The phenomenon of polytypism was first discovered by H. Baumhauer in 1912 [16] in silicon carbide crystals. It is not a unique property of silicon carbide but SiC is one of few compounds which crystallizes in several stable polytypes and the only compound discovered up to now which can form polytypes with very long unit cells. Indeed, silicon carbide with 594 bilayers in a unit cell was reported [17] and there is no signs that it is the polytype with the longest available period.

The distinction between crystal structures of different polytypes is most clearly seen in the (11$\bar{2}$0) crystal plane. In Fig. 4, this plane is coincidental with the plane of drawing. The dash-dotted lines denoted A, B and C are lying in this plane and their intersections with the (0001) plane mark possible locations of Si-C bonds directed along the c-axis. All Si atoms of one bilayer bound with C atoms in the next bilayer along the c-axis can be located only in the same positions. It cannot be that a part of them are in the A positions and the rest of them in positions B or C. Consequently, a SiC bilayer may be identified by the positions of Si atoms seen in the ($11\bar{2}0$) crystal plane. Then, a SiC polytype may be designated by the sequence of A, B and C symbols corresponding to SiC bilayers in a unit cell (so called ABC notation). Fig. 5 schematically shows the atoms layout in the ($11\bar{2}0$) crystal plane of four SiC polytypes with shortest translation vectors along the c-axis. Each polytype in this figure is signed using different notations. It is clear that the ABC notation is vivid and unambiguous but too cumbersome. Indeed, Fig. 5 shows also the SiC polytype with just only nine bilayers in the unit cell to demonstrate that the ABC notation is impractical for designation of long-period polytypes.

The compact alternative to the ABC notation was introduced by Lewis Ramsdell [18]. He designated SiC polytypes by the number of bilayers in the unit cell (n) followed by the letter C, H, or R, corresponding to the crystal symmetry which may be cubic, hexagonal, or rhombohedral, respectively. The SiC polytype with shortest period with only two bilayers in the unit cell is 2H-SiC in the Ramsdell notation. It has a wurtzite crystal structure. The next polytype is 3C-SiC. It has a face-centered cubic unit cell (with [111] direction coincided with the c-axis) and zinc blende crystal structure. All other SiC polytypes are hexagonal or rhombohedral.

There are few selection rules for n (the number of bilayers in the unit cell) in the Ramsdell notation. In the polytypes with rhombohedral lattice, the n number has to be a multiple of 3:

$$n = 3 \times m \tag{1}$$

where m is a positive number, and

$$n_+ - n_- = 3r \pm 1;$$

$$n_+ + n_- = m \tag{2}$$

where n_+ and n_- are numbers of bilayers with positive and negative pseudo-spins constituting 1/3 of the unit cell; and r is any integer. The lattice with hexagonal symmetry has to have even number of layers in the unit cell and stacking order of layers from $n/2$ to n has to be the reverse of the order of layers from 0 to $n/2$ [11]. It directly follows from these selection rules that there are no polytypes with 5 or 7 bilayers in the unit cell and that the shortest-period polytype with rhombohedral crystal symmetry is 9R-SiC. The Ramsdell notation is the most widely used one due to its compactness but it does not reflect an internal structure of a unit cell as the ABC notation.

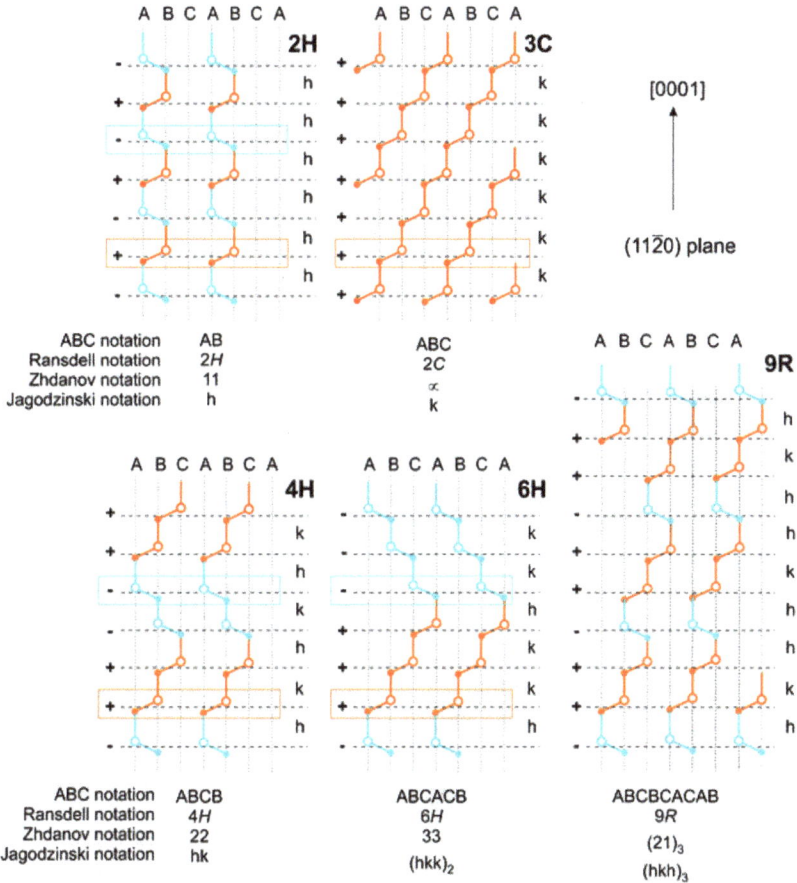

Figure 5. *Silicon (open circles) and carbon (closed circles) atom layout in the (11$\bar{2}$0) crystal plane of five SiC polytypes with short translation vectors along the c-axis. SiC bilayers are highlighted by rectangles. Their pseudo-spin is denoted in the column to the left. Dashed lines separate Si-C double layers with the same local symmetry denoted h or k in the right hand side. Each polytype is signed using different notations.*

A compact notation reflecting an internal structure of a unit cell was proposed by Zhdanov [19]. In this system, a polytype is designated by the sequence of numbers corresponding to the number of successive bilayers with the same pseudo-spin. Zhdanov notations for the first four SiC polytypes and for 9R-SiC are shown in Fig. 5. Polytype 2H is designated 11 in Zhdanov notation as every next successive bilayer changes its orientation. Polytype 3C is designated ∞ in Zhdanov notation as bilayers never change their orientation. The unit cell of polytype 6H comprises 3 successive layers of the same orientation in a row and then 3 layers with opposite orientation. It is designated 33 in Zhdanov notation. The sequence of bilayers' pseudo-spins in the unit cell of 9R-SiC is ++-++-++- and its designation in Zhdanov notation is $(21)_3$. In this notation, the relatively abundant long-period rhombohedral polytype 15R-SiC has a very compact designation $(23)_3$ which does reflect the internal crystal structure of its unit cell.

Another appropriate notation was proposed by Heinz Jagodzinski [20]. It is clearly seen in the Fig. 4 that Si and C atoms (highlighted by dotted rectangles) bounding two successive bilayers have different geometry of second-nearest neighbors depending on the orientation of stacked bilayers. These atoms occupy lattice sites with cubic or hexagonal local symmetry if they bound two bilayers with the same or opposite pseudo-spins, respectively. These layers of Si and C atoms with the same local symmetry (Si-C double layers) are separated by dashed lines in Fig. 5 and labelled "h" or "k" if they have hexagonal or cubic local symmetry, respectively. The Jagodzinski notation designates SiC polytypes by indication of local symmetry for each Si-C double layer in a unit cell. This notation system is less compact than the Ramsdell and Zhdanov notations, for example, it designates the polytype 15R as $(kkhkh)_3$. Nevertheless, the Jagodzinski notation is very useful because of it clearly reflects the polytype hexagonality (D) which is defined as:

$$D = n_h/(n_h + n_k) \qquad (3)$$

where n_h and n_k are the numbers of hexagonal and cubic double layers in a unit cell.

Finally, it should be noted that 3C-SiC is often called β-SiC and all other SiC polytypes have a common name α-SiC.

3.3   Stability, transformation and abundance of SiC polytypes

The most commonly observed polytypes are 3C, 2H, 4H, 6H, 8H and 15R. Rhombohedral polytypes with shorter periods, as 9R for example, where observed only as a stacking layer sequence in local nanoscale regions inside other more abundant polytypes [21, 22].

SiC polytypes are metastable since the stacking fault energy in silicon carbide is very low. They can be transformed into each other by thermal treatment [23]. Indeed, Krishna et al. reported transformation of small (about 1-2 mm size) 2H-SiC crystals in 3C-SiC after annealing in argon for 16 hours at temperatures from 1400 to 1800 °C [24]. Stability diagrams for the occurrence of SiC polytypes as a function of temperature can be found elsewhere [11]. As a rough guide, the most thermodynamically stable polytypes are 3C-SiC at temperatures below 1800 °C; 2H-SiC in the temperature range from 1800 to 2000 °C; and 4H-SiC at temperatures from 2000 to 2400 °C. Polytypes 6H, 15R and 8H-SiC can be crystallized at higher temperatures.

The most abundant naturally occurring terrestrial SiC polytype is 6H-SiC which is followed by 15R and 33R. Natural 3C-SiC is terrestrially extremely rare and natural 2H-SiC has never been reported [9].

SiC grown by the Acheson process has similar polytype distribution mostly containing 6H-SiC followed by 15R and 4H [11]. Both 3C and 2H-SiC were not recovered in significant amounts from this process.

The circumstellar silicon carbide is mainly 3C (about 80%) followed by 2H (about 2.7%). The rest 17% of interstellar SiC are crystal grains with variable proportion of 2H and 3C polytypes [7].

3.4   Chemical and physical properties of silicon carbide

The tough crystal structure of silicon carbide with strong chemical bonds defines its outstanding mechanical, thermal, chemical and electric properties. Some of them are shown in Table 1 side by side with silicon and diamond for comparison. All three materials are indirect band gap semiconductors with similar crystal structures but with different bond dissociation energies which define the distinction in their properties.

Silicon carbide is a very hard material - the third one after diamond and boron nitride in the Mohs scale. It remains solid up to the sublimation temperature of ~2500 °C and has very high Debye temperature (~850 °C) – the temperature corresponding to the maximum phonon frequency and limiting the maximum operating temperature of SiC as a semiconductor material. Silicon carbide has an excellent thermal conductivity comparable with that one of copper. It is resistant to radiation damages as a structural material and as a semiconductor [29].

Chemical inertness of silicon carbide is outstanding. It does not interact with any chemicals at room temperature. It can be etched only in molten alkalis at temperatures above 450 °C. Oxidation of silicon carbide starts to be noticeable only at temperatures above 900 °C. The etching rate of monocrystalline SiC in dry pure oxygen at 1200 °C is ranged from 2.5 to 20 nm/h depending on the etched surface orientation. For comparison, diamond burns away at the temperature above 800 °C and silicon etching rate in dry pure oxygen at 1200 °C is about 100 nm/h. It should be noted that, despite very low SiC oxidation rate, SiO_2 films can be grown on SiC thick enough for device applications. This is a great advantage of silicon carbide over other wide band semiconductors like diamond and III-V nitrides which do not have a native oxide.

Table 1. Selected physical properties of silicon carbide, silicon and diamond [10, 25-28].

Material		Si	SiC	C
Bond dissociation energy	eV	2.3	3.1	3.6
Bond length	nm	0.235	0.185	0.1545
Mohs Hardness		7	9.1 ÷ 9.4	10
Density	g/cm^3	2.33	3.21	3.52
Thermal conductivity at 300 K	W/cm/K	1.3	3.7 ÷ 4.9	20 ÷ 25
Debye temperature	°C	370	850	1600
Melting (decomposition) temperature	°C	1412	~2500 (sublimes)	~3600 (sublimes)
Relative dielectric constant		11.9	10.03	5.7
Band gap energy at 300 K	eV	1.12	3.23 *	5.47
Intrinsic carrier density at 300 K	cm^{-3}	$8.8 \cdot 10^9$	$1.7 \cdot 10^{-8}$ *	$\sim 10^{-27}$
Breakdown field at 300 K	V/cm	$(2 \div 4) \cdot 10^5$	$(1.6 \div 3) \cdot 10^6$ *	$10^6 \div 10^7$
Electron mobility at 300 K	cm^2/(V·s)	1400	900*	>2000
Shallow donor activation energy	meV	43 (Sb)	52 (N)*	600 (P)
Shallow acceptor activation energy	meV	45 (B)	190 (Al)*	370 (B)
Maximum diameter of native single crystal substrate	mm	450	200	10
Availability of native oxide		yes	yes	no

* Data for 4*H* polytype

Owing to its heat resistance, hardness, chemical inertness and affordable cost, silicon carbide finds numerous applications as an abrasive material and a component of refractory ceramic materials. Pure monocrystalline silicon carbide is transparent and colorless. It has a refractive index, dispersion and lustre index exceeding that of diamond and is used for making a very popular gemstone called moissanite.

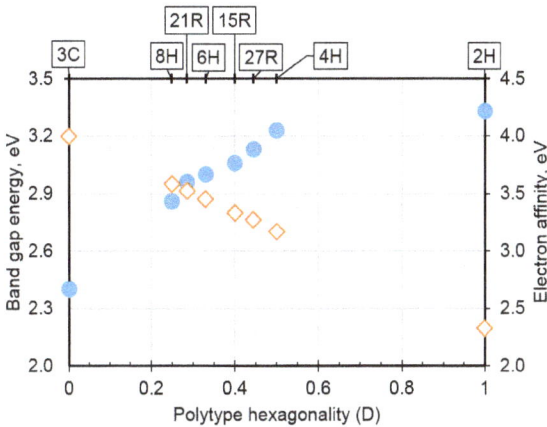

Figure 6.   *The band gap energy (solid circles) and electron affinity (open diamonds) of silicon carbide depending on the polytype hexagonality. The data shown in this figure are taken from [30, 31].*

3.5   Polytypism and electrical properties of silicon carbide

In the first approximation, electrical properties of silicon carbide are also defined by the high dissociation energy of Si-C bond. Indeed, as higher is this energy, more energy is required to ionize an atom and to generate an electron-hole (*e-h*) pair. This ionization energy roughly corresponds to the gap between conduction and valence energy bands (E_g). Silicon carbide is called a wide band gap semiconductor as its band gap is wider than that one of silicon.

As it was mentioned in Section 3.2, the local potential distribution in unit cells of different SiC polytypes is varying depending on their crystal structure, specifically on their hexagonality. This variation significantly influences all electrical properties of SiC polytypes including the band gap energy, electron affinity, ionisation energies of doping impurities, effective masses and mobilities of charge carriers, impact ionisation coefficients and others. This dependence is illustrated by Fig. 6 where the band gap energy and electron affinity are plotted versus the polytype hexagonality. The data shown in this figure were taken from [30, 31]. These are probably the most distinctive examples of electrical parameters depending almost linearly on polytypes hexagonality.

Other electrical parameters of SiC polytypes do not necessarily demonstrate linear or even monotonic dependence on their hexagonality. For example, the maximum electron

Materials Research Forum LLC
https://doi.org/10.21741/9781644900673-1

mobility both in *3C* and *4H*-SiC is higher than in *6H*-SiC. These three polytypes are the most widely used in SiC electronics and their selected electrical properties related to the semiconductor device processing are summarized in Table 2.

Comparison of electrical parameters shown in Table 2 leads to obvious conclusion that *4H*-SiC is the most suitable polytype for fabrication of power semiconductor devices due to its higher charge carriers mobilities and lower ionisation energies of donors and acceptors.

Table 2. Selected electrical properties of 3C, 4H and 6H-SiC [10, 26-28] [35].

Polytype		3*C*-SiC	6*H*-SiC	4*H*-SiC
Hexagonality		0	0.33	0.5
Band gap energy at 300 K	eV	2.36	3.08	3.23
Electron affinity	eV	4.0	3.45	3.17
Intrinsic carrier density at 300 K	cm^{-3}	$1.4 \cdot 10^{-1}$	$1.8 \cdot 10^{-7}$	$1.7 \cdot 10^{-8}$
Shallow donor (N) activation energy	meV	60-100	85-125	52
Shallow acceptor (Al) activation energy	meV	260	239	190
Maximum electron mobility at 300 K	cm^2/(V·s)	900	360	900
Maximum hole mobility at 300 K	cm^2/(V·s)	40	90	120
Maximum breakdown field at 300 K	V/cm	$\sim 1 \cdot 10^6$	$\sim 3 \cdot 10^6$	$\sim 3 \cdot 10^6$

3.6   Silicon carbide as a material for high temperature electronics

Thermal generation of electron-hole pairs in SiC is much lower than in silicon due to wider band gap and, hence, the intrinsic charge carrier density (n_i) at room temperature in *4H*-SiC is about 18 orders of magnitude lower than that one in Si. In fact, pure *4H*-SiC is a very good insulator at room temperature. It needs to be heated up to 500 °C to have the same intrinsic carrier density as silicon has at room temperature ($\sim 10^{10}$ cm^{-3}). To have the n_i value comparable with doping levels usually used in power semiconductor devices ($\sim 10^{14}$ cm^{-3}), SiC has to be heated up to 900 °C (for comparison, $n_i = 10^{14}$ cm^{-3} at about 200 °C in silicon). Hence, the maximum operating temperature of SiC semiconductor devices can reach SiC Debye temperature (850 °C). In practice, it is currently limited by thermal stability of dielectrics and metallisations used for fabrication of these devices.

The most startling examples of high temperature SiC devices were reported by the group of Philip Neudeck at NASA John Glenn Research Center. They demonstrated a short-term operation of packaged *4H*-SiC junction field effect transistor (JFET) logic integrated circuits (ICs) at ambient temperatures exceeding 800 °C in air [32]. They also demonstrated SiC lateral JFETs with operating time of 6000 hours at 500 °C which was limited by the thermal degradation of a metal stack used for the formation of ohmic contacts in these devices. [33, 34].

3.7 Silicon carbide as a material for high power electronics

Electron-hole pairs also can be generated by the impact ionization. At a sufficiently high electric field, some charge carriers can be accelerated to gain enough energy to be able to knock-out an electron from a shared orbital and to produce an *e-h* pair. Obviously, this energy has to be higher than E_g. This process is characterized by the electron (hole) impact ionization rate, which strongly depends on electric field, band gap energy, temperature and carrier scattering mechanisms in specific materials. At a critical electric field strength (F_{CR}), each electron (hole) generates at least one *e-h* pair during its travelling through a device structure. This generation leads to the infinite rise of charge carrier concentration and, hence, of electrical current (if it is not limited by an external circuit). A voltage drop which corresponds to F_{CR} in a structure with specific doping profile is called a breakdown voltage (V_B). Both F_{CR} and V_B are proportional to $(E_g)^{3/2}$ [25] and define the maximum blocking voltage of switching and rectifying devices made of specific material. The above reasoning is very simplified but provides a logical connection between the strength of chemical bonds in a particular semiconductor and the ability to block specific voltages by power devices made of this material.

Comparing electrical properties of 4*H*-SiC, silicon and diamond (given in Table 1), it can be noticed that the only disadvantage of silicon carbide is the low electron mobility. However, very high breakdown field and high thermal conductivity of silicon carbide have to afford substantial performance gain in many power devices, as compared to the conventional semiconductors. It is worth mentioning that diamond has even higher breakdown field and thermal conductivity but the donor activation energy in diamond is more than ten times higher than that one in silicon and silicon carbide making it barely useable for fabrication of power semiconductor devices.

To appreciate a potential efficiency gain of silicon carbide power devices over silicon counterparts, electrical characteristics and parameters of devices with the same blocking voltage ratings but made of SiC and Si are roughly estimated below. All devices of this kind have a thick low doped epitaxial layer which is completely depleted at reverse bias to block the applied voltage. This layer is doped by donor impurities providing electrons to conduct a current when the device is forward biased (in open state). In this particular example, we consider a unipolar device which does not have a charge carrier injection in the blocking layer at forward bias. It may be either Schottky diode, or Junction Barrier Schottky (JBS) diode, or metal-oxide-semiconductor field-effect transistor (MOSFET), or junction field-effect transistor (JFET). As far as Si and SiC have similar ionisation energies of donors (see Table 1), for the sake of simplicity it may be assumed that the electron density (n_e) is equal to the donor density (N_D) in blocking layer. Also we assume that the device termination efficiency is 100% and, hence, the maximum blocking voltage is equal to V_B and the maximum electric field is equal to F_{CR}.

The two most important parameters characterizing efficiency of power devices are (1) the on-resistance (R_{ON}) and (2) the charge which has to be removed from the blocking layer to switch it from the conducting state at forward bias to the insulating state at reverse bias (the reverse recovery charge, Q_{RR}). Both these parameters depend on the electron concentration in blocking layer, its thickness (d) and device area (S):

$$R_{ON} = \frac{d}{S \cdot q \cdot \mu_e \cdot n_e} = \frac{d}{S \cdot q \cdot \mu_e \cdot N_D} \tag{4}$$

$$Q_{RR} = q \cdot d \cdot S \cdot N_D \tag{5}$$

where q is the elementary charge and μ_e is the electron mobility.

As it is seen from Table 1, the critical field in 4H-SiC is about ten times higher than that one in silicon. Hence, a 4H-SiC device can have a blocking layer about ten times thinner than Si to support the same blocking voltage and this layer can be doped to higher level resulting in its lower on-resistance. More precisely, it can easily be shown that being designed in terms of minimization of specific on-resistance, blocking layer thickness and doping level are defined by the following expressions:

$$d = \frac{3V_B}{2F_{CR}} \tag{6}$$

$$N_D = \frac{4F_{CR}^2 \varepsilon \varepsilon_0}{9qV_B} \tag{7}$$

where ε_0 is the vacuum permittivity and ε is the relative permittivity.

Table 3 shows parameters and characteristics of unipolar power 4H-SiC and Si devices with optimized doping levels and thicknesses of blocking layers calculated using Eqs. 4-7. The area of the 4H-SiC blocking layer was chosen to keep its reverse recovery charge smaller than that one in Si. In this particular example, the 4H-SiC device has about ten times smaller area than the silicon one and the Q_{RR} value in the 4H-SiC device is for 22% lower than that one in the silicon device. At the same time, the on-resistance of 4H-SiC blocking layer is 56 times lower than that one for silicon. This very simplified consideration clearly shows the great potential of silicon carbide application in power semiconductor devices.

Table 3. Calculated parameters and characteristics of unipolar power 4H-SiC and Si devices.

Blocking voltage	V	650	
Forward current	A	50	
Semiconductor material parameters:			
Material		Si	4H-SiC
Electron mobility	cm^2/(V·s)	1400	900
Relative dielectric constant		11.7	9.7
Maximum electric field	V/cm	2.7×10^5	2.84×10^6
Typical contact resistivity	$\Omega\cdot$cm^2	6×10^{-6}	1×10^{-4}
Typical substrate resistivity	m$\Omega\cdot$cm	1	14
Typical substrate thickness	cm	0.04	0.04
Device structure parameters:			
Optimized blocking layer thickness	μm	36.1	3.4
Optimized blocking layer doping	cm^{-3}	3.2×10^{14}	3.0×10^{16}
Device area	cm^2	1	0.09
Calculated characteristics of blocking layer:			
Output capacitance at V_B	pF	28.7	27.1
Reverse recovery charge	μC	1.86	1.46
On-resistance of blocking layer	mΩ	50.0	0.9
Parasitic series resistance:			
Contact resistance	mΩ	0.006	1.11
Substrate resistance	mΩ	0.04	6.22
Calculated device characteristics:			
Total resistance	mΩ	50.0	8.23
Forward current density	A/cm^2	50	556
Dissipated power in on-state	W	125	21
Dissipated power density in on-state	W/cm^2	125	229

To fully realize this advantage of silicon carbide in practice, proper SiC crystal growth and device processing technologies are required. Indeed, the blocking layer in power devices is not a suspended piece of material. It has to be supported on a substrate and metallized to provide electrical contacts. Typical values of substrate thicknesses, contact resistivities and substrate resistivities are listed in Table 3. These parameters are responsible for parasitic series resistances (R_S) of simulated devices. The calculated R_S values are shown in Table 3 and it is clear that the series resistance in the Si device is negligible while in the 4H-SiC device, it is eight times larger than the resistance of blocking layer. The total resistance of the 4H-SiC device still remains six times smaller than that one for silicon but this particular estimations clearly show that further improvement in SiC device fabrication technology is highly desirable.

The last two lines in Table 3 compare total power and power density generated in Si and 4H-SiC devices at the same forward current. Two times higher power density in the 4H-SiC device is perfectly affordable as it can be efficiently dissipated owing to higher thermal conductivity and maximum operating temperature of silicon carbide.

4.   Silicon carbide in early radio technology

What is important in view of this book's subject-matter, the Acheson process enabled the industrial production of small silicon carbide crystals long before crystals of other materials, such as silicon or germanium, were available and this played a crucial role in the development of radio and electronics in general.

The early radio technology required the use of primitive diode-like devices to extract the signal from the tuned circuit. Early radios were devised before semiconductor or vacuum valve diodes appearance and mostly utilized so-called coherer devices, which were either made from magnetic metal grains pressure packed into small tins or from liquid mercury devices. These had very poor diode characteristics by modern standards and consequently had very low signal extraction efficiencies. It was known to that time that naturally occurring galena (PbS) crystals exhibit non-linear current-voltage (I-V) characteristics and this led to idea to use them as radio detectors to replace coherers. The patent on galena detector was granted to J. C. Bose in 1904 [36]. The device design was based on a mechanical contact of a small wire ("catwhisker") that made a point contact to the crystal surface while the other connection was a large area contact formed by a clamp or by a low-melting point alloy. The use of galena detectors was quite tricky as they required fine tuning of wire position and pressure and periodic re-application in order to maintain good electrical characteristics.

Many other materials were tested for use in radio detectors instead of galena (at least over 30,000 combinations of wires and crystals) [37] and in 1906, Henry Harrison Chase Dunwoody (1842-1933), retired US army general and a vice president of De Forest Wireless Telegraph Co., patented wireless telegraph system utilizing a "wave-responsive

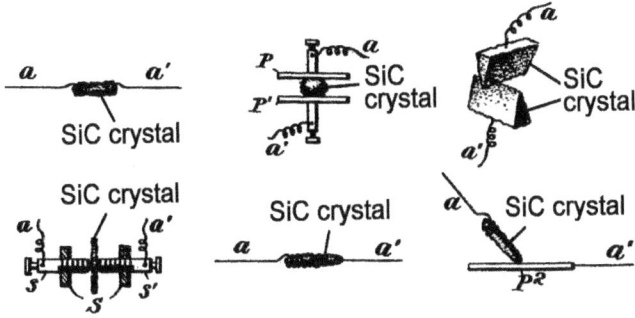

Figure 7. Some kinds of carborundum "wave-responsive devices" patented by General H.H.C. Dunwoody [38]. Public domain image. Source: United States Patent and Trademark Office, www.uspto.gov

device" made of carborundum [38]. Several ways of attaching the electrodes to the crystal covered by this patent are shown in Fig. 7. The main advantage of carborundum was its hardness which allowed to apply high pressure to contacts and get stable repeatable operation of these devices.

The first scientific paper describing electrical characteristics of carborundum detectors was published by George W. Pierce (1872-1956) in 1907 [39]. He investigated a "unilateral conductivity" of carborundum crystals clamped in silver jaws depending on the contact pressure, temperature and contact geometry. He found that some specimens demonstrated not only non-linear but also rectifying *I-V* characteristics as it is shown in Fig. 8. This rectifying behavior became more pronounced when one side of SiC crystal was "platinized" (covered by sputtered platinum). Pierce also demonstrated that the samples with rectifying characteristics (which are SiC point-contact Schottky diodes in modern terminology) could be

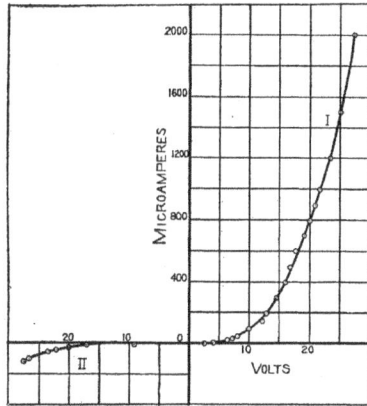

Figure 8. Rectifying I-V characteristics of carborundum detector [39]. Reprinted with permission. Copyright (1907) by the American Physical Society.

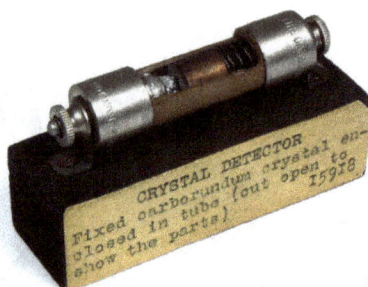

Figure 9.   Commercial packaged silicon carbide detector of the Carborundum Co., circa 1924. The fibre tube is sectioned to expose the SiC crystal soldered to the left contact and a brass plunger and spring connected to the opposite contact. Public domain image, Source: Museums Victoria,
https://collections.museumvictoria.com.au/items/409842

used as a radio detectors without additional voltage bias.

Significant improvements to the technology of metal contacts were made by G. W. Pickard [40] who pioneered the use of electroplated metal films as contacts to a range of semiconductor materials. The use of such contacts made electrical characteristics of the early carborundum detectors much more predictable and reliable. These detectors soon became the dominant radio detector device. Although they were less sensitive than made of galena, they were mechanically stable, did not require additional tuning and were sold adjusted and packaged in cartridges (see Fig. 9).

5.   Electroluminescence of silicon carbide

Following the widespread adoption of Acheson crystal silicon carbide diodes, it did not take long for electroluminescence to be discovered. In 1907, Captain Henry Round (an employee of the Marconi Company and an inventor of 117 patents) reported "a curious phenomenon" observed when a current was passed between two point contacts on a crystal of carborundum [41]. In his short "Note on Carborundum", Round reported that yellowish, green or blue light was observed in some crystals when a potential of 10 Volts was applied. The bright glow was observed at the negative electrode and only in the section facing to the positive electrode. Round correctly identified that this light emission was related to the electric field "produced by a junction of carborundum and another conductor." At that stage in the development of electronics, this light emission was

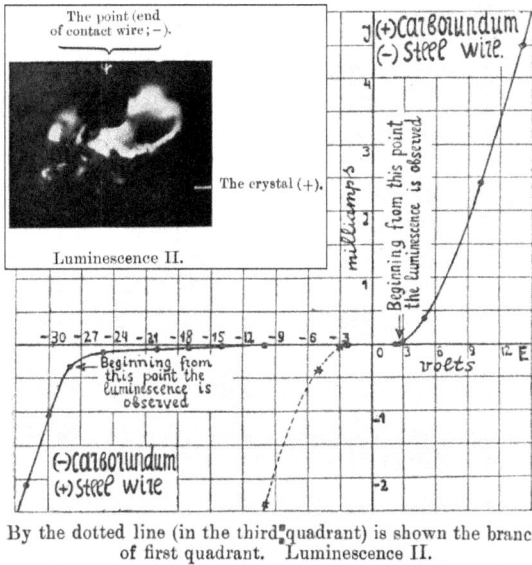

By the dotted line (in the third quadrant) is shown the branch of first quadrant. Luminescence II.

Figure 10. I-V characteristics of Lossev's SiC detector made of carborundum crystal with steel point-contact where starting points of electroluminescence are clearly identified. The SiC electroluminescence at forward bias is shown in the inset [44]. Reprinted with permission. Copyright (1928) by Taylor & Francis Ltd. https://www.tandfonline.com/doi/abs/10.1080/14786441108564683

probably considered too weak (in comparison with incandesced light bulb) for industrial use and no further studies were performed.

Electroluminescence from silicon carbide remained unexplored for nearly 15 years until it was re-discovered by Oleg Lossev (1903-1942), employee of the Nizhniy-Novgorod Radio Laboratory in Soviet Union [42, 43]. In fact, he reported two types of light emission from SiC – the emission that we would now call the "pre-breakdown light emission" under reverse bias and the electroluminescent emission seen previously by Round under forward bias [44].

Fig. 10 shows *I-V* characteristics of a SiC detector made of carborundum crystal with steel point-contact where the electroluminescence conditions are clearly identified. At the same time, Lossev mainly concentrated his efforts on more conventional use of silicon carbide and zinc oxide diodes as a part of tuned radio circuits. He discovered that *I-V* characteristics of some ZnO diodes have a sections with negative slope and designed an amplifying detector based on these diodes, which he named "crystadyne". However, electroluminescence was not completely neglected by him. Lossev realized that this light emission can be modulated to very high frequency and used for optical communication. He continued experiments to investigate the forward bias light emission and correctly identified that it occurred at the boundary between the top layer with "longer mean free path of electrons" and the body of SiC crystal with "shorter mean free path of electrons" (what we would now identify as layers of different doping polarity – although this concept was not known at that time). Lossev also attempted to use the emerging quantum theory to understand the effect and made some progress in connecting the light emission to a reverse photoelectric effect concept. He published 16 papers and obtained 4 patents on light emitting diodes (LEDs), photodiodes and optical recorders of high frequency signals [45].

Obviously, the electroluminescence observed by Lossev was an electroluminescence of SiC *p-n* junctions spontaneously formed in crystals grown by the Acheson method. In industrial terms, silicon carbide LEDs were still not a viable light source until much later.

6.   Silicon carbide varistors

In many countries, DC electrical grids were used in domestic houses in the 1920s and 1930s before AC became dominant. The DC grids were powered from massive neighborhood battery installations and were connected through low impedance connections to the domestic dwellings. To protect these power distribution systems as well as wired telecommunication lines against current surges which may be caused by lightning strikes, a new electrical component was developed. It was varistor, a non-linear resistor which resistance decreases with voltage increasing.

Early versions of varistors were made of copper oxide and copper. They were made obsolete by silicon carbide varistors developed in early 30s by R.O. Grisdale in Bell Laboratories [46]. SiC varistors were made by sintering a mixture of SiC granules and binding compound (graphite and clay). They had repeatable electric characteristics, better mechanical durability, higher voltage rating and higher operation temperature (up to 350 °C). It was found that the non-linear behavior of SiC varistors is caused by voltage dependent resistances of the contacts between SiC grains rather than by changing SiC bulk resistance. Various mechanisms of current conduction through the contacts between SiC crystallites are discussed elsewhere [47].

Although SiC varistors are now replaced in many applications by zinc oxide varistors which have higher nonlinearity and by transient voltage suppression diodes, they still remain highly demanded when high power rating is required and are produced commercially [48].

7. Lely platelets

On 23rd December 1947, John Bardeen and Walter Brattain working at Bell Labs in the US demonstrated the first transistor. This device and following commercial transistors were made of germanium. It has a relatively narrow band gap (E_g = 0.66 eV) which results in high intrinsic charge carrier concentration and low specific resistivity (about 60 $\Omega \cdot$cm at room temperature). Quite soon, it was realized that semiconductors with wider band gap are needed for solid state electronics to compete with vacuum tubes. To that time, silicon was widely used in the metallurgical industry and in 1937 Russell Ohl (discoverer of *p-n* junction) came up with the method for producing high purity monocrystalline silicon using crystal growth from a liquid phase [49]. It is not surprising that silicon became a material of choice for fabricating semiconductor devices for many years ahead due to its properties (E_g = 1.12 eV) and manufacturability.

Silicon carbide was also considered as a material with wider band gap for fabrication of semiconductor devices like recently invented transistors. It was already produced in a large scale and had a good track record for using in crystal detectors. But it was obvious that Acheson process was limited in both size and purity of SiC crystals and a new method needed to be developed for production of SiC wafers with "electronic" quality and usable size. Unfortunately, the crystal growth from a liquid phase was of no use for SiC – there was no suitable solvents and SiC has no appreciable "liquid" phase.

Silicon carbide vapor growth was a natural target and, in 1954, Jan Anthony Lely at Philips, Eindhoven (Netherlands) developed the method of producing high quality SiC crystals by sublimation that had potential for the semiconductor industry [50, 51]. The Lely method essentially used a high purity silicon carbide powder (2 and 3 in Fig. 11) as a raw material which was loaded in the graphite vessel (1 in Fig. 11) by that manner to leave a free sublimation space inside (23 in Fig. 11). The vessel was placed in the graphite heater (17 in Fig. 11) with water cooled copper electrodes (18 in Fig. 11) and heated under argon, hydrogen or carbon monoxide up to 2500 °C to bound the SiC liner and to reach the equilibrium vapor pressure inside the sublimation space. Sublimated SiC vapor (a disassociated mixture of silicon, carbon, silicon dicarbide and disilicon carbide) was redeposited "on other portions of the lining in the form of very pure silicon carbide crystals." The Lely furnace shown in Fig. 11 had the SiC load of 500 grams and the yield of useful crystals was from 30 to 200. These crystals (called nowadays Lely platelets) had low conductivity of *n* type due to residual nitrogen in protective gases. Highly conductive SiC crystals of both types with donor concentration up to 6×10^{19} cm^{-3} and acceptor concentration up to 2×10^{19} cm^{-3} were also grown by this method by adding nitrogen or AlCl$_3$ in the protective gas, respectively. Production of SiC crystals with *p-n* junctions was also demonstrated by changing the vapor pressure of the impurity gases.

Growth of silicon carbide wafer-like crystals with controlled doping by Lely method was a significant improvement over the Acheson process and attracted a great deal of interest. In the wake of these crystal growth developments, the US Air-force called perhaps the first large SiC conference in 1959 in Boston, USA [52]. Attended by nearly 500 delegates, it included William Shockley and many other distinguished figures. Held over 2 days, it included papers on growth (bulk and epi), material properties and characterization and devices. At the conference many excellent results were presented illustrating the competiveness of SiC technology at this stage despite the still inferior nature of the Lely grown crystals compared to those of silicon or germanium.

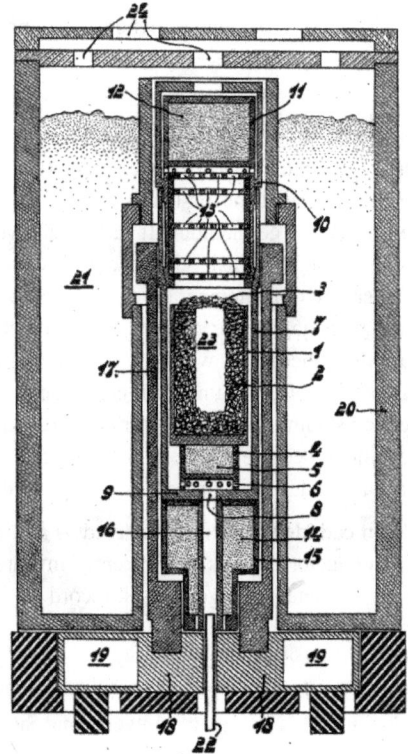

Figure 11. Schematic representation of sublimation furnace for manufacturing silicon carbide crystals [51]. Public domain image, source: United States Patent and Trademark Office, www.uspto.gov

The main disadvantage of the methods for growing SiC crystals based on the Lely technique was that the processes of crystal nucleation and growth were extremely difficult to control. Also, early Lely crystals had one nearly flat face and one resembling a fine staircase.

In the following two decades, extensive study of SiC material properties was performed and SiC processing technology was significantly improved. Many companies developed growth processes based on the Lely methods. A significant program existed in USSR with growth facility concentrated at Podolsk where numerous Lely furnaces were in operation. A number of significant developments were made in Westinghouse Electric

Corp. They had demonstrated growth of platelets with two flat sides at significant crystal sizes which was achieved by controlling the number of nucleation points of crystals by utilizing a graphite liner inside the growth zone [53]. Westinghouse also developed the in-situ doping technology initiated by Lely to enable them to control the electrical conductivity of the crystals. Few 6*H*-SiC Lely platelets with typical sizes and shapes are shown in Fig. 12.

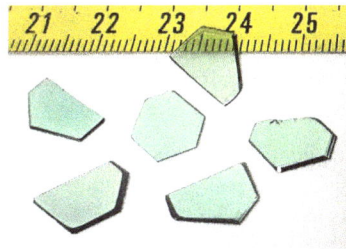

Figure 12. Low doped 6H-SiC wafers grown by the Lely method.

Further development of the Lely growth method was performed by the group of Yury A. Vodakov and Evgeniy N. Mokhov at the Ioffe Institute (Leningrad, USSR). They developed the method for growing long-period SiC polytypes by controlling the temperature and partial pressure of silicon vapors in the growth zone [54]. They demonstrated the reproducible growth of Lely crystals with rare polytypes 8*H*, 15*R*, 21*R*, 27*R* (shown in Fig. 13). These crystals are still commercially available from Nitride Crystals, Inc.

Lely platelets were used both for device research and small scale production of blue and yellow SiC LEDs for specific applications. At this stage, irregular shape of Lely platelets and their maximum sizes not exceeding few square centimeters became the limiting factors in developing industrial SiC electronics.

8.   Silicon carbide bulk crystal growth

The Lely method gave excellent quality crystals but there were many practical problems and it was uneconomic. The search was on for a method that could produce a more typical semiconductor "boule" and research programs existed in several countries. The breakthrough came from the research group at the Leningrad Electro-Technical Institute (LETI), USSR. In 1978, Yuri Tairov and Valeri Tsvetkov modified the Lely furnace by placing source material and a seed crystal in the opposite ends of the sublimation space [55, 56]. The graphite vessel was heated inductively with the temperature gradient about 30 °C to maintain a mass transfer from the source to the seed crystal. This SiC growth method was named as a modified Lely method (the name "LETI method" is also used). In contrast to the Lely method, the crystal growth is going in essentially nonequilibrium conditions in this method and can be carried out in a wide range of temperatures from 2000 to 2700 °C. Tairov and Tsvetkov studied carefully the influence of process parameters such as temperature, temperature gradients and Si/C ratio in the vapor phase [57]. They also investigated the role of the seed crystal and its preparation and were the first to understand the advantages of growing at a surface oriented at a small angled to the basal plane in SiC seed crystal.

The importance of the Tairov and Tsvetkov work cannot be overstated. The modified Lely method enabled to grow SiC boules with controlled polytype which could be sliced into wafers of standard sizes. It represented a major breakthrough on which all subsequent SiC developments have depended.

9. Silicon carbide epitaxial growth

Figure 13. Silicon carbide Lely crystals of rare long-period polytypes. Courtesy of Nitride Crystals, Inc. (http://www.nitride-crystals.com)

Availability of SiC substrates (even with standard sizes) was insufficient for commercial production of SiC devices. The second component of crystal growth, SiC epitaxy needed to be developed. The first attempt was done at the same time when Tairov and Tsvetkov were working on the modified Lely method. The group of Yury Vodakov and Evgeniy Mokhov at Ioffe Institute proposed to use a SiC wafer as a source material in a sublimation growth process [58]. The source and seed wafers were placed face to face separated by a small gap from 0.6 to 6 mm but not exceeding 0.2 of the maximum wafer's size. This growth chamber geometry reduced the loss of sublimated vapors in lateral direction and maintained required partial pressure of silicon close to the equilibrium pressure even at reduced pressure of inert gases in the chamber [59]. This method made it possible to grow epitaxial layers up to 100 μm thick at relatively low temperatures (about 1600 °C). The grown epitaxial layers were of very high crystal quality with dislocation densities below 100 cm^{-2} and can be doped up to the impurity solubility limit. This method (named as a "sublimation sandwich method") is still in use as a laboratory growth method of high quality epitaxial SiC layers but was never commercialized due to difficulties in control of layer thickness and doping profile of epitaxial layers.

The key breakthrough in SiC epitaxy came from Kyoto University, Japan, where the group of Shigehiro Nishino and Hiroyuki Matsunami was working on silicon carbide growth by chemical vapor deposition (CVD) [60]. The first layers were grown on 6H-SiC substrates at temperatures from 1500 to 1750 °C using C_3H_8 and $SiCl_4$ as a source gases and hydrogen as a carrier gas. Although CVD has advantages over the sublimation sandwich method in the precise control of layer thickness and doping, epitaxial films grown by this technique had mosaic structure and mixed polytypes and were not useful for fabrication of SiC devices due to their low crystalline quality [61].

In early 1980s, the same group developed the growth of cubic silicon carbide on silicon substrates by chemical vapor deposition [62]. They used a horizontal cold-wall reactor with graphite RF-heated susceptor. The process was carried out at atmospheric pressure using propane and silane as a source gases and hydrogen as a carrier gas. A 20 nm thick polycrystalline SiC buffer layer was grown *in-situ* before growing silicon carbide. Large area heteroepitaxial 3*C*-SiC layers with uniform thickness up to 34 μm were grown but they were strained and of insufficient crystalline quality due to the difference in thermal expansion coefficients ($2.6 \cdot 10^{-6} °C^{-1}$ for Si vs. $4.67 \cdot 10^{-6} °C^{-1}$ for SiC) and 20% lattice mismatch of SiC layer and Si substrate [63].

In 1987, K. Shibahara, S. Nishino and H. Matsunami discovered that introduction of Si(100) substrate inclination towards (011) effectively eliminates antiphase domains and improves crystalline quality of 3*C*-SiC films [64]. They applied this technique ("step-flow growth on off-oriented substrates") in CVD of 6*H*-SiC on native substrates and demonstrated an epitaxial growth of SiC layers with high crystalline quality inheriting the substrate polytype [65]. The process was carried out in a horizontal cold-wall reactor with graphite RF-heated susceptor at atmospheric pressure and relatively low temperature of 1400-1500 °C. [66]. SiH_4 (1% in H_2) and C_3H_8 (1% in H_2) were used as source gases. Nitrogen and trimethylaluminum (TMA) were used for *n* and *p* type doping, respectively. SiC epitaxial layers where grown on (0001)Si face of 6*H*-SiC substrates off-oriented to the basal plane at the angle of 1.5 - 6° in [11$\overline{2}$0] direction. The substrate surface was etched *in-situ* by hydrogen chloride before CVD growth. The relationship between polytypes of the grown layers and the magnitude and direction of off-orientation angle between the growth surface and (0001) plane in 6*H*-SiC substrate is shown in Fig. 14a. The model of step-flow growth across the terraces on off oriented 6*H*-SiC (0001) substrate is shown in Fig. 14b. The comprehensive overview of step-flow growth technique (also called step-controlled epitaxy) can be found elsewhere [67].

The CVD growth of silicon carbide was further improved in 1993 when the hot-wall reactor was introduced by the group at Linköping University, Sweden, led by Prof Erik Janzén [68]. This reactor comprised an RF-heated susceptor cut from one single block of graphite with a rectangular hole to place substrates. It was insulated from the water cooled quartz walls by high purity graphite felt. This type of reactor made it possible to grow very thick epitaxial SiC layers at higher temperatures with low background doping

Materials Research Forum LLC
https://doi.org/10.21741/9781644900673-1

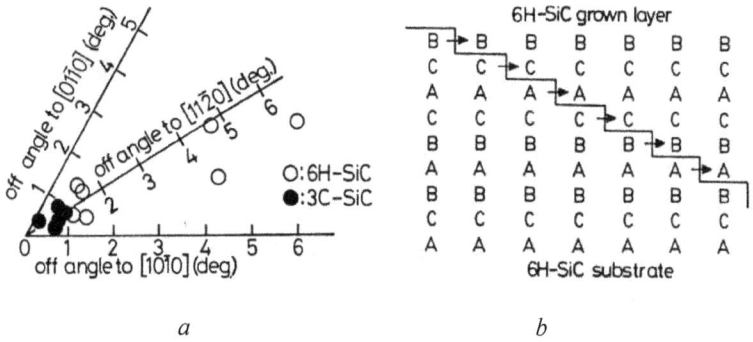

Figure 14. (a) *Polytypes of the grown layers depending on the magnitude and direction of off-orientation angle between the growth surface and (0001) plane in 6H-SiC substrate.* (b) *Model of homoepitaxial growth on off oriented 6H-SiC (0001) substrate* [65]. *Reprinted with permission. Copyright 1987 the Japan Society of Applied Physics.*

(below 2×10^{14} cm^{-3}) [69]. This silicon carbide epitaxial growth technique is called high-temperature CVD (HTCVD) and is indispensable in fabrication of high power and high voltage SiC devices.

10. Emergence of industrial silicon carbide electronics

To the end of 1980s, availability of SiC substrates and epitaxial growth techniques enabled to investigate silicon carbide electrical properties in details, develop core SiC device processing technologies (plasma etching, doping by ion implantation, formation of ohmic and Schottky contacts, oxidation), and demonstrate all basic semiconductor devices made of silicon carbide including *p-i-n* and Schottky diodes, MOSFETs, JFETs, Bipolar Junction Transistors (BJTs), LEDs [70-73]. Irregular shaped Lely platelets were still used for these researches and small scale production of LEDs as the modified Lely method still was not introduced in the industry.

10.1 Foundation of Cree Research, Inc. and the first commercial blue LEDs

The breakthrough came from the North Carolina State University. There was a long-standing program in silicon carbide led by Dr Robert Davis. His team started working on SiC in the early 80s and made rapid progress in establishing production of SiC wafers using sublimation methods derived from the work of Tairov and Tsvetkov but with

Advancing Silicon Carbide Electronics Technology II   Materials Research Forum LLC
Materials Research Foundations **69** (2020) 1-62   https://doi.org/10.21741/9781644900673-1

numerous improvements. In 1987, the North Carolina group filed US and Japanese patents that were, on approval, to play a crucial underpinning role in future developments [74]. The patent highlighted the role of technological issues such as the polytype of the source material, using source materials that were not single crystals, controlling the surface area and particle size distribution of the source materials and controlling the thermal gradient between source and seed continuously to maintain optimal conditions. These developments (and subsequent further refinements) enabled the growth of high quality SiC substrates of increasing area. In 1987, the group demonstrated $6H$-SiC bulk growth capability in the range of 10 mm diameter. To this time, the group also had developed SiC epitaxy and was able to demonstrate some good devices.

In the summer of 1987, based on these results, they decided to form Cree Research, Inc. to commercialize silicon carbide electronics. The company started with selling blue LEDs in 1988. The overview of development, fabrication, characterization of $6H$-SiC blue LEDs, green LEDs and UV photodiodes at Cree Research, Inc. can be found elsewhere [75]. Since that time up to now, Cree dominates silicon carbide market and is the leader in silicon carbide technology research and development.

10.2   Industrial SiC wafer growth

In 1991, Cree Research, Inc. released to the market the first commercial $6H$-SiC wafers with standard 1 inch diameter. The overview of SiC bulk growth in Cree Research, Inc. can be found elsewhere [76, 77].

In general, silicon carbide wafers grown by the modified Lely method have lower crystalline quality than that one of Lely platelets [78]. The most common crystallographic defects in SiC wafers are micropipes, $3C$ polytype inclusions, threading edge dislocations (TED), threading screw dislocations (TSD), stacking faults (SF), grain boundaries. The most distinctive defects in SiC wafers are micropipes which are not found in Lely platelets [79]. They are the screw dislocations with large Burgers vectors propagating along the c-axis through the whole wafer and subsequently grown epitaxial layers. They have a hollow cores with diameters from about 0.1 μm to 5 μm and even a single micropipe found in an active device area is detrimental to the performance of high power and high voltage devices. SiC wafers fabricated in early 1990s had micropipe densities (MPD) from 100 to 1000 cm^{-2}. It was not a big issue since the SiC wafers were used for fabrication of low voltage blue LEDs and for the demonstration of the technological feasibility to produce high power devices. However commercialization of SiC power devices required high yield fabrication of high voltage devices with active areas exceeding 0.1 cm^2 which was not achievable on wafers with MPD > 100 cm^{-2}. This demand of high crystal quality SiC wafers stimulated extensive research in optimization of modified Lely growth method in the second half of 1990s and early 2000s. Cree, Inc. reported SiC wafers with diameters of 25 mm and free of micropipes in 2001 [77]. In 2004, D. Nakamura *et al.* reported the 'repeated a-face' (RAF) growth process [80] which allowed them to grow 4H-SiC wafers, 2.0 inches in diameter, having zero MPD and total dislocation density of 75 cm^{-2}. In 2007, Cree, Inc. demonstrated 4 inch 4H-SiC wafers with zero MPD.

Nowadays, single crystal SiC substrates at sizes up to 150 mm diameter and MPD < 1 cm^{-2} are manufactured commercially. Currently, 150 mm 4H-SiC substrates are the standard choice for high volume production of MOSFETs and Schottky diodes although 200 mm SiC wafers have been already demonstrated and are under development. 6H-SiC and 4H-SiC wafers with various levels of p and n type doping and wafer diameters ranged from 2 to 6 inch are available on the market. Alongside this volume production of SiC wafers with p- and n-type doping, there is a steady industrial development of semi-insulating (SI) substrates (including the use of CVD based crystal growth) although at far smaller volume. These SI substrates are predominantly aimed at development of high power SiC RF devices as well as at using as substrates for GaN RF devices and for graphene growth. Commercial silicon carbide substrates with typical doping levels are shown in Fig. 15.

The global production of the silicon carbide wafers reached 453,000 pcs in 2017 [81] and continues to increase rapidly. The main SiC wafer suppliers are Wolfspeed (a Cree company), SiCrystal, TankeBlue Semiconductor Co., Dow Corning, II-VI Inc., Norstel AB. Some of these companies offer also a growth of epitaxial structures with bespoke doping profiles (Wolfspeed, Dow Corning, Norstel AB).

10.3   Preconditions and demands for SiC power electronics

In 1994, SiC LEDs were made obsolete by much brighter LEDs based on III-V nitride semiconductors which had more than 1000 times higher external quantum efficiency due to direct band gap structure. Green and blue LEDs with InGaN/AlGaN epitaxial structures grown on sapphire substrates were demonstrated by Shuji Nakamura [82] and commercialized by Nichia Chemical Industries in 1994. Six months later, Cree, Inc. released to the market blue LEDs with nitride device structures grown on SiC substrates. The advantage of using SiC substrates is originated from closer SiC lattice constant matching with GaN and its higher thermal conductivity than that one of sapphire. Furthermore, electrical conductance of SiC made it available to design LEDs with the vertical current flow geometry and, hence, lower resistance while sapphire is a very good insulator. Currently, SiC substrates are used for fabrication of highly efficient and high power nitride LEDs. It is worth noting that only using SiC substrates in large quantities

Figure 15. Commercial 3 inch silicon carbide wafers. From the left to the right: Semi-Insulating 6H-SiC wafer with resistivity more than 10^6 $\Omega \cdot cm$; moderately doped 4H-SiC wafer with n-type of conductivity and resistivity of 0.032 $\Omega \cdot cm$; heavily doped 4H-SiC wafer with n-type of conductivity and resistivity of 0.018 $\Omega \cdot cm$.

for mass production of SiC and III-V nitride based LEDs was the main driving force of SiC wafer size increasing and dramatic improvement of their crystal quality in 1990s.

With commercialization of LEDs based on III-V nitrides, it became apparent that the most feasible field of application for SiC devices is power electronics. Indeed, in the late 1990's and the first years of the twenty first century, new measures were introduced to tackle climate change and reduce the carbon footprint. They included not only increasing energy generation from renewable sources but also reduction of energy losses during electric power transmission, conversion and use. This stimulated a great demand for highly efficient power electronics to use in solar and wind power generation, lighting, railway traction, power factor correction, induction heating, uninterrupted power supply and many other applications [83]. Furthermore, in 2000, Toyota released to the worldwide market the first mass-produced hybrid electrical vehicle (HEV) which required efficient, compact and low weight electrical motor driver and battery charger. Almost all these applications needed semiconductor power devices with blocking voltages ranking from 600 V to 6.5 kV. The market in this voltage range was dominated by silicon super junction MOSFETs and insulated gate bipolar transistors (IGBTs). To that point in time, silicon power devices were under development for more than 50 years and approaching their performance limitations. By this reason, the interest to silicon carbide as a new material for high power and high voltage semiconductor devices resurrected due to its outstanding material properties and relatively matured growth and processing technology.

Materials Research Forum LLC
https://doi.org/10.21741/9781644900673-1

10.4 4*H*-SiC polytype as a material for power electronics

While SiC blue LEDs where produced on 6*H*-SiC substrates, highly efficient SiC power devices have to be made of 4*H*-SiC polytype since it has an electron mobility which is almost isotropic [84] and about three times higher than that one in 6*H*-SiC [85]. In 1994, Cree Research, Inc. released to the market the first 4*H*-SiC wafers with standard diameter of 1 3/16 inch (~30 mm) and in 2001, they made commercially available 3 inch substrates of both 6*H* and 4*H* polytypes. Currently, Wolfspeed (a Cree Company) offers only 4*H*-SiC wafers (of 100 and 150 mm diameters) as a standard commercial product.

To the middle of 1990s, the 4*H*-SiC polytype was adequately investigated and almost all its material parameters were measured [35]. The only parameters which remained not sufficiently understood were electron and hole impact ionisation rates. The precise dependence of these rates on electric field is required for design of power devices, specifically to calculate the avalanche breakdown voltage (V_B) which corresponds to the maximum blocking voltage (V_{BL}) in power devises. To that point in time, impact ionisation rates were measured only in 6*H*-SiC polytype [86], [87]. It was also known that *p-n* junctions formed on (0001) plane of 6*H*-SiC had negative temperature coefficients of V_B [88] making this polytype barely usable for fabrication of power devices with continuous mode of operation and recoverable breakdown. Only in 1997, ionisation rates were measured in 4*H*-SiC by A. Konstantinov *et al.* [89, 90]. They measured ionisation rates in p^+-*n* structures grown by CVD techniques on commercial 4*H*-SiC substrates off-oriented from the basal plane by 3.5°. They also shown that, in contrast to 6*H*-SiC, the *p-n* junctions formed on (0001) plane of 4*H*-SiC have positive temperature coefficients of V_B [91]. This crucial advance in 4*H*-SiC characterization confirmed that this polytype is the most suitable for fabrication of SiC power devices. Using ionisation rates measured by Konstantinov *et al.* in analytical models and numerical simulations made accurate design of SiC devices available. Solid lines in Fig. 16 show avalanche breakdown voltages of 4*H*-SiC p^+-n^--n^+ structures calculated for different blocking layer thicknesses using impact ionisation rates reported by A. Konstantinov *et al.* [89]. For comparison, blue dashed lines in this figure show breakdown voltages of Si p^+-n^--n^+ junctions calculated using expressions from [25].

10.5 4*H*-SiC unipolar power devices

As it was discussed in Section 3.6 of this chapter, SiC unipolar power devices have a great advantage over their silicon counterparts due to significantly lower conduction and switching losses. These devices include JBS and Schottky diodes, JFETs and MOSFETs. Furthermore, they also outperform silicon and silicon carbide bipolar devices when used in power applications with voltage ranking from 600 V to 6.5 kV [92]. By this reason and

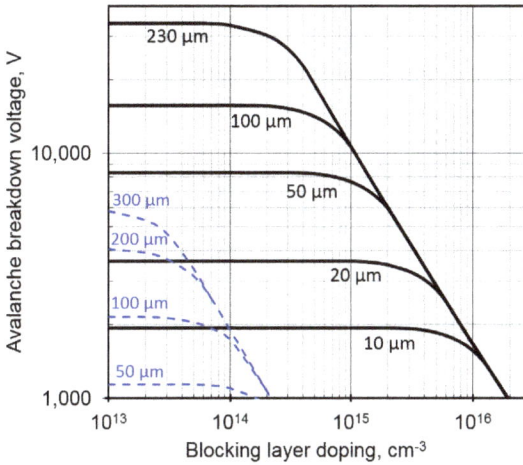

Figure 16. Avalanche breakdown voltage of Si (dashed lines) and 4H-SiC (solid lines) p^+-n^--n^+ structures as a function of n^--layer doping level with the n^--layer thickness as a parameter.

in view of perspective high volume market for these devices, development of 4*H*-SiC unipolar devices became a mainstream in silicon carbide research in the end of 1990s and the first decade of twenty first century.

10.5.1 4*H*-SiC power Schottky diodes

The first high voltage SiC Schottky diodes with breakdown voltages above 400 V were reported by the group of B. Javant Baliga from North Carolina State University in 1992 [93]. These were Pt/6*H*-SiC diodes on 10 µm thick epilayers doped to 4×10^{16} cm^{-3}. As it was explained in Section 10.4, 4*H*-SiC polytype is more suitable for fabrication of power devices. Furthermore, it was obvious that efficient device edge termination needs to be developed for further increasing of blocking voltages. In 1994, Akira Itoh, Tsunenobu Kimoto, and Hiroyuki Matsunami from Kyoto University investigated different metals (Au, Ti and Ni) as Schottky contacts to 4*H*-SiC and demonstrated low leakage Ti/4*H*-SiC rectifiers with blocking voltages up to 800 V on 10 µm thick epilayers doped to 5×10^{15} cm^{-3} [94]. One year later, they demonstrated small area Ti/4*H*-SiC Schottky rectifiers on 10 µm thick epilayers doped to 7×10^{15} cm^{-3} with almost 100% efficient edge termination by boron implantation and blocking voltages up to 1750 V [95].

Up to now, the JBS diodes with highest blocking voltages were demonstrated by Y. Jiang *et al.* [96]. The JBS diodes with Ni Schottky contacts and the edge termination by floating field rings were fabricated on a 4-inch 4H-SiC wafer with 100 μm thick epitaxial layer doped to $2.7{\times}10^{14}$ cm^{-3}. Diodes with 0.1 cm^2 active area were characterized at forward currents up to 4.5 A at 200 °C and demonstrated reverse blocking voltages exceeding 8 kV at temperatures up to 125 °C.

In 2001, Infineon released to the market the first SiC Schottky diodes with blocking voltages up to 600 V [97, 98]. Today, SiC Schottky and JBS diodes are commercially available from many suppliers and diodes with maximum rating of 1700 V/50 A are offered by Wolfspeed [99].

10.5.2 4H-SiC power JFETs

With successful commercialization of SiC Schottky and JBS diodes, it became apparent that SiC switching devices were also needed to fully realize the potential of silicon carbide as a material for high power electronics.

The wider bandgap of SiC than that one of Si means that the built-in voltage is higher and the space charge region (SCR) is wider in SiC p^+-n junction than that ones in Si. This makes it possible to fabricate normally-off SiC JFETs which have a sufficiently thin channel to be punched-through by SCR (completely depleted) at zero gate bias. As far as the channel is physically located inside the semiconductor, JFETs are free of problems arising from the oxide reliability and dielectric strength. Furthermore, electrons in a JFET's channel should ideally have a low-field mobility very close to its value in a bulk crystal since there is no surface or SiC/SiO$_2$ interface carrier scattering in these devices [100].

The first 4H-SiC JFETs were reported by P. Ivanov *et al.* in 1993 [101]. That were normally-on devices with buried p^+-gate and n-channel layer (1.2 μm thick; $1.8{\times}10^{17}$cm^{-3}) grown by the sublimation epitaxy on 4H-SiC substrates cut from a single crystal boule grown by the LETI method. The substrate surface was oriented parallel to the c-axis. The channel height was defined by the reactive ion etching of the top n-layer. Devices with the gate length of 9 μm and channel width of 0.7 mm operated at temperatures up to 400 °C and were characterized at currents up to 80 mA and voltages up to 60 V. The electron low field mobility of 340 cm^2/(V·s) was extracted from device characteristics at room temperature. The electron mobility in JFETs with epitaxial channels oriented parallel the basal plane was reported by P. Sannuti *et al.* in 2005 [102]. They measured channel mobility of 398 cm^2/(V·s) in lateral JFETs with channel doping of $1{\times}10^{17}$cm^{-3}. Both these mobility values (along and perpendicular to the c-axis) are very well compared to the bulk electron mobility in $4H$-SiC at this doping level (~ 400 cm^2/(V·s)).

The first power $4H$-SiC JFETs were reported by H. Mitlehner *et al.* in 1999 [103]. That were vertical JFETs (VJFETs) with planar channel which schematic cross-section is shown in Fig. 17a. These devices were fabricated by selective Al implantation into thick $4H$-SiC drift layer to form a p^+ well buried gate followed by the second epitaxy to grow 2.5 μm high channel with 7×10^{15}cm^{-3} doping. Then, the top p^+ gate was formed by the second Al implantation. Fabricated VJFETs with a 15 μm thick blocking layer doped to 4.5×10^{15} cm^{-3}, a channel width of 32 cm and an active area of 4.1 mm^2 operated at currents up to 5 A with specific on-resistance resistance $R_{ON\text{-}SP} < 15$ mΩ·cm^2. They were capable to block 1800 V at the gate-source voltage (V_{GS}) of -20 V. These devices were used in a cascode circuit with low voltage Si MOSFET to operate as normally-off switches [104].

The first $4H$-SiC power VJFETs with vertical channel were reported by H. Onose *et al.* in 2001 [105]. The schematic cross-section of these devices is shown in Fig. 17d. The gate p^+-n junctions were formed by deep ion implantation (about 2 μm) into 20 μm thick epitaxial layer doped to 2.5×10^{15} cm^{-3}. Fabricated VJFETs demonstrated normally-off operation but required $V_{GS} = -50$ V to block 2000 V. These devices had relatively high $R_{ON\text{-}SP}$ (~ 70 mΩ·cm^2) due to low channel mobility caused by defects resulted from the very high energy implantation (up to 1.3 MeV) needed to form deep p^+ gate regions.

Deep high energy implantation was eliminated from SiC VJFET processing by J. Zhao *et al.* in 2003 [106]. They reported $4H$-SiC trenched and implanted VJFETs (TI-VJFETs) with gate p^+-regions formed by ion implantation in sidewalls of deep trenches etched in a drift epitaxial layer. The schematic cross-section of these devices is shown in Fig. 17e. The shallow low energy ion implantation in the sidewalls has enabled to form long channels with precisely controlled uniform opening width in these devices. Fabricated TI-VJFETs with 9.6 μm thick $4H$-SiC drift layer doped to 6.5×10^{15}cm^{-3} and long vertical channel with opening of 0.63 μm demonstrated normally-off operation with $R_{ON\text{-}SP} =$ 3.6 mΩ·cm^2 and maximum $V_{BL} = 1726$ V at $V_{GS} = 0$ V. The electron channel mobility of 561 cm^2/(V·s) was extracted from device characteristics. This channel mobility value probably is the highest reported for SiC JFETs although it is noticeably lower than the bulk electron mobility in $4H$-SiC at the same doping level (~ 850 cm^2/(V·s)) [85].

TI-VJETs were intensively developed and in 2008, Y. Li *et al.* reported normally-on devices with $V_{BL} = 1650$ V and record $R_{ON\text{-}SP} = 1.88$ mΩ·cm^2 [107]. At the same time, normally-on and normally-off SiC TI-VJFETs were brought to the market by SemiSouth Laboratories, Inc. The normally-off TI-VJFETs designed for 800 V applications had

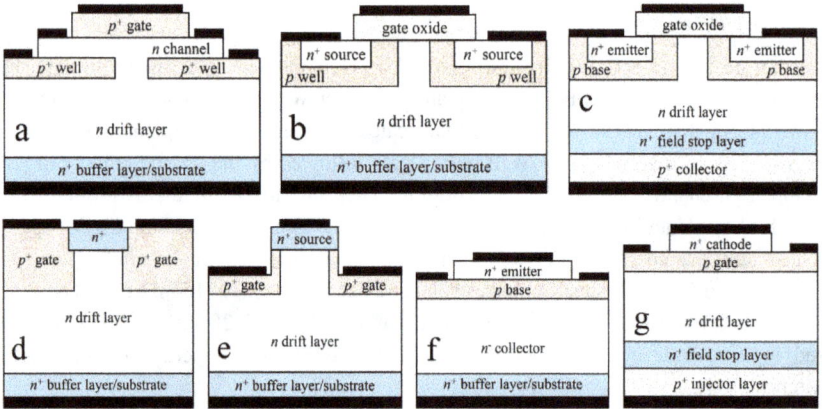

Figure 17.   Schematic cross-sections of different SiC switching devices (not in scale). (a) VJFET with planar channel; (b) DMOSFET; (c) IGBT; (d) VJFET with implanted gate; (e) TI-VJFET; (f) BJT; (g) GTO thyrystor.

R_{ON-SP} < 2.9 mΩ·cm^2 and < 6.6 mΩ·cm^2 when measured at 25 °C and 200 °C, respectively; devices designed for 1200 V applications had R_{ON-SP} < 4.3 mΩ·cm^2 at 25 °C and < 12.8 mΩ·cm^2 at 200 °C. [108].

SiC JFETs demonstrated very low on-resistance but their commercialization was hampered by the unconventional and very complex device fabrication which is not compatible with Double Diffused Metal Oxide Semiconductor (DMOS) processing, the major power FET technology in silicon electronics. Characteristics of SiC JFETs are very sensible to the channel doping level which needs to be strictly controlled and device processing needs to be tuned individually for each epi-wafer depending on its doping. SiC JFETs require specially designed gate driving circuit since the blocking voltage depends on the gate-source bias. Nevertheless, normally-on SiC TI-VJFETs rated up to 1700 V/8 A/400 mΩ and 1200 V/120 A/9 mΩ are currently commercially available from United Silicon Carbide, Inc. [109].

10.5.3 4*H*-SiC power MOSFETs

SiC MOSFETs have many inherent advantages over JFETs. They have the highest industry acceptance due to simple power and circuit designs. MOSFETs are voltage controlled devices and do not require any special gate voltage shape unlike in the case of JFETs.

The first detailed study of Al/SiO$_2$/SiC MOS structures was reported by A. Suzuki *et al.* in 1982. [110] SiO$_2$ layers were grown on the C face of 6*H*–SiC Lely platelets at 850 – 1100 °C in wet oxide. The accumulation, depletion and inversion regions were clearly observed. The minimum surface-state density was found to be about 2×10^{12} cm^{-2}eV^{-1} and the oxide breakdown strength of 2×10^6 V/cm.

For many years, successful development of SiC power MOSFETs was hampered by low quality of silicon oxide either grown or deposited on silicon carbide. High density of interfacial states near the SiC conduction band edge and charge trapping in silicon oxide resulted in electrons Coulomb scattering at the interface leading to low channel mobility and unacceptable high on-state resistivity. The field effect mobility of electrons in inversed layers was typically measured not exceeding 20 cm^2/(V·s) in 6*H*-SiC and about 5 cm^2/(V·s) in 4*H*-SiC devices. Furthermore, fabrication of SiC MOSFETs was incompatible with DMOS technology due to very low diffusion coefficient of doping impurities in silicon carbide at any reasonable temperature.

In 1997, the first high-voltage SiC MOSFET with vertical structure (shown in Fig. 17b) adopting DMOS process was reported by Shenoy *et al.* [111]. Both n^+ source regions and *p*-type wells were formed by ion implantation (instead of double diffusion used in silicon DMOS process) in 10 μm thick 6*H*-SiC drift layer doped to 6.5×10^{15} cm^{-3}. Both implants were activated simultaneously by annealing at 1600 °C for 30 min in an argon ambient. 60 nm thick gate oxide was grown at 1150 °C in wet oxygen. Fabricated devices shown threshold voltages between 8 and 10 V and were capable to block voltages up to 760 V. Channel mobilities up to 26 cm/(V·s) were measured in these double implanted MOSFETs (DIMOSFETs).

A satisfactory solution of the low channel mobility problem in SiC MOSFETs also became from the silicon technology. Since 1980, thermal nitridation of silicon dioxide by annealing in nitrogen containing gases was used in silicon MOS technology to produce thin gate insulators [112]. It was well known that the nitridation effectively reduces the interface trap density in silicon MOS devices [113]. In 1997, the group headed by Sima Dimitrijev at Griffith University, Australia, reported that post oxidation annealing in nitric oxide (NO) significantly reduced the interface defect density near the conduction band edge on 6*H*-SiC [114]. They grew 3.5 nm thick SiO$_2$ on commercial 6*H*–SiC wafers with nitrogen doping up to 4.8×10^{17} cm^{-3} and annealed it in NO at 1100 °C for 5 min. The authors found that the total trap density at the SiO$_2$/6*H*–SiC was reduced from 1.6×10^{10} cm^{-2} to 8×10^9 cm^{-2} by this annealing.

In 2001, the post oxidation annealing in nitric oxide was employed in SiC MOSFETs processing for the first time by Chung *et al.* [115]. The authors reported a substantial increase of inversion layer electron mobility in lateral 4H-SiC MOSFETs fabricated on commercial epiwafers with p-type doping of 4×10^{16} cm^{-3}. The 40 nm thick gate oxide layer was grown by a combination of thermal oxidations in dry and wet oxygen and then annealed in NO at 1175 °C for 2 hours. The increase of inversion layer electron mobility from 5 to 37 cm^2/(V·s) was discovered in NO annealed devices. This was one of the most important milestones in the development and commercialization of SiC power MOSFETs as NO annealing became a standard processing step in fabrication of these devices.

In 2005, the next important breakthrough in improvement of SiO$_2$/4H-SiC interface was made by G. Gudjónsson *et al.* [116]. The authors applied annealing in nitrogen ambient in fabrication of 4H-SiC DIMOSFET with ion implanted gate channels and extracted the peak field effect mobility of 100 cm^2/(V·s) in transistors with Al concentration of 1×10^{17} cm^{-3} and 51 cm^2/(V·s) in transistors with Al concentration of 5×10^{17} cm^{-3}.

Although the field effect mobility in SiC MOSFETs remained to be a much smaller fraction of the bulk mobility than it is in Si devices (about 10% for SiC compared to 50% for Si), it was high enough to fabricate SiC MOSFETs with blocking voltages exceeding 600 V and with channel resistances much lower than the resistance of the drift layer. The effect of channel mobility on DIMOSFET device performance depending on blocking voltage is shown in Fig. 10 of Chapter 4 in the first volume of this book [92].

In 2011, Cree, Inc. launched the first commercial silicon carbide power MOSFET [117] . Today, SiC MOSFETs rated up to 1700 V/72 A (with on-resistance of 45 mΩ) are commercially available from Wolfspeed [99]. The SiC power MOSFETs with record breakdown voltages were reported by John Palmour in 2014 [118]. 4H-SiC DIMOSFETs with die size of 64 mm^2 exhibited $R_{ON\text{-}SP}$ = 208 mΩ·cm^2 and V_B = 15.5 kV. The fabricated devices were tested at currents up to 10 A and were capable to switch efficiently at frequency of 20 kHz and above. Detailed reviews of current status of SiC power MOSFETs, their integration with SiC Schottky diodes and potential for their future development can be found elsewhere [119] [120].

10.6 Development of 4H-SiC power bipolar devices

Bipolar semiconductor devices can be more efficient at higher blocking voltages than the unipolar ones. In these devices, minor carriers are injected in the thick drift layer allowing the charge carrier density to exceed the drift layer doping level for several times (so called conductivity modulation). For sufficiently high blocking voltage ratings defined by the thickness of a drift layer (V_{BL}> 600 V for Si and V_{BL}> 6000 V for SiC), reduction of a drift layer resistivity resulted from the conductivity modulation compensates addition resistance of a p^+ layer and a p-n junction built-in voltage drop (in devices with the odd number of p-n junctions) and bipolar devices become more efficient than the unipolar ones. The main condition for the effective conductivity modulation is that the injected minority carriers have to have the lifetime long enough to drift through the full length of a blocking layer (about 40 μm for V_{BL}= 6000 V in a 4H-SiC p-i-n junction). That was the first pitfall on the way to SiC power bipolar devices because of the first SiC epitaxial layers grown in early 1990s suffered from very low minority carrier lifetimes.

10.6.1 Minority carrier lifetime enhancement

Key lifetime limiting factors in epitaxial layers grown by different methods in early 1990s were compensating dopant species; unintentional metallic impurities; thickness and doping uniformity of thick epitaxial layers [121]. The maximum minority carrier (hole) lifetime of 105 ns was measured in 10 μm thick low doped (2×10^{16} cm^{-3}) n-type 6H-SiC epitaxial layer. This lifetime corresponded to the diffusion length of about 6-8 μm [122]. The minority carrier (electron) lifetime in low doped p-type 4H-SiC epitaxial layer was even lower and not exceeding 80 ns [123].

The first significant improvement of epitaxial layers crystal quality and substantial increase of a minority carriers lifetime happened in the second half of 1990s with development of HTCVD growth method [68]. Minority carrier (hole) lifetimes changing from 0.6 to 3.8 μs at temperatures from 20 to 277 °C and corresponding to 16 and 22 μm diffusion length were measured in 50 μm thick low doped n-type 4H-SiC layer grown by this method [124]. In 2001, the highest minority carrier lifetime of 2 μs measured at RT in 40 μm thick 4H–SiC n-type (10^{16} cm^{-3}) epilayers grown on a p-type substrate was reported by P. Grivickas in 2001 [125]. This lifetime corresponded to the diffusion length of about 30 μm (for ambipolar diffusion coefficient of 4 cm^{2}/s) [126] but still remained too low for conductivity modulation of thick drift layers required for the development of high voltage bipolar SiC devices (100 ÷ 200 μm).

Almost miraculous solution of this problem came from Japan in 2007. To that time, it was known that the point defects with energy levels located for 0.67 eV below the conduction band edge (so-called Z$_{1/2}$ centers) are the main lifetime killers in thick 4H-SiC epitaxial layers [127]. They act as recombination centers resulting in reduction of the carrier lifetime. The typical concentration of these centers was about 1×10^{13} cm^{-3}. In 2006, Liutauras Storasta and Hidekazu Tsuchida from Central Research Institute of Electric Power Industry (CRIEPI), Nagasaka, reported the two times increase of carrier lifetime in 30 μm thick n-type 4H-SiC epitaxial layers after carbon ion implantation into the shallow surface layer (~ 250 nm) and subsequent annealing at 1600 °C for 10 minutes [128]. They also observed by deep level transient spectroscopy that the traps' density reduced from 3×10^{13} cm^{-3} to below the detection limit (< 5×10^{11} cm^{-3}) in the material underneath the implanted layer for the depth of 4 µm. The authors suggested that the $Z_{1/2}$ centers are associated with carbon vacancies and carbon atoms from the implanted layer diffuse through these vacancies during the post-implantation annealing into the semiconductor bulk and effectively "repair" these defects. It is worth to note that this incredibly fast carbon diffusion for a huge depth is going through the vacancies and the concentration of diffused atoms cannot exceed the concentration of pre-existing vacancies. For comparison, the diffusion depth of carbon impurities after annealing at 1600 °C for 10 minutes calculated with use of "conventional" diffusion coefficient [129], [130] is only about 40 nm.

In 2009, Toru Hiyoshi and Tsunenobu Kimoto from Kyoto University proposed to replace the implantation and subsequent annealing by a single processing step of thermal oxidation [131]. In the case, the carbon interstitial atoms are generated at the SiO_2/SiC interface during the oxidation. The authors demonstrated the reduction of $Z_{1/2}$ centres concentration below the detection limit for a depth of about 50 µm from the surface in a 100 µm thick n-type 4H-SiC epilayer after thermal oxidation at 1300 °C for 5 hours and carrier lifetimes increasing from 0.73 µs to 1.62 µs after thermal oxidation at 1300 °C for 10 hours.

Since the publication of Toru Hiyoshi and Tsunenobu Kimoto in 2009 [131], the lifetime enhancement thermal oxidation has become a standard step in processing SiC power devices. In 2012, K. Kawahara *et al.* reported the minority carriers lifetime increase from 0.6 to 6.5 µs in a n-type 4H-SiC layer (95 µm thick, 2×10^{15} cm^{-3}) after oxidation at 1400 °C for 16.5 hours [132]. Recently, S. Ryu *et al.* reported carrier lifetimes ranged from 15 µs to 20 µs (corresponding to the diffusion length of 90 µm at ambipolar diffusion coefficient of 4 cm^2/s) in a n-type 4H-SiC layer (140 µm thick, 2×10^{14} cm^{-3}) measured after the lifetime enhancement thermal oxidation at 1450 °C for 5 hours [133].

10.6.2 Suppressing stacking faults

The second huge obstacle on the way to the SiC power devices was identified in 2000. H. Lendenmann *et al.* reported at the European Conference on Silicon Carbide and Related Materials (ECSCRM-2000) that the voltage drop in 4H-SiC p-n junction diodes anomalously increased during their operation at a forward bias [134]. It was observed that the triangular planar defects with edges along <11$\bar{2}$0> directions appeared and expanded in SiC epilayers concurrently with the degradation of I-V characteristics. These defects were interpreted as stacking faults (SF) lying in basal planes of silicon carbide [135] [136]. It was found that the energy of electron-hole recombination is high enough to induce the stacking faults nucleation and expansion. This effect is more noticeable in high voltage SiC devices where the drift layer is thick. Since the carrier recombination is a fundamental process in bipolar semiconductor devices and cannot be avoided, the development of SiC power devices seemed desperate. The news was so shocking that some companies abandoned their SiC research and development programs.

Tremendous efforts were spent to overcome this problem. First, it was found that stacking faults originate from basal plane dislocations (BPDs) presented in a SiC drift layer and in a substrate near the interface with the epitaxial layer. In 2004, T. Ohno *et al.* showed that the propagation of BPDs from a 4*H*-SiC substrate to the epilayer can be suppressed by optimization of the epitaxial process (C/Si ratio, growth temperature and growth rate) and demonstrated the 4*H*-SiC layers with BPD density well below 10^3 cm^{-2} [137].

In 2005, J. Sumakeris *et al.* from Cree reported that BPDs can be converted to threading edge dislocations (TEDs) which do not act as nucleation sites for SFs by growing SiC epilayers on the Si face of SiC substrates etched in molten potassium hydroxide (KOH) [138]. This etching is highly anisotropic and produces the BPD etch pits to enable conversion of the BPDs into TEDs by promoting lateral growth at the narrow sector of BPD etch pits. One year later, Cree demonstrated the 4*H*-SiC layers with BPD density well below 10 cm^{-2} grown on KOH etched 4*H*-SiC substrates [139].

Further improvement of SiC epitaxial process was reported by W. Chen and M. Capano from Perdue University in 2005 [140]. They demonstrated that growing of epitaxial layers on low off-cut SiC substrates significantly enhances BPD conversion into TEDs. A BPD density of 2.6 cm^{-2} was achieved in 20 μm thick epitaxial layers grown on 4° off-cut angle 4*H*-SiC substrates (without KOH pre-etching) which is 100 times lower than the BPD density measured in epi-layers grown on the 8° off-angle 4*H*-SiC substrates. Since then, the 4° off-cut angle 4*H*-SiC substrates became a new standard in commercial SiC wafers.

Stacking faults nucleation sites can be introduced also by post-growth device processing. An example of a SF generation and propagation from a defect in a mesa-structure sidewall formed by reactive ion etching is shown in Fig. 19, Chapter 4 of this Volume. This source of SFs can be effectively reduced by optimization of device processing. Indeed, Y. Bu *et al.* reported bipolar degradation free 6.5 kV *p-i-n* diodes with active area of 0.22 cm^2 and 65 μm thick 4*H*-SiC drift layer (doped to 1×10^{15} cm^{-3}) grown on 4° off-cut commercial substrates [141]. To suppress the BPDs generation the Al implanted dose in junction termination extension was kept below 1×10^{15} cm^{-2} and damaged layer after mesa etching was removed by sacrificial oxidation at 1350 °C for 300 min.

10.6.3 Advanced 4*H*-SiC power bipolar devices

Resolving the problems with stacking faults formation and low minority carrier lifetime paved the way to successful development of 4*H*-SiC power bipolar devices. In 2012, H. Niwa *et al.* [142] reported a SiC *p-i-n* diode with a breakdown voltage of 21.7 kV and 186 μm thick drift layer doped to 2.3×10^{14} cm^{-3}. It was the highest breakdown voltage among any semiconductor devices reported to the time of publication. Differential on-resistance and voltage drop at 50 A/cm^2 were 63.4 mΩ·cm^2 and 9.3 V, respectively. The on-resistance of a Schottky barrier diode using the same epilayer was 592 mΩ·cm^2, indicating effective conductivity modulation of the drift layer.

A BJT schematic cross-section is shown in Fig. 17f. The first SiC BJTs were reported by W. v. Münch and P. Hoeckin in 1978 [73]. These devices were fabricated on *n*-type 6*H*-SiC Lely platelets with doping level of 5×10^{18} cm^{-3}. The base Al doped *p*-type layer (0.8 μm, 4×10^{17} cm^{-3}) was grown directly on the substrate. Devices with an emitter area of 200×200 μm^2 were characterized at currents up to 0.8 mA and capable to block up to 50 V. The current gain was measured from 4 to 8 corresponding to a minority carrier lifetime of 5 nsec and an electron mobility of 140 cm^2/(V·s).

The first high voltage 4*H*-SiC BJT was reported in 2000 [143]. The fabricated devices with a 10 μm thick drift layer doped to 1.2×10^{16} cm^{-3} and collector area of about 1.6×1.6 mm^2 demonstrated a current gain of 9 at 2.7 A and blocking voltages up to 800 V.

4*H*-SiC BJTs with relatively large active area (4.3 mm^2) were reported by M. Domeij *et al.* in 2012 [144]. These devices demonstrated a current gain above 100 at collector currents of 15 A (350 A/cm^2), specific on-resistance less than 3 mΩ·cm^2, maximum collector current of 30 A and breakdown voltage of 1850 V. Small area 4*H*-SiC BJTs with the record blocking voltages of 21 kV (current gain of 63, R_{ON-SP} = 321 mΩ·cm^2) were reported in 2012 [145]. Small area 4*H*-SiC BJTs with the record current gain of 139 and blocking voltages up to 15.8 kV were reported in 2018 [146].

Gate turn-off (GTO) thyristors (schematic cross-section shown in Fig. 17g) do not have a gate oxide layer, and do not demonstrate a current saturation behaviour. These two features enable 4*H*-SiC GTO thyristors to operate at very high junction temperatures and at very high currents with low conduction losses. Recently, GTO thyristors utilizing 140 μm thick drift *n*-type 4*H*-SiC layers doped to 2×10^{14} cm^{-3} were reported by Ryu *et al.* [133]. The fabricated devices with the chip size of 1 cm^2 and 0.465 cm^2 active area demonstrated a room temperature forward voltage drop of 5.18 V at current densities of 100 A/cm^2 and leakage currents below 0.17 μA at 15 kV blocking voltages.

A schematic cross-section of IGBT is shown in Fig. 17c. 4*H*-SiC IGBTs with 230 μm thick *n*-type drift layers doped to 2.5×10^{14} cm^{-3}, chip size of 0.81 cm^2 and active area of 0.28 cm^2 were reported by E. Van Brunt *et al.* in 2015 [147]. Prior to the device fabrication, a lifetime enhancement thermal oxidation at 1300 °C for 15 hours was performed to increase the ambipolar lifetime from less than 2 to more than 10 μs. The fabricated IGBTs exhibit an on-state voltage of 11.8 V at a forward current of 20 A and gate bias of 20 V. They demonstrated leakage current of 10 μA at $V_{BL} = 27.5$ kV which is the highest reported blocking voltage among any semiconductor switching devices.

10.7 Emergence of automobile SiC power electronics

In 2008, a great impact on development of SiC power electronics was given by the introduction to the market of the first mass-produced electrical vehicles (EVs) - Tesla Motors launched its first all-electric car. There are two parts of the electrical powertrain in these vehicles which crucially affect their performance. They are a battery charger and an inverter which converts DC power from a battery pack into AC power for a motor. The efficiency of power conversion by these units is very important since the on-board stored energy in EVs is limited by the battery capacitance. The first electric cars (and majority of EVs today) had inverters with conversion efficiency ranged from 80 to 95% and based on silicon IGBTs. These inverters, even at the efficiency of 95%, dissipate too much energy and require liquid cooling. They are heavier and larger than the electrical motors powered by them. Replacing silicon IGBTs with SiC MOSFETs could increase the inverter's conversion efficiency up to 99% [148] with significant reduction of its weight and size. This potential application of SiC power devices in the high volume automotive market gave impetus to the intensification of research in design and technology of SiC devices and the first commercial silicon carbide power MOSFET was launched by Cree, Inc. in 2011 [117].

In 2017, Tesla launched the Model 3, the first electric car with inverters based on SiC MOSFETs. To the date of this publication, its weekly production reached 6000 cars each of which using 48 silicon carbide MOSFETs (rated at 650 V/100 A) produced by STMicroelectronics fab in Catania (Italy) [149]. Nowadays, the market of SiC power devices is growing very rapidly and SiC industrial sector now shows considerable diversity with successful companies operating in different modes. Some, such as Cree/Wolfspeed and ROHM are strongly integrated with activities from wafer growth, device fabrication and even power module production. Other companies are focused purely on wafer growth whilst others manufacture devices or even provide foundry services. The industry has become strongly international in nature with significant activities in more than 10 countries over the world. This suggests that the SiC power electronics will continue to prosper as an industrial technology over the next few decades.

An additional driving force of further development of SiC technology is a great potential of silicon carbide as a material for high temperature and high frequency electronics, which is still not realized and awaiting for convincing demonstration of SiC advantageous over conventional semiconductors for these applications.

Conclusion

Silicon carbide technology has more than hundred years of history from its discovery to use of modern advanced SiC semiconductor devices in automobile electronics. This chapter briefly describes silicon carbide properties and highlights the main milestones in development of silicon carbide electronics.

Nowadays, with great progress in growth and epitaxy and the widespread commercial availability of high quality SiC epitaxial structures, development of appropriate device processing is stimulated by growing demand for SiC devices with improving performance and reliability and high cost efficiency of their commercial production. The crucially selected core technologies of SiC device processing are presented in the following chapters of this book.

References

[1]   J. J. Berzelius, "Untersuchungen über die Flussspathsäure und deren merkwürdigsten Verbindungen," *Annalen der Physik und der physikalischen Chemie,* vol. 77, no. 6, pp. 169-230, 1824. https://doi.org/10.1002/andp.18240770603

[2]   G. Pensl, F. Ciobanu, T. Frank, M. Krieger, S. Reshanov, F. Schmid, M. Weidner, "SiC MATERIAL PROPERTIES," *Sic Materials and Devices*, WORLD SCIENTIFIC, 2006, pp. 1-41.

[3]   E. G. Acheson, "Carborundum: Its history, manufacture and uses," *Journal of the Franklin Institute,* vol. 136, no. 3, pp. 194-203, 1893. https://doi.org/10.1016/0016-0032(93)90311-h

[4]   E. G. Acheson, "Carborundum: Its history, manufacture and uses," *Journal of the Franklin Institute,* vol. 136, no. 4, pp. 279-289, 1893. https://doi.org/https://doi.org/10.1016/0016-0032(93)90369-6

[5]   E. G. Acheson, *PRODUCTION 0F ARTIFICIAL CRYSTALLINE GARBONACEOUS MATERIALS*, US Patent 492,767, 1893.

[6]   M. H. Moissan, "Etude du siliciure de carbone de la meteorite Canyon Diablo,"
       Comptes rendus hebdomadaires des séances de l'Académie des sciences., vol. 140,
       pp. 405–406, 1905. https://doi.org/10.5962/bhl.part.29049

[7]   T. L. Daulton, T. J. Bernatowicz, R. S. Lewis, S. Messenger, F. J. Stadermann, S.
       Amari, "Polytype distribution of circumstellar silicon carbide," *Geochimica et
       Cosmochimica Acta,* vol. 67, no. 24, pp. 4743-4767, 2003.
       https://doi.org/10.1016/s0016-7037(03)00272-2

[8]   F. V. Kaminskiy, V. J. Bukin, S. V. Potapov, N. G. Arkus, V. G. Ivanova,
       "Discoveries of silicon carbide under natural conditions and their genetic
       importance," *International Geology Review,* vol. 11, no. 5, pp. 561-569, 1969.
       https://doi.org/10.1080/00206816909475090

[9]   A. A. Shiryaev, W. L. Griffin, E. Stoyanov, "Moissanite (SiC) from kimberlites:
       Polytypes, trace elements, inclusions and speculations on origin," *Lithos,* vol. 122,
       no. 3, pp. 152-164, 2011.
       https://doi.org/https://doi.org/10.1016/j.lithos.2010.12.011

[10]   S. Adachi, "Properties of Group-IV, III-V and II-VI Semiconductors," John Wiley
       & Sons, Ltd, 2005.

[11]   A. R. Verma, P. Krishna, *Polymorphism and Polytypism in Crystals*, New York
       Wiley, 1966.

[12]   F. Bechstedt, P. Kackell, A. Zywietz, K. Karch, B. Adolph, K. Tenelsen,
       J. Furthmuller, "Polytypism and Properties of Silicon Carbide," *physica status
       solidi (b),* vol. 202, no. 1, pp. 35-62, 1997.
       https://doi.org/10.1002/1521-3951(199707)202:1<35::aid-pssb35>3.0.co;2-8

[13]   W. van Haeringen, P. A. Bobbert, W. H. Backes, "On the Band Gap Variation in
       SiC Polytypes," *physica status solidi (b),* vol. 202, no. 1, pp. 63-79, 1997.
           https://doi.org/10.1002/1521-3951(199707)202:1<63::Aid-
       pssb63>3.0.Co;2-e

[14]   W. R. L. Lambrecht, S. Limpijumnong, S. N. Rashkeev, B. Segall, "Electronic
       Band Structure of SiC Polytypes: A Discussion of Theory and Experiment,"
       physica status solidi (b), vol. 202, no. 1, pp. 5-33, 1997.
       https://doi.org/10.1002/1521-3951(199707)202:1

[15]   A. Lebedev, Y. Tairov, "Polytypism in SiC: Theory and experiment," *Journal of
       Crystal Growth,* vol. 401, pp. 392-396, 2014.
       https://doi.org/https://doi.org/10.1016/j.jcrysgro.2014.01.021

[16] H. Baumhauer, "VII. Über die Krystalle des Carborundums," *Zeitschrift für Kristallographie - Crystalline Materials*, vol. 50, no. 1-6, pp. 33-39, 1912. https://doi.org/10.1524/zkri.1912.50.1.33

[17] G. Honjo, S. Miyake, T. Tomita, "Silicon carbide of 594 layers," *Acta Crystallographica*, vol. 3, no. 5, pp. 396-397, 1950. https://doi.org/10.1107/s0365110x50001105

[18] L. S. Ramsdell, "Studies on silicon carbide," *American Mineralogist,* vol. 32, pp. 64-82, 1947.

[19] A. L. Ortiz, F. Sanchez-Bajo, F. L. Cumbrera, F. Guiberteau, "The prolific polytypism of silicon carbide," *Journal of Applied Crystallography,* vol. 46, no. 1, pp. 242-247, 2013. doi:10.1107/S0021889812049151

[20] H. Jagodzinski, "Eindimensionale Fehlordnung in Kristallen und ihr Einfluss auf die Rontgeninterferenzen. I. Berechnung des Fehlordnungsgrades aus den Rontgenintensitaten," *Acta Crystallographica,* vol. 2, no. 4, pp. 201-207, 1949. doi:10.1107/S0365110X49000552

[21] U. Kaiser, A. Chuvilin, V. Kyznetsov, Y. Butenko, "Evidence for 9R-SiC?," *Microscopy and Microanalysis,* vol. 7, no. 04, pp. 368-369, 2001. https://doi.org/10.1017/s1431927601010364

[22] D. S. Korolev, A. A. Nikolskaya, N. O. Krivulin, A. I. Belov, A. N. Mikhaylov, D. A. Pavlov, D. I. Tetelbaum, N. A. Sobolev, M. Kumar, "Formation of hexagonal 9R silicon polytype by ion implantation," *Technical Physics Letters,* vol. 43, no. 8, pp. 767-769, 2017. https://doi.org/10.1134/s1063785017080211

[23] N. W. Jepps, T. F. Page, "Polytypic transformations in silicon carbide," *Progress in Crystal Growth and Characterization,* vol. 7, no. 1-4, pp. 259-307, 1983. https://doi.org/10.1016/0146-3535(83)90034-5

[24] P. Krishna, R. C. Marshall, C. E. Ryan, "The discovery of a 2H-3C solid state transformation in silicon carbide single crystals," *Journal of Crystal Growth,* vol. 8, no. 1, pp. 129-131, 1971. https://doi.org/https://doi.org/10.1016/0022-0248(71)90033-9

[25] S. M. Sze, *Physics of Semiconductor Devices*, Second ed., New York: Wiley, 1981, pp. 868.

[26] Y. A. Goldberg, M. E. Levinshtein, S. L. Rumyantsev, "Silicon Carbide," in: *Properties of Advanced Semiconductor Materials: GaN, AlN, InN, BN, SiC, SiGe,*

M. E. Levinshtein, S. L. Rumyantsev and M. S. Shur, eds., New York: John Wiley & Sons, Inc. , 2001.

[27] W. J. Choyke, G. Pensl, "Physical Properties of SiC," *MRS Bulletin,* vol. 22, no. 3, pp. 25-29, 1997. https://doi.org/10.1557/s0883769400032723

[28] A. A. Lebedev, "Deep level centers in silicon carbide: A review," *Semiconductors,* vol. 33, no. 2, pp. 107-130, 1999. https://doi.org/10.1134/1.1187657

[29] A. A. Lebedev ed. "Radiation Effects in Silicon Carbide," *Materials Research Foundations*, Millersville: Materials Research Forum LLC, 2017, p. 171.

[30] M. J. Bozack, "Surface Studies on SiC as Related to Contacts," *physica status solidi (b),* vol. 202, no. 1, pp. 549-580, 1997. https://doi.org/10.1002/1521-3951(199707)202:1<549::aid-pssb549>3.0.co;2-6

[31] S. Y. Davydov, "On the electron affinity of silicon carbide polytypes," *Semiconductors,* vol. 41, no. 6, pp. 696-698, 2007. https://doi.org/10.1134/s1063782607060152

[32] P. G. Neudeck, D. J. Spry, L. Chen, N. F. Prokop, M. J. Krasowski, "Demonstration of 4H-SiC Digital Integrated Circuits Above 800 °C," *IEEE Electron Device Letters,* vol. 38, no. 8, pp. 1082-1085, 2017. https://doi.org/10.1109/led.2017.2719280

[33] P. G. Neudeck, D. J. Spry, C. Liang-Yu, G. M. Beheim, R. S. Okojie, C. W. Chang, R. D. Meredith, T. L. Ferrier, L. J. Evans, M. J. Krasowski, N. F. Prokop, "Stable Electrical Operation of 6H-SiC JFETs and ICs for Thousands of Hours at 500C," *Electron Device Letters, IEEE,* vol. 29, no. 5, pp. 456-459, 2008.

[34] P. G. Neudeck, S. L. Garverick, D. J. Spry, L.-Y. Chen, G. M. Beheim, M. J. Krasowski, M. Mehregany, "Extreme temperature 6H-SiC JFET integrated circuit technology," *physica status solidi (a),* vol. 206, no. 10, pp. 2329-2345, 2009.

[35] G. L. Harris ed. "Properties of Silicon Carbide," London, United Kingdom: INSPEC, the Institution of Electrical Engineers, 1995, p. 295.

[36] J. C. Bose, *Detector for electrical disturbances*, US Patent 755,840, 1904.

[37] T. H. Lee, "The (pre-) history of the integrated circuit: a random walk," *IEEE Solid-State Circuits Society Newsletter,* vol. 12, no. 2, pp. 16-22, 2007. https://doi.org/10.1109/N-SSC.2007.4785573

[38] H. H. C. Dunwoody, *Wireless telegraph system*, US Patent 837616, 1906.

[39]　G. W. Pierce, "Crystal Rectifiers for Electric Currents and Electric Oscillations. Part I. Carborundum," *Physical Review (Series I), vol.* 25, no. 1, pp. 31-60, 1907. https://doi.org/10.1103/physrevseriesi.25.31

[40]　G. W. Pickard, *Oscillation detector and rectifier*, US Patent 912,613, 1909.

[41]　H. J. Round, "A Note on Carborundum," *Electrical World*, 1907, p. 309.

[42]　O. V. Lossev, "Behavior of contact detectors; the effect of temperature on the generating contacts (in Russian) " *Telegrafia i telefonia bez provodov (TiTbp)*, no. 18, pp. 45-62, 1923.

[43]　O. V. Lossev, "Oscillating Crystals," *The Wireless World and Radio Review*, no. 271, pp. 93-96, 1924.

[44]　O. V. Lossev, "Luminous carborundum detector and detection effect and oscillations with crystals," *The London, Edinburgh, and Dublin Philosophical Magazine and Journal of Science,* vol. 6, no. 39, pp. 1024-1044, 1928. https://doi.org/10.1080/14786441108564683

[45]　E. E. Loebner, "Subhistories of the light emitting diode," *IEEE Transactions on Electron Devices,* vol. 23, no. 7, pp. 675-699, 1976. https://doi.org/10.1109/t-ed.1976.18472

[46]　R. O. Grisdale, "Silicon carbide varistor," *Bell Laboratories Record,* vol. 19, no. 10, pp. 46-51, 1940.

[47]　J. Mitchell, J. Shewchun, "High-Current Characteristics of Silicon Carbide Varistors," *Journal of Applied Physics,* vol. 42, no. 2, pp. 889-892, 1971. https://doi.org/10.1063/1.1660124

[48]　*HVR International Product Catalogue*: Information on www.hvrint.de.

[49]　M. Riordan, L. Hoddeson, "The origins of the pn junction," *IEEE Spectrum,* vol. 34, no. 6, pp. 46-51, 1997. https://doi.org/10.1109/6.591664

[50]　J. A. Lely, "Darstellung von Einkristallen von Silicium Carbid und Beherrschung von Art und Menge der eingebauten Verunreinigungen," *Berichte der Deutschen Keramischen Gesellschaft,* vol. 8, pp. 229, 1955.

[51]　J. A. Lely, SUBLIMATION PROCESS FOR MANUFACTURING SILICON CARBIDE CRYSTALS, US Patent 2,854,364, 1958.

[52]　J. R. O'Connor, C. E. Smiltens eds., "Silicon Carbide, A High Temperature Semiconductor," New York: Pergamon, 1960, p. 521.

[53] H. C. Chang, L. J. Kroko, APPARATUS FOR AND PREPARATION OF SILICON CARBIDE SINGLE CRYSTALS, US Patent 3,275,415, 1966.

[54] Y. A. Vodakov, E. N. Mokhov, A. D. Roenkov, D. T. Saidbekov, "Effect of crystallographic orientation on the polytype stabilization and transformation of silicon carbide," Physica Status Solidi (a), vol. 51, no. 1, pp. 209-215, 1979. https://doi.org/10.1002/pssa.2210510123

[55] Y. M. Tairov, V. F. Tsvetkov, "Investigation of growth processes of ingots of silicon carbide single crystals," Journal of Crystal Growth, vol. 43, no. 2, pp. 209-212, 1978. https://doi.org/10.1016/0022-0248(78)90169-0

[56] Y. M. Tairov, V. F. Tsvetkov, "General principles of growing large-size single crystals of various silicon carbide polytypes," Journal of Crystal Growth, vol. 52, pp. 146-150, 1981. https://doi.org/10.1016/0022-0248(81)90184-6

[57] Y. M. Tairov, V. F. Tsvetkov, "Progress in controlling the growth of polytypic crystals," Progress in Crystal Growth and Characterization, vol. 7, no. 1-4, pp. 111-162, 1983. https://doi.org/10.1016/0146-3535(83)90031-x

[58] Y. A. Vodakov, E. N. Mokhov, M. G. Ramm, A. D. Roenkov, "Epitaxial growth of silicon carbide layers by sublimation „sandwich method" (I) growth kinetics in vacuum," Kristall und Technik, vol. 14, no. 6, pp. 729-740, 1979. https://doi.org/10.1002/crat.19790140618

[59] Y. A. Vodakov, E. N. Mokhov, Method for epitaxial production of semiconductor silicon carbide utilizing a close-space sublimation deposition technique, US Patent 4,147,572, 1979.

[60] H. Matsunami, S. Nishino, M. Odaka, T. Tanaka, "Epitaxial growth of α-SiC layers by chemical vapor deposition technique," Journal of Crystal Growth, vol. 31, pp. 72-75, 1975. https://doi.org/10.1016/0022-0248(75)90113-x

[61] S. Nishino, H. Matsunami, T. Tanaka, "Growth and morphology of 6H-SiC epitaxial layers by CVD," Journal of Crystal Growth, vol. 45, pp. 144-149, 1978. https://doi.org/10.1016/0022-0248(78)90426-8

[62] H. Matsunami, S. Nishino, H. Ono, "IVA-8 heteroepitaxial growth of cubic silicon carbide on foreign substrates," IEEE Transactions on Electron Devices, vol. 28, no. 10, pp. 1235-1236, 1981. https://doi.org/10.1109/t-ed.1981.20556

[63] S. Nishino, J. A. Powell, H. A. Will, "Production of large-area single-crystal wafers of cubic SiC for semiconductor devices," Applied Physics Letters, vol. 42, no. 5, pp. 460-462, 1983. https://doi.org/10.1063/1.93970

[64]   K. Shibahara, S. Nishino, H. Matsunami, "Antiphase-domain-free growth of cubic
       SiC on Si(100)," Applied Physics Letters, vol. 50, no. 26, pp. 1888-1890, 1987.
       https://doi.org/10.1063/1.97676

[65]   N. Kuroda, K. Shibahara, W. Yoo, S. Nishino, H. Matsunami, "Step-Controlled
       VPE Growth of SiC Single Crystals at Low Temperatures," in Extended Abstracts
       of the 1987 Conference on Solid State Devices and Materials, 1987, pp. 227-230.

[66]   H. Matsunami, T. Kimoto, "Step-controlled epitaxial growth of SiC: High quality
       homoepitaxy," Materials Science and Engineering: R: Reports, vol. 20, no. 3, pp.
       125-166, 1997. https://doi.org/http://dx.doi.org/10.1016/S0927-796X(97)00005-3

[67]   T. Kimoto, A. Itoh, H. Matsunami, "Step-Controlled Epitaxial Growth of High-
       Quality SiC Layers," physica status solidi (b), vol. 202, no. 1, pp. 247-262, 1997.
       https://doi.org/10.1002/1521-3951(199707)202:1<247::aid-pssb247>3.0.co;2-q

[68]   O. Kordina, C. Hallin, R. C. Glass, E. Janzen, "A novel hot-wall CVD reactor for
       SiC epitaxy," Inst. Phys. Conf. Ser., vol. 137, pp. 41-44, 1994.

[69]   O. Kordina, A. Henry, J. P. Bergman, N. T. Son, W. M. Chen, C. Hallin, E.
       Janzén, "High quality 4H-SiC epitaxial layers grown by chemical vapor
       deposition," Applied Physics Letters, vol. 66, no. 11, pp. 1373-1375, 1995.
       https://doi.org/10.1063/1.113205

[70]   P. A. Ivanov, V. E. Chelnokov, "Recent developments in SiC single-crystal
       electronics," Semiconductor Science and Technology, vol. 7, no. 7, pp. 863-880,
       1992. https://doi.org/10.1088/0268-1242/7/7/001

[71]   J. W. Palmour, J. A. Edmond, H. S. Kong, C. H. Carter, "6H-silicon carbide
       devices and applications," Physica B: Condensed Matter, vol. 185, no. 1-4, pp.
       461-465, 1993. https://doi.org/10.1016/0921-4526(93)90278-e

[72]   J. W. Palmour, J. A. Edmond, H. S. Kong, J. C. H. Carter, "Vertical power devices
       in silicon carbide," Silicon Carbide and Related Materials, Inst. of Phys. Conf.
       Series vol. 137, pp. 499-502, 1994.

[73]   W. v. Münch, P. Hoeck, "Silicon carbide bipolar transistor," Solid-State
       Electronics, vol. 21, no. 2, pp. 479-480, 1978.
       https://doi.org/10.1016/0038-1101(78)90283-6

[74]   R. F. Davis, C. H. Carter, C. E. Hunter, Sublimation of silicon carbide to produce
       large, device quality single crystals of silicon carbide, US Patent 4,866,005, 1989.

[75]   J. Edmond, H. Kong, A. Suvorov, D. Waltz, J. C. Carter, "6H-Silicon Carbide
       Light Emitting Diodes and UV Photodiodes," physica status solidi (a), vol. 162,

no. 1, pp. 481-491, 1997.
https://doi.org/10.1002/1521-396x(199707)162:1<481::aid-pssa481>3.0.co;2-o

[76] R. C. Glass, D. Henshall, V. F. Tsvetkov, J. C. H. Carter, "SiC Seeded Crystal
      Growth," physica status solidi (b), vol. 202, no. 1, pp. 149-162, 1997.
      https://doi.org/10.1002/1521-3951(199707)202:1<149::aid-pssb149>3.0.co;2-m

[77] S. G. Müller, R. C. Glass, H. M. Hobgood, V. F. Tsvetkov, M. Brady, D. Henshall,
      D. Malta, R. Singh, J. Palmour, C. H. Carter, "Progress in the industrial production
      of SiC substrates for semiconductor devices," Materials Science and Engineering:
      B, vol. 80, no. 1-3, pp. 327-331, 2001. https://doi.org/10.1016/s0921-
      5107(00)00658-9

[78] M. Tuominen, R. Yakimova, R. C. Glass, T. Tuomi, E. Janzén, "Crystalline
      imperfections in 4H SiC grown with a seeded Lely method," Journal of Crystal
      Growth, vol. 144, no. 3-4, pp. 267-276, 1994.
      https://doi.org/10.1016/0022-0248(94)90466-9

[79] J. Heindl, H. P. Strunk, V. D. Heydemann, G. Pensl, "Micropipes: Hollow Tubes
      in Silicon Carbide," physica status solidi (a), vol. 162, no. 1, pp. 251-262, 1997.
      https://doi.org/10.1002/1521-396x(199707)162:1<251::aid-pssa251>3.0.co;2-7

[80] D. Nakamura, I. Gunjishima, S. Yamaguchi, T. Ito, A. Okamoto, H. Kondo, S.
      Onda, K. Takatori, "Ultrahigh-quality silicon carbide single crystals," Nature, vol.
      430, no. 7003, pp. 1009-1012, 2004. https://doi.org/10.1038/nature02810

[81] Information on https://www.reportsanddata.com/press-release/global-silicon-
      carbide-wafer-market

[82] S. Nakamura, T. Mukai, M. Senoh, "Candela-class high-brightness InGaN/AlGaN
      double-heterostructure blue-light-emitting diodes," Applied Physics Letters, vol.
      64, no. 13, pp. 1687-1689, 1994. https://doi.org/10.1063/1.111832

[83] X. She, A. Q. Huang, O. Lucia, B. Ozpineci, "Review of Silicon Carbide Power
      Devices and Their Applications," IEEE Transactions on Industrial Electronics, vol.
      64, no. 10, pp. 8193-8205, 2017. https://doi.org/10.1109/tie.2017.2652401

[84] M. Schadt, G. Pensl, R. P. Devaty, W. J. Choyke, R. Stein, D. Stephani,
      "Anisotropy of the electron Hall mobility in 4H, 6H, and 15R silicon carbide,"
      Applied Physics Letters, vol. 65, no. 24, pp. 3120-3122, 1994.
      https://doi.org/doi:http://dx.doi.org/10.1063/1.112455

[85]   W. J. Schaffer, H. S. Kong, G. H. Negley, J. Palmour, "Hall effect and CV measurements on epitaxial 6H- and 4H-SiC," Inst. Phys. Conf. Ser., vol. 137, pp. 155-159, 1994.

[86]   A. P. Dmitriev, A. O. Konstantinov, D. Litvin, V. I. Sankin, "Impact ionization and superlattice in 6H-SiC," Soviet physics. Semiconductors, vol. 17, pp. 686-689, 1983.

[87]   A. O. Konstantinov, "Influence of temperature on impact ionization and avalanche breakdown in silicon carbide," Soviet Phys. Semicond, vol. 23, no. 1, pp. 31-35, 1989. [in Russian: А. О. Константинов, Температурная зависимость ударной ионизации и лавинного пробоя в карбиде кремния, ФТП 23, (1989) с. 52-57].

[88]   K. V. Vassilevski, V. A. Dmitriev, A. V. Zorenko, "Silicon carbide diode operating at avalanche breakdown current density of 60 kA/cm^2," Journal of Applied Physics, vol. 74, no. 12, pp. 7612-7614, 1993. https://doi.org/10.1063/1.354963

[89]   A. O. Konstantinov, Q. Wahab, N. Nordell, U. Lindefelt, "Ionization rates and critical fields in 4H silicon carbide," Applied Physics Letters, vol. 71, no. 1, pp. 90-92, 1997. https://doi.org/doi:http://dx.doi.org/10.1063/1.119478

[90]   A. O. Konstantinov, Q. Wahab, N. Nordell, U. Lindefelt, "Study of avalanche breakdown and impact ionization in 4H silicon carbide," Journal of Electronic Materials, vol. 27, no. 4, pp. 335-341, 1998. https://doi.org/10.1007/s11664-998-0411-x

[91]   A. O. Konstantinov, N. Nordell, Q. Wahab, U. Lindefelt, "Temperature dependence of avalanche breakdown for epitaxial diodes in 4H silicon carbide," Applied Physics Letters, vol. 73, pp. 1850-1852, 1998. https://doi.org/10.1063/1.122303

[92]   M. Bakowski, "Status and Prospects of SiC Power Devices," in: Advancing Silicon Carbide Electronics Technology I, K. Zekentes and K. Vasilevskiy, eds., Millersville: Materials Research Forum LLC, 2018, pp. 191-236.

[93]   M. Bhatnagar, P. K. McLarty, B. J. Baliga, "Silicon-carbide high-voltage (400 V) Schottky barrier diodes," IEEE Electron Device Letters, vol. 13, no. 10, pp. 501-503, 1992. https://doi.org/10.1109/55.192814

[94]   A. Itoh, T. Kimoto, H. Matsunami, "High performance of high-voltage 4H-SiC Schottky barrier diodes," IEEE Electron Device Letters, vol. 16, no. 6, pp. 280-282, 1995. https://doi.org/10.1109/55.790735

[95]   A. Itoh, T. Kimoto, H. Matsunami, "Excellent reverse blocking characteristics of high-voltage 4H-SiC Schottky rectifiers with boron-implanted edge termination," IEEE Electron Device Letters, vol. 17, no. 3, pp. 139-141, 1996. https://doi.org/10.1109/55.485193

[96]   Y. Jiang, W. Sung, X. Song, H. Ke, S. Liu, B. J. Baliga, A. Q. Huang, E. Van Brunt, "10kV SiC MPS diodes for high temperature applications," in 28th International Symposium on Power Semiconductor Devices and ICs (ISPSD), pp. 43-46, 2016.

[97]   "CoolSiC™ Automotive Discrete Schottky Diodes " Infenion Application Note; information on www.infineon.com

[98]   M. Holz, G. Hultsch, T. Scherg, R. Rupp, "Reliability considerations for recent Infineon SiC diode releases," Microelectronics Reliability, vol. 47, no. 9, pp. 1741-1745, 2007. https://doi.org/https://doi.org/10.1016/j.microrel.2007.07.031

[99]   Information on http://www.wolfspeed.com

[100] D. Stephani, P. Friedrichs, "SILICON CARBIDE JUNCTION FIELD EFFECT TRANSISTORS," International Journal of High Speed Electronics and Systems, vol. 16, no. 03, pp. 825-854, 2006. https://doi.org/10.1142/s012915640600403x

[101] P. A. Ivanov, N. S. Savkina, V. N. Panteleev, V. E. Chelnokov, "Junction field-effect transistors based on 4H-silicon carbide," Institute of Physics Conference Series, 137. pp. 593-595, 1994.

[102] P. Sannuti, X. Li, F. Yan, K. Sheng, J. H. Zhao, "Channel electron mobility in 4H-SiC lateral junction field effect transistors," Solid-State Electronics, vol. 49, no. 12, pp. 1900-1904, 2005. https://doi.org/10.1016/j.sse.2005.10.027

[103] H. Mitlehner, W. Bartsch, K. O. Dohnke, P. Friedrichs, R. Kaltschmidt, U. Weinert, B. Weis, D. Stephani, "Dynamic characteristics of high voltage 4H-SiC vertical JFETs," in 11th International Symposium on Power Semiconductor Devices and ICs. ISPSD'99 Proceedings 1999, pp. 339-342.

[104] P. Friedrichs, H. Mitlehner, K. O. Dohnke, D. Peters, R. Schorner, U. Weinert, E. Baudelot, D. Stephani, "SiC power devices with low on-resistance for fast switching applications," 12th International Symposium on Power Semiconductor Devices & ICs. Proceedings, pp. 213-216, 2000. https://doi.org/10.1109/ISPSD.2000.856809

[105] H. Onose, A. Watanabe, T. Someya, Y. Kobayashi, "2 kV 4H-SiC Junction FETs," Materials Science Forum, vol. 389-393, pp. 1227-1230, 2002. doi:10.4028/www.scientific.net/msf.389-393.1227

[106] J. H. Zhao, K. Tone, X. Li, P. Alexandrov, L. Fursin, M. Weiner, "3.6 mΩ·cm^2, 1726 V 4H-SiC normally-off trenched-and-implanted vertical JFETs," in ISPSD '03. IEEE 15th International Symposium on Power Semiconductor Devices and ICs, 2003.

[107] Y. Li, P. Alexandrov, J. H. Zhao, "1.88-mΩ·cm^2 1650-V Normally on 4H-SiC TI-VJFET," IEEE Transactions on Electron Devices, vol. 55, no. 8, pp. 1880-1886, 2008. https://doi.org/10.1109/ted.2008.926678

[108] I. Sankin, D. C. Sheridan, W. Draper, V. Bondarenko, R. Kelley, M. S. Mazzola, J. B. Casady, "Normally-Off SiC VJFETs for 800 V and 1200 V Power Switching Applications," Proceedings of 20th International Symposium on Power Semiconductor Devices and IC's, 18-22 May 2008, pp. 260-262, https://doi.org/10.1109/ISPSD.2008.4538948

[109] Information on https://unitedsic.com

[110] A. Suzuki, A. Ashida, N. Furui, K. Mameno, H. Matsunami, "Thermal Oxidation of SiC and Electrical Properties of Al–SiO2–SiC MOS Structure," Japanese Journal of Applied Physics, vol. 21, no. Part 1, No. 4, pp. 579-585, 1982. https://doi.org/10.1143/jjap.21.579

[111] J. N. Shenoy, J. A. Cooper, M. R. Melloch, "High-voltage double-implanted power MOSFET's in 6H-SiC," IEEE Electron Device Letters, vol. 18, no. 3, pp. 93-95, 1997. https://doi.org/10.1109/55.556091

[112] T. Ito, "Direct Thermal Nitridation of Silicon Dioxide Films in Anhydrous Ammonia Gas," Journal of The Electrochemical Society, vol. 127, no. 9, pp. 2053, 1980. https://doi.org/10.1149/1.2130065

[113] H. Hwang, W. Ting, B. Maiti, D. L. Kwong, J. Lee, "Electrical characteristics of ultrathin oxynitride gate dielectric prepared by rapid thermal oxidation of Si in N$_2$O," Applied Physics Letters, vol. 57, no. 10, pp. 1010-1011, 1990. https://doi.org/10.1063/1.103550

[114] H.-f. Li, S. Dimitrijev, H. B. Harrison, D. Sweatman, "Interfacial characteristics of N$_2$O and NO nitrided SiO$_2$ grown on SiC by rapid thermal processing," Applied Physics Letters, vol. 70, no. 15, pp. 2028-2030, 1997. https://doi.org/10.1063/1.118773

[115] G. Y. Chung, C. C. Tin, J. R. Williams, K. McDonald, R. K. Chanana, R. A. Weller, S. T. Pantelides, L. C. Feldman, O. W. Holland, M. K. Das, J. W. Palmour, "Improved inversion channel mobility for 4H-SiC MOSFETs following high temperature anneals in nitric oxide," IEEE Electron Device Letters, vol. 22, no. 4, pp. 176-178, 2001. https://doi.org/10.1109/55.915604

[116] G. Gudjonsson, H. O. Olafsson, F. Allerstam, P. A. Nilsson, E. O. Sveinbjornsson, H. Zirath, T. Rodle, R. Jos, "High field-effect mobility in n-channel Si face 4H-SiC MOSFETs with gate oxide grown on aluminum ion-implanted material," Electron Device Letters, IEEE, vol. 26, no. 2, pp. 96-98, 2005. https://doi.org/10.1109/LED.2004.841191

[117] "Cree Launches Industry's First Commercial Silicon Carbide Power MOSFET; Destined to Replace Silicon Devices in High-Voltage Power Electronics," JANUARY 17, 2011; information on https://www.cree.com

[118] J. W. Palmour, L. Cheng, V. Pala, E. V. Brunt, D. J. Lichtenwalner, G. Y. Wang, J. Richmond, M. O'Loughlin, S. Ryu, S. T. Allen, A. A. Burk, C. Scozzie, "Silicon carbide power MOSFETs: Breakthrough performance from 900 V up to 15 kV," in IEEE 26th International Symposium on Power Semiconductor Devices & IC's (ISPSD), 2014, https://doi.org/10.1109/ispsd.2014.6855980

[119] G. Liu, "Silicon carbide: A unique platform for metal-oxide-semiconductor physics," Applied Physics Reviews, vol. 2, pp. 021307, 2015. https://doi.org/10.1063/1.4922748

[120] S. Dimitrijev, "SiC power MOSFETs: The current status and the potential for future development," in IEEE 30th International Conference on Microelectronics (MIEL), 9-11 Oct. 2017, pp. 29-34, https://doi.org/10.1109/MIEL.2017.8190064

[121] R. Singh, "HIGH POWER SIC PIN RECTIFIERS," International Journal of High Speed Electronics and Systems, vol. 15, no. 04, pp. 867-898, 2005. https://doi.org/10.1142/S0129156405003442

[122] N. Ramungul, V. Khemka, T. P. Chow, M. Ghezzo, J. W. Kretchmer, "Carrier Lifetime Extraction from a 6H-SiC High Voltage p-i-n Rectifier Reverse Recovery Waveform," Materials Science Forum, vol. 264-268, pp. 1065-1068, 1998. https://doi.org/10.4028/www.scientific.net/msf.264-268.1065

[123] M. E. Levinshtein, J. W. Palmour, S. L. Rumyantsev, R. Singh, "Forward current-voltage characteristics of silicon carbide thyristors and diodes at high current densities," Semiconductor Science and Technology, vol. 13, no. 9, pp. 1006-1010, 1998. https://doi.org/10.1088/0268-1242/13/9/007

[124] P. A. Ivanov, M. E. Levinshtein, K. G. Irvine, O. Kordina, J. W. Palmour, S. L. Rumyantsev, R. Singh, "High hole lifetime (3.8 [micro sign]s) in 4H-SiC diodes with 5.5 kV blocking voltage," Electronics Letters, vol. 35, no. 16, pp. 1382, 1999. https://doi.org/10.1049/el:19990897

[125] P. Grivickas, A. Galeckas, J. Linnros, M. Syväjärvi, R. Yakimova, V. Grivickas, J. A. Tellefsen, "Carrier lifetime investigation in 4H–SiC grown by CVD and sublimation epitaxy," Materials Science in Semiconductor Processing, vol. 4, no. 1-3, pp. 191-194, 2001. https://doi.org/10.1016/s1369-8001(00)00133-5

[126] P. Grivickas, J. Linnros, V. Grivickas, "Free Carrier Diffusion Measurements in Epitaxial 4H-SiC with a Fourier Transient Grating Technique: Injection Dependence," Materials Science Forum, vol. 338-342, pp. 671-674, 2000. https://doi.org/10.4028/www.scientific.net/msf.338-342.671

[127] J. Zhang, L. Storasta, J. P. Bergman, N. T. Son, E. Janzén, "Electrically active defects in n-type 4H–silicon carbide grown in a vertical hot-wall reactor," Journal of Applied Physics, vol. 93, no. 8, pp. 4708-4714, 2003. https://doi.org/10.1063/1.1543240

[128] L. Storasta, H. Tsuchida, "Reduction of traps and improvement of carrier lifetime in 4H-SiC epilayers by ion implantation," Applied Physics Letters, vol. 90, no. 6, pp. 062116, 2007. https://doi.org/10.1063/1.2472530

[129] Y. M. Tairov, V. F. Tsvetkov, "Semiconductor Compounds AIVBIV," in: Handbook on electrotechnical materials, Y. V. Koritskii, V. V. Pasynkov and B. M. Tareev, eds., Leningrad: Energomashizdat, 1988, p. 728. [in Russian]

[130] R. N. Ghoshtagore, R. L. Coble, "Self-Diffusion in Silicon Carbide," Physical Review, vol. 143, no. 2, pp. 623-626, 1966.

[131] T. Hiyoshi, T. Kimoto, "Reduction of Deep Levels and Improvement of Carrier Lifetime in n-Type 4H-SiC by Thermal Oxidation," Applied Physics Express, vol. 2, pp. 041101, 2009. https://doi.org/10.1143/apex.2.041101

[132] K. Kawahara, J. Suda, T. Kimoto, "Analytical model for reduction of deep levels in SiC by thermal oxidation," Journal of Applied Physics, vol. 111, no. 5, pp. 053710, 2012. https://doi.org/10.1063/1.3692766

[133] S. H. Ryu, D. J. Lichtenwalner, M. O'Loughlin, C. Capell, J. Richmond, E. van Brunt, C. Jonas, Y. Lemma, A. Burk, B. Hull, M. McCain, S. Sabri, H. O'Brien, A. Ogunniyi, A. Lelis, J. Casady, D. Grider, S. Allen, J. W. Palmour, "15 kV n-GTOs

Materials Research Forum LLC
https://doi.org/10.21741/9781644900673-1

in 4H-SiC," Materials Science Forum, vol. 963, pp. 651-654, 2019.
https://doi.org/10.4028/www.scientific.net/MSF.963.651

[134] H. Lendenmann, F. Dahlquist, N. Johansson, R. Söderholm, P. Å. Nilsson,
P. Bergman, P. Skytt, "Long Term Operation of 4.5 kV PiN and 2.5 kV JBS
Diodes," Materials Science Forum, vol. 353-356, pp. 727-730, 2001.
https://doi.org/10.4028/www.scientific.net/msf.353-356.727

[135] P. Bergman, H. Lendenmann, P. Å. Nilsson, U. Lindefelt, P. Skytt, "Crystal
Defects as Source of Anomalous Forward Voltage Increase of 4H-SiC Diodes,"
Materials Science Forum, vol. 353-356, pp. 299-302, 2001.
https://doi.org/10.4028/www.scientific.net/msf.353-356.299

[136] J. Q. Liu, M. Skowronski, C. Hallin, R. Söderholm, H. Lendenmann, "Structure of
recombination-induced stacking faults in high-voltage SiC p–n junctions," Applied
Physics Letters, vol. 80, no. 5, pp. 749-751, 2002.
https://doi.org/10.1063/1.1446212

[137] T. Ohno, H. Yamaguchi, S. Kuroda, K. Kojima, T. Suzuki, K. Arai, "Influence of
growth conditions on basal plane dislocation in 4H-SiC epitaxial layer," Journal of
Crystal Growth, vol. 271, no. 1-2, pp. 1-7, 2004.
https://doi.org/10.1016/j.jcrysgro.2004.04.044

[138] J. J. Sumakeris, J. R. Jenny, A. R. Powell, "Bulk Crystal Growth, Epitaxy, and
Defect Reduction in Silicon Carbide Materials for Microwave and Power
Devices," MRS Bulletin, vol. 30, no. 4, pp. 280-286, 2005.
https://doi.org/10.1557/mrs2005.74

[139] J. J. Sumakeris, P. Bergman, M. K. Das, C. Hallin, B. A. Hull, E. Janzén,
H. Lendenmann, M. J. O'Loughlin, M. J. Paisley, S. Y. Ha, M. Skowronski,
J. W. Palmour, C. H. Carter Jr, "Techniques for Minimizing the Basal Plane
Dislocation Density in SiC Epilayers to Reduce V_f Drift in SiC Bipolar Power
Devices," Materials Science Forum, vol. 527-529, pp. 141-146,
2006.https://doi.org/10.4028/www.scientific.net/msf.527-529.141

[140] W. Chen, M. A. Capano, "Growth and characterization of 4H-SiC epilayers on
substrates with different off-cut angles," Journal of Applied Physics, vol. 98, no.
11, pp. 114907, 2005. https://doi.org/10.1063/1.2137442

[141] Y. Bu, H. Yoshimoto, N. Watanabe, A. Shima, "Fabrication of 4H-SiC PiN diodes
without bipolar degradation by improved device processes," Journal of Applied
Physics, vol. 122, no. 24, pp. 244504, 2017. https://doi.org/10.1063/1.5001370

[142] H. Niwa, J. Suda, T. Kimoto, "21.7 kV 4H-SiC PiN Diode with a Space-Modulated Junction Termination Extension," Applied Physics Express, vol. 5, no. 6, pp. 064001, 2012. https://doi.org/10.1143/apex.5.064001

[143] Y. Luo, L. Fursin, J. H. Zhao, "Demonstration of 4H-SiC power bipolar junction transistors," Electronics Letters, vol. 36, no. 17, pp. 1496, 2000. https://doi.org/10.1049/el:20001059

[144] M. Domeij, A. Konstantinov, A. Lindgren, C. Zaring, K. Gumaelius, M. Reimark, "Large Area 1200 V SiC BJTs with β>100 and ρ_{ON} < 3 m$\Omega\cdot$cm^2," Materials Science Forum, vol. 717-720, pp. 1123-1126, 2012. https://doi.org/10.4028/www.scientific.net/msf.717-720.1123

[145] H. Miyake, T. Okuda, H. Niwa, T. Kimoto, J. Suda, "21-kV SiC BJTs With Space-Modulated Junction Termination Extension," IEEE Electron Device Letters, vol. 33, no. 11, pp. 1598-1600, 2012. https://doi.org/10.1109/LED.2012.2215004

[146] A. Salemi, H. Elahipanah, K. Jacobs, C.-M. Zetterling, M. Ostling, "15 kV-Class Implantation-Free 4H-SiC BJTs With Record High Current Gain," IEEE Electron Device Letters, vol. 39, no. 1, pp. 63-66, 2018. https://doi.org/10.1109/led.2017.2774139

[147] E. van Brunt, L. Cheng, M. J. O'Loughlin, J. Richmond, V. Pala, J. W. Palmour, C. W. Tipton, C. Scozzie, "27 kV, 20 A 4H-SiC n-IGBTs," Materials Science Forum, vol. 821-823, pp. 847-850, 2015. https://doi.org/10.4028/www.scientific.net/msf.821-823.847

[148] N. Zabihi, A. Mumtaz, T. Logan, T. Daranagama, R. A. McMahon, "SiC Power Devices for Applications in Hybrid and Electric Vehicles," Materials Science Forum, vol. 963, pp. 869-872, 2019. https://doi.org/10.4028/www.scientific.net/MSF.963.869

[149] "IS TESLA'S PRODUCTION CREATING A SIC MOSFET SHORTAGE?," 2019; information on https://www.pntpower.com/is-teslas-production-creating-a-sic-mosfet-shortage

Advancing Silicon Carbide Electronics Technology II
Materials Research Foundations **69** (2020) 63-106

Materials Research Forum LLC
https://doi.org/10.21741/9781644900673-2

CHAPTER 2

Dielectrics in Silicon Carbide Devices: Technology and Application

Anthony O'Neill[1]*, Oliver Vavasour[2], Stephen Russell[2], Faiz Arith[1], Jesus Urresti[1], Peter Gammon[2]

[1]School of Engineering, Newcastle University, Newcastle upon Tyne, United Kingdom

[2]School of Engineering, University of Warwick, Coventry CV4 7AL, United Kingdom

* anthony.oneill@newcastle.ac.uk

Abstract

Formation of dielectric layers on SiC is a key feature of device processing technology. Achieving high mobility SiC MOSFETs is dependent on solving challenges within gate stack formation, where the dielectric plays a central role. Dielectrics also play a key role in surface passivation of SiC devices. This chapter reviews the main dielectrics that are used in SiC devices. The most commonly used dielectrics in electronic devices are SiO_2 and Si_3N_4 and so these are introduced first, followed by high-κ dielectrics (i.e. dielectrics with higher permittivity than Si_3N_4). The methods of dielectric deposition are discussed before focusing on SiC thermal oxidation. Different parameters of the oxidation process and post-oxidation annealing, which have an impact on oxide quality and the formation of residual carbon in the SiO_2/SiC interface, are evaluated. Efforts to improve electron mobility in SiC MOSFETs using a variety of dielectric layer formation techniques are reviewed, indicating where progress has been made. Issues surrounding SiC surface passivation by dielectrics are also discussed.

Keywords

Silicon Oxide, High-κ Dielectrics, MOSFET, Gate Oxide, Surface Passivation, Post Oxidation Annealing, Field Effect Mobility, Silicon Nitride

Contents

List of used symbols and abbreviations..65

1. Introduction...68

1.1   Interface-trapped charge effects and requirements 68

1.2   Near-interface trap effects 69

1.3   SiC MOS interface requirements 70

2.   Dielectrics in SiC device processing **71**

2.1   Silicon dioxide in SiC devices 71

2.2   Silicon nitride in SiC devices 73

2.3   High-κ dielectrics in SiC devices 74

3.   Methods of dielectric deposition used in SiC device processing **75**

3.1   Plasma enhanced chemical vapor deposition of dielectric on SiC 76

3.2   Deposition of silicon oxide films using TEOS 77

3.3   Atomic layer deposition of gate dielectrics in SiC devices 78

3.4   Densification of dielectrics deposited on SiC 80

3.5   Deposition methods conclusion 80

4.   Thermal oxidation of SiC **81**

4.1   SiC oxidation rates and a modified Deal-Grove model 81

4.2   Interface traps introduced during thermal oxidation of silicon carbide 83

4.3   High temperature oxidation 85

4.4   Low temperature oxidation 86

4.5   Post oxidation annealing 88

4.6   Thermal oxidation conclusion 89

5.   Other methods to improve channel mobility **89**

5.1   Sodium enhanced oxidation 89

5.2   Counter doped channel regions 90

5.3   Alternative SiC crystal faces 90

6.   Surface passivation by dielectrics **91**

7.   Summary **92**

Acknowledgements **93**

References **94**

List of used symbols and abbreviations

A	fitting parameter in Deal-Grove model of oxidation;
B	fitting parameter in Deal-Grove model of oxidation;
C_{dep}	depletion capacitance per unit area;
C_{GC}	gate-to-channel capacitance per unit area;
C_{it}	capacitive term per unit area associated with D_{it};
C_{ox}	gate oxide capacitance per unit area;
D_{it}	density of interface traps;
d_{ox}	SiO_2 thickness;
E_C	conduction band minimum energy;
F_{eff}	effective electric field;
E_g	energy bandgap;
E_i	intrinsic Fermi level;
F_{ox}	dielectric strength;
E_V	valence band maximum energy;
$F_{1/2}$	Fermi-Dirac integral (of order 1/2);
g_d^i	intrinsic channel conductance;
g_m^i	intrinsic transconductance;
I_D	drain current;
L	gate length of MOSFET;
n	ideality factor in definition of the subthreshold;
N_s	density of charge carriers on the surface;
N_{depl}	depletion charge density close to an inverted SiC surface;
q	elementary charge;
Q_{it}	density of interface-trapped charge;
Q_{mobile}	density of mobile charge in the MOSFET channel;
Q_n	channel charge density;
S	subthreshold slope;

T	temperature;
t	time;
V_{DS}	source-drain voltage;
V_{GS}	gate-source voltage;
V_t	threshold voltage;
W	gate width of MOSFET;
β	current gain of BJT;
ε_r	relative permittivity (often referred to as dielectric constant);
η	weighting function for N_S in definition of F_{eff} depending on substrate orientation;
κ	relative permittivity (often referred to as dielectric constant);
μ_c	charge carrier mobility contribution from Coulombic scattering;
μ_{eff}	effective mobility;
μ_{FE}	field effect mobility;
μ_i	charge carrier mobility contribution from scattering at the SiC/SiO_2 interface (roughness);
μ_{inv}	inversion layer mobility;
μ_p	charge carrier mobility contribution from phonon scattering;
χ_{ox}	oxide electron affinity;
ψ_S	surface potential;
3D	three-dimensional;
ALD	atomic layer deposition;
BJT	bipolar junction transistors;
BSG	borosilicate glass;
BTS	bias temperature stress;
CMOS	complementary metal oxide semiconductor field effect transistor;
CNL	charge neutrality level;
CVD	chemical vapor deposition;

DFT	density functional theory;
DLTS	deep level transient spectroscopy;
EELS	electron energy loss spectroscopy;
EOT	effective oxide thickness;
FINFET	fin field effect transistor;
HRTEM	high resolution transmission electron microscopy;
IGBT	insulated gate bipolar transistor;
LPCVD	low pressure chemical vapor deposition;
MIS	metal insulator semiconductor;
MOS	metal oxide semiconductor;
MOSCAP	metal oxide semiconductor capacitor;
MOSFET	metal oxide semiconductor field effect transistor;
NIT	near interface trap;
ONO	oxide/nitride/oxide;
PDA	post deposition annealing;
PECVD	plasma enhanced chemical vapor deposition;
POA	post oxidation annealing;
PSG	phosphosilicate glass;
RF	radio frequency;
SCM	scanning capacitance microscopy;
SIMS	secondary ion mass spectroscopy;
TDDB	time-dependent dielectric breakdown;
TEOS	tetraethyl orthosilicate;
TMA	trimethylaluminium;
UMOSFET	U metal oxide semiconductor field effect transistor.
XPS	x-ray photoelectron spectroscopy.

Advancing Silicon Carbide Electronics Technology II   Materials Research Forum LLC
Materials Research Foundations **69** (2020) 63-106   https://doi.org/10.21741/9781644900673-2

1. Introduction

Dielectrics are critically important materials used widely in semiconductor devices. They are used for electrical isolation between conductive elements, such as metal interconnections and semiconductors, as well as for passivation and protection of free semiconductor surfaces. Another important application of dielectrics is their use as the insulating layer in metal-insulator-semiconductor (MIS) capacitors, usually referred to as metal-oxide-semiconductor (MOS) capacitors (MOSCAPs) regardless of dielectric, and MOS field effect transistors (MOSFETs) [1]. The choice of an appropriate dielectric for specific application depends on its electrical properties, such as energy bandgap (E_g), critical electric field (F_{ox}, also known as 'dielectric strength'), relative permittivity (also referred to as dielectric constant and designated here as ε_r or κ) and electron affinity (χ_{ox}), as well as on its chemical and mechanical compatibility with semiconductor device design and processing. These dielectric properties greatly influence the effective operation of semiconductor devices, such as their ability to withstand high electric field, high current density and high temperature. Different dielectrics applicable in SiC devices are discussed in this chapter.

The structural quality of a dielectric layer on a semiconductor is extremely important. For instance, in MOSFETs, the dielectric layer acts as an insulator between the gate contact and the channel, which is a thin (< 5 nm) semiconductor inversion layer at the semiconductor/dielectric interface, along which the flow of electrons is controlled by the gate voltage. The deposition or growth of a gate dielectric layer gives rise to defects located inside the dielectric and close to the dielectric/semiconductor interface. These defects can deteriorate device performance and reliability by reducing the dielectric strength and increasing gate leakage current.

1.1 Interface-trapped charge effects and requirements

The most important type of defect in MOS structures is interface-trapped charge. These defects are physically located at the dielectric-semiconductor interface and can include dangling bonds, Si-Si bonds, C-C bonds and various disorderly complexes created at the boundary between dielectric and semiconductor. These defects act as electrically available states, and electrons and holes can be captured and emitted. The density of interface traps charge per unit area and unit energy is given the symbol D_{it}.

When the interface traps are charged, the charge can decrease the channel conductivity due to Coulomb scattering of electrons. Furthermore, the active trapping and release of charge can further affect device performance. The capture and emission of electrons depends on the applied gate voltage, temperature and the initial trap state, and causes

Advancing Silicon Carbide Electronics Technology II   Materials Research Forum LLC
Materials Research Foundations **69** (2020) 63-106   https://doi.org/10.21741/9781644900673-2

instability in MOSFET operation and drift of its parameters. A semiconductor/dielectric interface always has some trapped charge, and it must be kept below certain limit. This limit can be estimated by recalling that the inversion layer surface charge density, qN_s, is given by:

$$qN_s = C_{ox} (V_{GS} - V_t) \tag{1}$$

where C_{ox} is the gate oxide capacitance per unit area, V_{GS} is the gate voltage, V_t is the threshold voltage, q is the elementary charge and N_s is the charge carriers surface density in the MOSFET channel. If the silicon dioxide (SiO_2) thickness (t_{ox}) is 100 nm and the gate overdrive voltage ($V_{GS} - V_t$) is 5 V then from Eq. 1 it is found that $N_s \sim 10^{12}$ cm^{-2}. From Eq. 1, N_s increases linearly as V_{GS} increases or as t_{ox} decreases. Ideally, D_{it} needs to be lower than 10^{12} cm^{-2}, in the range 10^{10} - 10^{11} cm^{-2}, and the bulk density of electrically active defects in the SiO_2 below 10^{15} cm^{-3} or below 4×10^{-5} at.%.

1.2   Near-interface trap effects

In addition to defect states physically located at the interface, charge can also become trapped in the dielectric. Some states will have constant, static occupancy, independent of gate voltage, and this trapped charge will contribute to shifts in flatband voltage for MOSCAPs and threshold voltage for MOSFETs. Some states, however, are capable of changing their occupancy. The nature, origin and effect of oxide states varies between material systems, a variety of descriptions are used and there is some disagreement among researchers regarding the topic.

For the $4H$-SiC/SiO$_2$ system, the preferred nomenclature in the SiC/SiO$_2$ system is "near-interface traps" (NITs), although the term "border traps" is occasionally used, and they are observed to have an energy level close to the SiC conduction band edge [2]. Modelling using Density Functional Theory (DFT) has showed that CO molecules, C interstitials and C pairs can grow within the oxide, implicating them as the source of NITs [3]. Experiments using a thermally stimulated current technique have experimentally confirmed the presence of traps close to the conduction band edge of $4H$-SiC [4]. C-V and deep level transient spectroscopy (DLTS) measurements have found these traps centered 0.1 eV below the conduction band edge in $4H$-SiC compared to 0.5 eV below the conduction band edge in $6H$-SiC [5, 6].

NITs in the SiC/SiO$_2$ system have a very short time constant, trapping and releasing charge at least as fast as conventional interface-trapped charge. As such, their effect on device structures is generally lumped with conventional interface-trapped charge. NITs are not as widely studied as conventional interface-trapped charge, measured using C-V

techniques, but are important to improving channel mobility and have shown some response to changes in fabrication processes [7].

1.3 SiC MOS interface requirements

In general, trap states and trapped charge must be minimized in device structures. The main impediment to SiC devices is interface charge: both conventional interface-trapped charge, located physically at the interface, and near-interface traps, located sufficiently close to the interface to be electrically active. These affect the mobility of electrons in the channel, as detailed below in Section 2.1, but channel mobility is not the sole requirement of the SiC dielectric system.

The threshold voltage of MOSFETs can be influenced by fixed charge in the oxide and, for power applications, the target for threshold voltage is ~5 V to allow for easy operation but prevent accidental turn-on. Any alternative dielectric process must not drift further from this target than the incumbent process. Moreover, threshold voltage can be further influenced by trap states in the oxide, leading to instability over time and operation cycles. In addition to threshold voltage effects, DC leakage and dielectric breakdown are undesirable and should be controlled. In particular, deteriorating leakage performance over time and operation cycles has been observed in SiC MOS structures using time-dependent dielectric breakdown (TDDB) techniques [8, 9]. Leakage and leakage stability over time must also not be compromised by any alternative dielectric process.

These are specific cases of a general requirement: the dielectric must offer high reliability in addition to low D_{it} and high channel mobility. Many dielectric processes discussed below offer improved channel mobility at the expense of reliability. The current incumbent technology is based on thermal oxidation – SiC is the only compound semiconductor that can be thermally oxidized to produce a SiO_2 layer, but the presence of carbon (C) atoms results in significant deterioration of the oxide quality, for all types of defect and trap state, in comparison to that grown on silicon (Si). For this reason, formation of SiO_2 layers on SiC requires tailored processes, unique from Si, such as post oxidation annealing (POA) or ultrathin interfacial oxides. Achieving a high-quality dielectric layer, with high structural quality and low interfacial charge, remains one of the main challenges in the development of SiC electronics. Different methods of dielectric deposition on SiC are reviewed in this chapter, alongside thermal oxidation.

2.　Dielectrics in SiC device processing

2.1　Silicon dioxide in SiC devices

The success of Si electronics is in no small part due to the fact that a high structural quality SiO_2 can be readily grown on the Si surface by thermal oxidation. SiO_2 is used as the dielectric in SiC MOSFETs, as well as being a surface passivation layer and as a sacrificial layer in device fabrication. Unfortunately, the SiC/SiO_2 interface is poor compared with Si/SiO_2 and contains high levels of charged defects.

Electron mobility in the MOSFET channel is affected by several scattering processes, and by trapping processes. The scattering processes can be represented by Matthiesen's rule:

$$\frac{1}{\mu_{inv}} = \frac{1}{\mu_c} + \frac{1}{\mu_p} + \frac{1}{\mu_i} \tag{2}$$

The first term μ_c corresponds to Coulomb scattering resulting from carrier-carrier interaction, electron scattering by fixed charges at the interface traps and remote scattering by charged defects in the SiO_2 or SiC. The second term μ_p corresponds with phonon scattering and is material dependent. The final term μ_i corresponds with scattering due to the SiC/SiO_2 interface roughness in MOSFETs and dominates mobility at large gate voltages, V_{GS} [10]. These are combined to give the mobility for the inversion layer μ_{inv}.

Experimentally, field-effect mobility, μ_{FE}, is commonly used as a figure of merit for SiC MOSFETs. It can be calculated from MOSFET electrical measurements by [11]:

$$\mu_{FE} = \frac{L\, g_m^i}{W C_{ox} V_{DS}} \tag{3}$$

In Eq. 3, g_m^i is the intrinsic transconductance [12], L is the gate length, and W is the gate width, V_{DS} is the source-drain voltage and C_{ox} is the gate oxide capacitance, measured using a split C-V configuration. μ_{FE} is dependent on the electric field in the channel and, as a result, it is often plotted as a function of effective electric field, F_{eff}, defined by:

$$F_{eff} = \frac{q}{\varepsilon_0 \varepsilon_{SiC}} (N_{depl} + \eta N_s) \tag{4}$$

In Eq. 4, ε_{SiC} is the relative permittivity of SiC, N_{depl} is the depletion charge density close to the SiC surface, N_S is the inversion charge density, and η is the weighting function for N_S that depends on substrate orientation. For Si MOSFETs, $\eta = 1/2$ for electrons on the (100) face, while $\eta = 11/32$ is obtained theoretically [10]. Ohashi *et al.* [13] have shown that $\eta = 1/3$ is a better fit to mobility data in C face SiC MOSFETs and this is expected to be the case for Si face SiC MOSFETs too.

Effective mobility, μ_{eff}, is more commonly used as a figure of merit for Si MOSFETs. A plot of μ_{eff} versus F_{eff} is known as a "universal mobility curve" because it is independent of substrate impurity concentration or bias. Effective mobility can be calculated from MOSFET measurements by:

$$\mu_{eff} = \frac{L\, g_d^i}{W Q_n} \tag{5}$$

In Eq. 5, g_d^i is the channel conductance of the intrinsic device (e.g. when source/drain parasitics are removed). In an ideal MOSFET, free of interface charge, Q_n is equal to the density of mobile charge in the channel, Q_{mobile}, which is calculated from the gate-to-channel capacitance per unit area, C_{GC}, according to [11]:

$$Q_{mobile} = \int_{-\infty}^{V_{GS}} C_{GC} dV_{GS} \tag{6}$$

In the case of SiC MOSFETs, not all channel charge is mobile and so

$$Q_n = Q_{mobile} + Q_{it} \tag{7}$$

where Q_{it} is density of charge trapped at the interface. Trapped charge can be calculated from the measured D_{it} by:

$$Q_{it} = q \int_{E_i}^{E_c} D_{it}\, F_{1/2}(E) dE \tag{8}$$

In Eq. 7, E_C is the conduction band minimum, E_i is the intrinsic Fermi energy (i.e. the function integrates from E_i to the conduction band edge) and $F_{1/2}$ is the Fermi-Dirac integral (of order 1/2). When substituted into Eq. 4, this can be summarized as:

$$\mu_{eff} = \frac{L\, g_d^i}{W(Q_{mobile} + Q_{it})} \tag{9}$$

Interface defects in SiC include impurities, features and decorations from bulk defect propagation. They may contribute to several mobility degradation processes: to Coulomb scattering as fixed or occupied charge states, to surface roughness scattering as points or regions of displacement and/or to trapping processes as variable-occupancy states. For the purposes of this chapter, the effect of interface degradation is considered as a lumped process, without distinguishing between different mobility degradation mechanisms.

4H-SiC, with a standard gate dielectric formed by thermal oxidation at around 1100 °C and without post oxidation annealing (POA) yields channel mobility values below 10 cm^2/(V·s) [14], compared to typical Si values > 200 cm^2/(V·s) [10]. On the other hand, bulk electron mobilities are of comparable magnitude in Si and SiC, being 900 cm^2/(V·s) for 4H-SiC and 1450 cm^2/(V·s) for Si. Therefore, the low electron mobility in $4H$-SiC MOSFETs is a result of poor interface quality between SiC and SiO_2. A large quantity of research over the last 20 years has looked at improving the electron mobility in SiC MOSFETs, for example, by reducing D_{it}.

SiO_2 can also be deposited by a variety of methods, as discussed in Section 3. Some combination of thermally grown and deposited SiO_2 can also be found in applications.

2.2   Silicon nitride in SiC devices

Silicon nitride (Si_3N_4) was investigated in the 1960s as an alternative dielectric to SiO_2 for the scaling of Si devices, because it has a higher permittivity. However, issues with interface quality and charge injection into the nitride made it an unsuitable choice for standard MOSFET technology. The oxide-nitride-oxide (ONO) structure was subsequently proposed for use in SiC power device technology, where it is desirable to employ a gate dielectric with a higher ε_r than SiO_2, as SiC devices need to withstand high fields during operation compared with their silicon counterparts.

Lipkin and Palmour [1] have compared the reliability of a range of dielectrics (SiO_2, Si_3N_4, ONO, AlN) used on SiC for high voltage applications. While silicon oxides exhibited dielectric breakdowns at ~ 10 MV/cm, nitrides exhibited leakage current breakdown fields at ~ 5 MV/cm, which is related to the smaller electron affinity step using nitrides (1 to 2 eV) compared with oxides (~3 eV). ONO layers showed a combination of leakage and dielectric breakdown, with leakage current above 6.5 MV/cm. The breakdown field measured in MOSFETs is similar for oxide and ONO gate dielectrics, but the ONO thickness can be larger than the oxide and have the same capacitance, i.e. the same "effective oxide thickness" (EOT), thanks to the higher dielectric constant in the nitride.

The reliability performance of ONO was demonstrated to be equivalent to SiO_2 by time-dependent dielectric breakdown (TDDB) testing using ONO in a MOS capacitor, which consists of a bottom thermal oxide, LPCVD grown nitride and a pyrogenically oxidized top oxide with an estimated equivalent SiO_2 thickness of 40 nm [15]. Thermal oxide has been demonstrated to be superior to deposited oxide for the bottom oxide layer, but variations in nitride and top oxide formation have not been investigated. Further work using ONO structures showed the impact of underlying defects on the quality of thermal oxides and how the ONO structure can mitigate their impact [16]. ONO gate stacks have been incorporated into a normally-off MOSFET [17]. The ONO stack has been consistently shown to be a stable alternative to thermal oxidation and it can even be stressed reliably with a large negative bias, showing dielectric breakdown field strength of −19.6 MV/cm [18].

2.3 High-κ dielectrics in SiC devices

According to Gauss's law, the product of permittivity and electric field normal to the dielectric/semiconductor interface has to be continuous. Therefore, for a given electric field in a semiconductor, the electric field in the dielectric can be lowered by replacing SiO_2 or Si_3N_4 with a dielectric that has a higher permittivity (κ). These high-κ dielectrics, such as HfO_2, Al_2O_3, $BaTiO_3$, TiO_2 and Ta_2O_5, have been of interest to the Si technology community because of their very high permittivities, and could potentially be of benefit to SiC technology in this regard. However, other material properties of high-κ dielectrics can mean that they are susceptible to leakage current and degradation, even at low electric fields, which detracts from their utility [1].

High-κ dielectrics have been used in CMOS logic technologies since the 45 nm node was introduced in 2007 [19]. To achieve the ever-higher drive currents required by Moore's Law, these MOSFETs required ever-thinner gate oxides, which increases gate leakage by quantum mechanical tunneling. This leakage current can be minimized by using thicker layers of high-κ dielectrics that maintain the same capacitance and EOT, while increasing the gate oxide potential barrier to electrons. However, it is necessary to incorporate an ultrathin thermal SiO_2 layer prior to high-κ dielectric deposition, in order to reduce interface states to an acceptable level. The choice of high-κ dielectric also depends on there being a large enough electron affinity difference between the channel semiconductor and the gate dielectric to create a potential barrier to electrons in the gate dielectric and thereby minimize tunnel current.

SiC has a larger band gap than Si, so the potential barrier in the conduction band at the SiO_2/SiC interface is smaller. Consequently a larger tunneling current occurs than would be observed in an equivalent Si MOS gate stack with the same gate oxide thickness [8]. HfO_2, which has been used as a preferred high-κ material for Si technology, is not suitable for the SiC system due to a small electron affinity step (barrier height) at the HfO_2/SiC interface as shown in Fig. 1 [20, 21]. Therefore other high-κ materials that offer a wider bandgap and more favorable electron affinity step have been considered e.g. Al_2O_3 [21-25], $LaSiO_x$ [26] and AlN [27].

Among those materials, Al_2O_3 offers good performance in terms of gate leakage reliability and MOSFET channel mobility, but is not without problems. Tanner *et al.* [28] claimed that a deposited Al_2O_3 layer was crystallized by post deposition annealing (PDA) at 1100 °C. This reduced the electric field needed to generate a leakage current of $1\times10^{-6}\,A\cdot cm^{-2}$ from $5\,MV\cdot cm^{-1}$ to $1\,MV\cdot cm^{-1}$. Additionally, Al_2O_3 suffers from trapping and de-trapping of electrons, a mechanism due to oxygen vacancies [29]. This results in hysteresis in MOSFET I_D-V_{GS} characteristics, which can be mitigated by

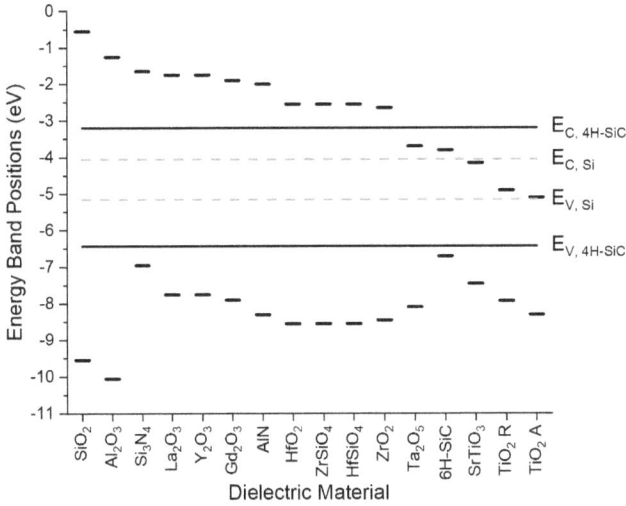

Figure 1.   Energy bandgaps of dielectrics compared with Si and SiC.

annealing during processing. Despite these issues, the maturity of Al_2O_3 technology and its integration with semiconductors results in it being commonly used in MOS structures.

For the formation of a good MOSFET gate stack, it is useful if there is a thin SiO_2 layer between the channel and dielectric layer to screen the remote scattering. The high-κ oxide layer should also be thin, so that the defect density is low, and so there are fewer traps that may otherwise contribute to I_D-V_{GS} hysteresis [29]. High-κ dielectrics must be deposited, and atomic layer deposition (ALD) is the preferred fabrication method for high-κ dielectrics, as discussed in Section 3.3.

3.   Methods of dielectric deposition used in SiC device processing

Thermal oxidation of SiC produces a relatively poor SiC/SiO_2 interface in comparison with Si/SiO_2. Excess C is produced during oxidation that cannot incorporate into stoichiometric SiC or SiO_2 and so gives rise to C related defects close to the interface. These lead to a high density of interface states and hence a low channel mobility in SiC MOSFETs. One potential approach to mitigate against this is to deposit an oxide rather than consuming Si and C in thermal oxidation. SiO_2 may be deposited either by evaporation or by sputtering but these methods can only offer a "line-of-sight" deposited

Materials Research Forum LLC
https://doi.org/10.21741/9781644900673-2

film and so the layers are non-conformal [30]. The chemical vapor deposition (CVD) methods discussed below offer conformal coating, making them more common for fabrication of gate stacks and passivation dielectrics in SiC devices.

3.1 Plasma enhanced chemical vapor deposition of dielectric on SiC

CVD is the formation of a solid film on a substrate by the chemical reaction of vapor phase reactants that have the correct constituents. The reactants are introduced into a reaction chamber at a specific temperature to achieve required chemical reactions. Typical CVD processes for commonly used dielectrics in SiC technology, such as SiO_2, make use of silane (SiH_4) through the following reaction:

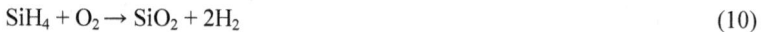

$$SiH_4 + O_2 \rightarrow SiO_2 + 2H_2 \tag{10}$$

Unlike conventional CVD, which relies on thermal energy to initiate and maintain chemical reactions, plasma enhanced CVD (PECVD) uses an RF induced glow discharge to provide energy to the reacting chemical species [30]. As a result, it enables higher deposition rates at lower temperatures. Desirable properties of PECVD films include reasonable electrical parameters, good long-term reliability, low pinhole count, good adhesion, good step coverage and conformity with underlying surfaces.

A glow discharge or plasma is created by an RF field applied to a low pressure gas, which creates free electrons. These electrons gain enough energy in the applied electric field that, when they collide with reactant gas molecules, the gas molecules decompose. These energetic chemical species are then adsorbed onto the surface where a film builds up. Improved film quality compared with thermal CVD processes is achieved because the chemical species (radicals) form stronger bonds to the surface and can more easily migrate along the surface. A potential problem with the use of these higher energy chemical species is poorer stoichiometry in the deposited films.

Some of the earliest studies on 6H-SiC claimed CVD SiO_2 quality to be as good as thermally grown oxide [31]. As SiC technology progressed and the problems with thermally grown gate oxides became clearer, PECVD deposited films, along with a suitable post oxidation anneal, were pursued and shown to give satisfactory interface state densities $\sim 10^{11}$ $1/(cm^2 \cdot eV)$ at 0.2 eV below the conduction band edge, resulting in lateral MOSFETs with mobilities over 50 $cm^2/(V \cdot s)$ [32-34]. For the CVD process, the most common implementation is PECVD with a deposition temperature of approximately 400 °C and with silane and O_2 as silicon and oxygen precursors, although other precursors (e.g. disilane, N_2O) can be used.

Masato *et al.* [35] compared an oxide thermally grown in N_2O to a PECVD deposited SiO_2 with post oxidation N_2O and N_2 anneal. The deposited oxide demonstrated a marginally better mobility at 26 cm^2/(V·s), compared to 20 cm^2/(V·s) for the thermally grown oxide. Moreover, the deposited oxide showed significant improvement in time-dependent dielectric breakdown testing, withstanding on average 70 C/cm^2 of charge injected into the gate prior to breakdown, compared to 27.5 C/cm^2 for the thermal oxide [35].

Another advantage of PECVD-deposited oxides is their thickness uniformity compared to thermal oxides for three-dimensional (3D) structures in SiC. For example, a trench in SiC, needed in the fabrication of UMOSFETs, will reveal different crystal planes. The SiC oxidation rate is different for different crystal planes, which will result in different thermally grown oxide thicknesses, while a conformally deposited PECVD oxide is of uniform thickness. PECVD has proved beneficial in the fabrication of 3D (or FINFET style) gate structures in 4*H*-SiC, which have 16× improved drain current density (compared to a planar device) due to the utilization of crystal planes with higher mobility [36].

3.2   Deposition of silicon oxide films using TEOS

Silane used for SiO_2 deposition is pyrophoric – it ignites spontaneously on contact with air. A safer alternative precursor for SiO_2 deposition is tetraethyl orthosilicate (TEOS), $Si(OC_2H_5)_4$, which is a colorless liquid that degrades in water:

$$Si(OC_2H_5)_4 + 2H_2O \rightarrow SiO_2 + 4C_2H_5OH \tag{11}$$

TEOS is relatively inert and liquid at room temperature. Its vapor can be supplied to the reaction chamber by a bubbler and N_2 carrier gas or by direct liquid injection. TEOS can be deposited via low pressure CVD (LPCVD) or PECVD. The chemical reaction for SiO_2 deposition using TEOS at temperatures above 600 °C is:

$$Si(OC_2H_5)_4 \text{ (liquid) } \rightarrow SiO_2 \text{ (solid) } +2C_2H_4 \text{ (gas) } +2H_2O \text{ (gas)} \tag{12}$$

In the PECVD process, TEOS oxide can be deposited at lower temperatures (below 450 °C):

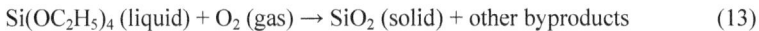

$$Si(OC_2H_5)_4 \text{ (liquid) } + O_2 \text{ (gas) } \rightarrow SiO_2 \text{ (solid) } + \text{other byproducts} \tag{13}$$

Generally, deposited TEOS oxides have been used to passivate the SiC surface, or create field oxides, given the method has a high deposition rate and is well-suited to thick oxide films. It is not generally used as a gate oxide, as its breakdown electric field strength has been shown [37] to be 60% of the equivalent oxide developed via a PECVD/silane process and the TEOS oxide also had 2-3× greater interface trap density. Another study [38, 39] reported an 8× improvement in channel mobility using a PECVD/TEOS oxide compared to counterparts with gate oxide formed by thermal oxidation (from 5 up to 40 $cm^2/(V \cdot s)$). However, these devices highlighted a key problem with TEOS layers as a gate dielectric, which is increased gate leakage. This is likely due to the quality of the TEOS dielectric, which can include pin holes, defects or trapped charge originating from dangling bonds. This often makes a densification process necessary after the TEOS oxide deposition to improve the layer's stoichiometry [40].

Deposition of a gate oxide from TEOS has been combined with oxide doping by phosphorous to passivate the near-interface traps (phosphidation treatments are discussed in more detail in section 4.5). Lateral 4*H*-SiC MOSFETs fabricated by using this technique have demonstrated a channel mobility of 80 $cm^2/(V \cdot s)$ [41].

3.3   Atomic layer deposition of gate dielectrics in SiC devices

Atomic Layer Deposition (ALD) is another method of depositing dielectrics on various substrates by utilizing a vapor phase technique. Owing to its sequential chemical process, the deposited layer thickness can be controlled to Angstrom level of accuracy with excellent conformality, even in high aspect ratio structures [42]. Deposition of high-κ dielectrics such as Al_2O_3 [43, 44] and SiO_2 [45] on SiC substrates have been reported. A schematic of a process for depositing Al_2O_3 on 4*H*-SiC is shown in Fig. 2. Initially, the 4*H*-SiC surface is terminated with a hydroxyl (OH) group following exposure to air as shown in Fig. 2a. Once the wafer has been inserted in the ALD chamber, a first chemical precursor, trimethylaluminium (TMA), is introduced. An ALD precursor will adsorb to and react with the 4*H*-SiC surface, producing a single uniform monolayer on the surface as depicted in Fig. 2b. Then the TMA and any by-product elements are pumped away before introducing water vapor in the chamber as shown in Fig. 2c. Now the water vapor reacts and replaces the CH_3 groups with OH groups. A further purge removes the remaining H_2O and CH_4 produced, leaving the surface terminated with OH groups. The cycle can then be repeated to grow a film of Al_2O_3 of desired thickness. During this process, the chamber pressure is kept at 600 mTorr with a temperature below 300 °C to prevent the surface from being oxidized. The precursors are transported to the reaction chamber by vapor draw with N_2 carrier gas. For ALD SiO_2 deposition on 4*H*-SiC, Yang *et al.* [45] used 3-Aminopropyltriethoxysilane, H_2O and O_3 as precursors, but ALD deposition of SiO_2 is not as mature as CVD processes and other precursors, such as bis(tert-butylamino)silane and bis(diethylamino)silane can be used.

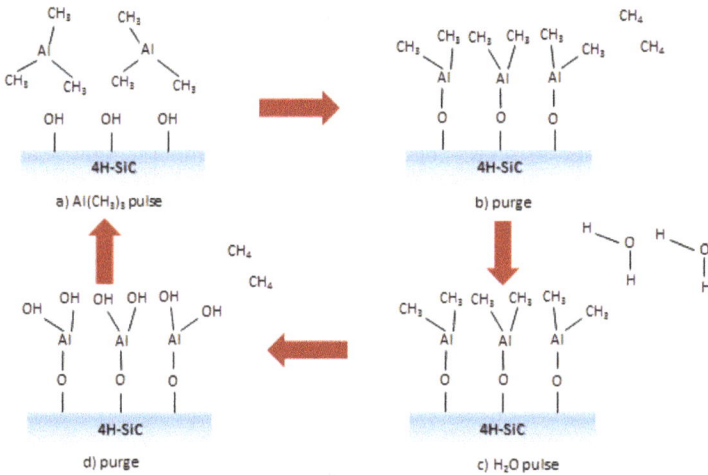

Figure 2. *ALD reaction cycle showing the growth of Al$_2$O$_3$ using TMA and water as precursors, with CH$_4$ as a by-product.*

Both ALD Al$_2$O$_3$ and ALD SiO$_2$ have been used as a gate oxide for 4*H*-SiC MOSFET fabrication. In 2009, Lichtenwalner *et al.* [23] reported a mobility of more than 100 cm^2/(V·s) in 4*H*-SiC MOSFETs using, as a gate dielectric, an ALD deposited Al$_2$O$_3$ with following post deposition annealing (PDA) at 400 °C for 30 s. Prior to the deposition of Al$_2$O$_3$, the samples were annealed in NO at high temperature for a short time to grow a SiO$_2$ layer and to control D_{it}.

Yang *et al.* [45] deposited 30 nm of SiO$_2$ by ALD and subsequently performed PDA in a nitrous oxide (N$_2$O) ambient. The highest electron mobility of 26 cm^2/(V·s) was achieved by performing PDA at 1100 °C for 40 s. The gate oxide could withstand effective fields up to 6 MV/cm within a leakage current range of 1×10^{-7} A/cm^2. This value of maximum electric field is small compared to the thermally grown SiO$_2$, which can typically withstand up to 10 MV/cm. In other work, Yang *et al.* [26] inserted 1 nm of lanthanum silicate (LaSiO$_x$) between ALD deposited SiO$_2$ and 4*H*-SiC to form a gate stack. Peak mobility of 132.6 cm^2/(V·s) was found, with 3× larger current carrying capability compared to gate oxide without La$_2$O$_3$ but no F_{ox} data was given.

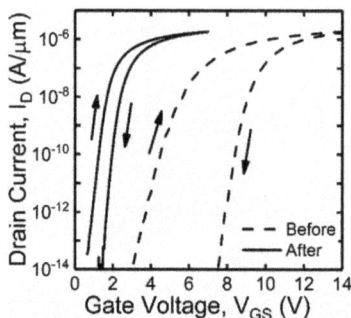

Figure 3.   Before and after densification process of Al_2O_3 gate dielectric in MOSFETs. Reprinted, with permission, from [43]. © 2018 IEEE.

3.4   Densification of dielectrics deposited on SiC

As-deposited dielectrics are in general not lattice matched to the SiC substrate. They typically contain many defects and are more loosely packed than in their ideal crystalline state. In electrical terms, this means an uncontrolled increase in charge traps, which adversely affects electronic devices. Annealing such deposited dielectrics results in their densification and thereby improves their properties. The higher temperature of PDA allows the migration of constituent ions and the reduction of defect states in order to densify the dielectric.

Al_2O_3 is known to have quite a high defect concentration [29]. In particular the O vacancy has 5 stable charge states (+2, +1, 0, -1, -2) in the energy band gap, corresponding with four possible charge state transitions. If Al_2O_3 is used in the gate stack of a SiC MOSFET, the change in charge state of O vacancies changes the level of trapped charge in the oxide and so shifts the threshold voltage. As V_{GS} is swept in either direction, charge state transitions occur at different points, giving rise to a hysteresis in the I_D-V_{GS} curve, as shown in Fig. 3. Annealing performed immediately after the oxide deposition alleviates this problem [43].

3.5   Deposition methods conclusion

Deposition methods are of particular interest because no underlying substrate material is consumed and no carbon is incorporated into the oxide and interface. PECVD using silane offers high deposition rates, LPCVD and PECVD using TEOS can offer high deposition rates and good passivation, but require densification, and ALD offers excellent conformal coverage and fine control for low oxide thickness. Deposited dielectrics benefit from PDA to improve oxide quality and/or interfacial layer formation to control D_{it}. In particular, ALD deposited Al_2O_3 contains high concentrations of oxygen vacancies as-deposited, and requires post-deposition annealing to control hysteresis – annealing processes are discussed in more detail in Section 4.5. Deposited dielectrics are not used commercially as gate oxides: PECVD and LPCVD are used for SiC field oxides and surface passivation (discussed in more detail in Section 6), whereas ALD is currently limited to academic research.

4.   Thermal oxidation of SiC

SiC is the only compound semiconductor in which thermal oxidation results in the formation of a native oxide SiO_2. During the SiC thermal oxidation process, the volume of grown SiO_2 is equal to 2.16× the volume of SiC consumed, which is similar to the case of Si oxidation. In this section, we report first on the mechanics of SiO_2 growth on SiC by thermal oxidation and show the resulting oxidation rates. The problems and issues with these insulating layers are then discussed, given the presence of significant trapped C close to the interface [46].

4.1   SiC oxidation rates and a modified Deal-Grove model

For Si, the kinetics of thermal oxidation was modeled by Deal and Grove in 1965 [47] and this model is still an accepted approximation of ultimate oxide thickness, given process temperatures and times. The model assumes an initial layer of SiO_2 and that oxidation occurs at the Si/SiO_2 interface. Oxide growth is limited by the inward movement of the oxidant rather than the outward movement of Si. As laid out in the original work [47], the following three steps make up the kinetics of Si thermal oxidation:

1. a flux of oxidant species arrives at the oxide film's outer surface;

2. it is transported across the oxide film towards the semiconductor;

3. it arrives at the semiconductor surface and reacts to form new SiO_2.

However, when considering the specific case of SiC, this simple case no longer holds true and one must factor in the removal of excess C. Given that:

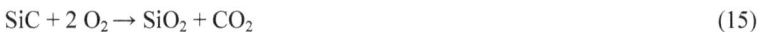

$$SiC + 3/2\ O_2 \rightarrow SiO_2 + CO \qquad (14)$$

$$SiC + 2\ O_2 \rightarrow SiO_2 + CO_2 \qquad (15)$$

two more steps can be added to the above as follows [48]:

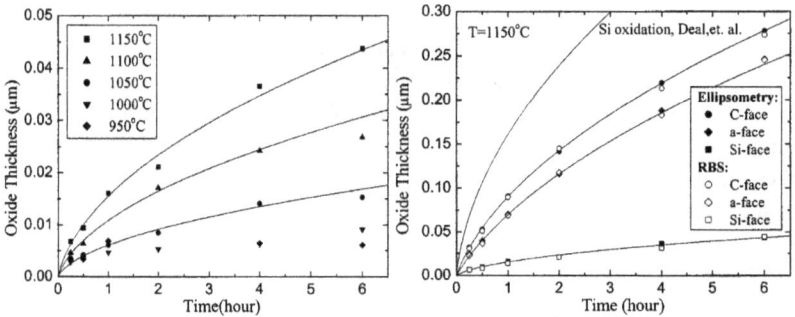

Figure 4.   Left, oxide thickness as a function of time and temperature on the (0001) Si face of 4H-SiC. Right, oxide thickness as a function of time and SiC orientation on 4H-SiC, with Si oxidation as a reference. Reproduced from [48] with permission from AIP publishing.

4. out-diffusion of product gases (e.g. CO) through the oxide film;

5. removal of product gases away from the oxide surface.

In 2004, Song *et al.* [48], produced a modified Deal-Grove SiC oxidation model using these five steps as their starting point. The oxide thickness using this model is predicted as:

$$d_{ox}^2 + Ad_{ox} = Bt \tag{16}$$

where d_{ox} is the resulting SiO_2 thickness after an oxidation time t, B is a parabolic rate constant and B/A is a linear rate constant – these are derived in full in [48]. Subsequently, this model has been further improved [49] to better account for the early stages of oxidation. However, using Eq. 16, it is possible to approximate the oxidation rate for a number of temperatures, times and for different SiC crystal faces, as shown in Fig. 4. Oxidation on the Si face (0001) is slower than both the vertical a-plane (11$\bar{2}$0) by 6-10×, and the comparatively unused C face (000$\bar{1}$). The vast majority of SiC devices use the long-established Si face, however, both the C face and the a-plane are of interest due to reports of high mobility on these surfaces, which shall be further discussed in Section 4.6.3. The a-plane can be exploited in SiC trench MOSFET architectures, while the C face of SiC has been used in the development of SiC IGBTs (insulated gate bipolar transistors) [50].

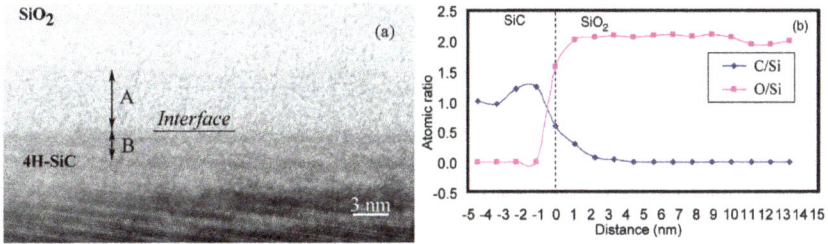

Figure 5.   (a) High resolution transmission electron microscope (HRTEM) image showing transition layers appearing on both sides of the SiC/SiO₂ interface and extending for the depth of 4.8 nm in to SiO₂ and 3.3 nm in to SiC; (b) Profiles of the C/Si and O/Si ratios as measured using EELS. Note the increased carbon content on both sides of the interface which correlates with the interfacial transition layers displaying a distinct contrast in the HRTEM image. Reproduced from [52], with the permission of AIP Publishing.

4.2   Interface traps introduced during thermal oxidation of silicon carbide

As described in Eq. 14 and Eq. 15, unlike Si thermal oxidation, C is released during SiC thermal oxidation, for example as CO or CO_2 gases. However, in reality, a small amount of C remains close to the SiO_2/SiC interface according to the following reaction:

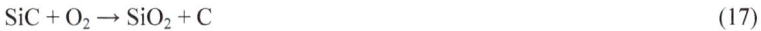

$$SiC + O_2 \rightarrow SiO_2 + C \tag{17}$$

What happens to the C remains the subject of some debate, though a wealth of evidence links the presence of carbon clusters to high interface state densities and hence low channel mobilities.

High resolution transmission electron microscopy (HRTEM) in Fig. 5a shows the existence of nm-scale transition layers at the SiC/SiO_2 interface [46, 51, 52]. For a 62 nm thermally grown SiO_2 layer, electron energy loss spectroscopy (EELS) data in Fig. 5b confirm a non-stoichiometric C/Si ratio extending 4 nm into the SiO_2 and 4 nm into the SiC. This can mean an excess of C or a deficit of Si in the SiC, or both. Specific defect identities cannot be assigned from this data but C interstitials, ternary SiO_xC_y phases and amorphous SiC are possible explanations. Graphitic features were detected using Raman spectroscopy on 4*H*-SiC surfaces following oxidation and etch [53], which could correspond to C clusters in SiC. Recently, rather than trying to mitigate the effects of excess C resulting from thermal oxidation of SiC to form a MOSFET gate stack, Arith *et al.* have minimized the formation of excess C by minimizing thermal oxidation of SiC [43]. They grew less than 1 nm of SiO_2 by low temperature (600 °C) oxidation. In this way a field effect mobility of $125 \text{ cm}^2/(\text{V·s})$ was obtained in $4H$-SiC MOSFETs, indicating a high quality SiC/SiO$_2$ interface.

The C that resides at the interface between the SiC and the SiO$_2$ introduces trap states, with energy levels within the SiC bandgap. So-called "deep" interface traps have energy levels close to the middle of the SiC bandgap, which results in a net positive or negative charge at the interface. The polarity of deep interface traps depend on their energy relative to both the Fermi level and the charge neutrality level (CNL) of SiC, which has been simulated to be 1.71-1.85 eV above the valence band for $4H$-SiC [54]. Gap state energies lower than the CNL, but greater than the Fermi level will act as ionized donors resulting in a net positive charge. Gap state energies greater than the CNL, but lower the Fermi level act as ionized acceptors resulting in a net negative charge.

Deep level traps are considered to have a major detrimental impact on the carrier lifetime of $4H$-SiC, particularly the $Z_{1/2}$ ($E_C - 0.65$ eV) and EH$_{6/7}$ centers ($E_C - 1.55$ eV) [55]. These can be largely reduced or suppressed via thermal oxidation at 1150–1300 °C, minimizing their impact on the SiC channel. However, this process has been shown, via deep level transient spectroscopy (DLTS) studies on oxidized $4H$-SiC, to result in a significant increase in the C interstitials related to the HK0 centre ($E_V + 0.78$ eV) [55]. This hypothesis was reinforced by theoretical work identifying the HK0 as a di-interstitial of C that forms in the substrate after oxidation [56] that cannot be readily passivated by post oxidation annealing (POA). This deep level trap, it is concluded, is the likely cause of poor channel mobility of electrons in SiC MOSFETs.

In addition to the electrical action of residual carbon, it also results in increased interface roughness. As carbon tends to form clusters, rather than being distributed uniformly across the interface, interface roughness increases and mobility is degraded by interface roughness scattering as well as Coulomb scattering and carrier trapping. As discussed in Section 2.1, these effects are considered as a lumped process in this chapter.

Many approaches have been investigated to mitigate the effects of the presence of C. For example, re-oxidation where oxygen is supplied at a lower temperature after oxide growth to remove C clusters, i.e. CO molecules combine with oxygen and are moved away in the form of CO$_2$ [57, 58]. Argon annealing at 600 °C has also shown to be effective at reducing the presence of C clusters in the SiC/SiO$_2$ interface [59]. Other methods include the use of N containing gases, like N$_2$, NO and N$_2$O, either during oxidation or POA [60].

Similar to silicon technology, wet oxidation can give a much higher oxidation rate than dry oxidation. The increased oxidation rate significantly reduces control over the oxide thickness and uniformity. In addition, wet oxidation generates a high quantity of stacking faults and surface pits, causing deterioration in SiO_2/SiC interface quality and oxide breakdown strength [61, 62]. Due to these issues, wet oxidation is not commonly used in SiC.

4.3 High temperature oxidation

In Si device technology, gate oxidation is typically performed at a temperature between 800 °C and 1000 °C. Thermal oxidation of SiC initially mimicked Si technology, but with an upper temperature limit of 1200 °C due to the slower rate of oxidation. Oxidation at 1300 °C was suggested to offer a decrease in C content and led to a lower D_{it} [63]. MOS capacitors with thermal oxide grown at 1400 °C also showed a similar result [64]. The reason suggested was a higher oxidation rate of C than Si at these elevated temperatures, leading to faster removal of carbon atoms from the interface.

There has been evidence to suggest that using even higher temperatures, up to 1600 °C, may offer further benefit. This can be achieved using a conventional dry oxidation process in a tube furnace, itself made from SiC, rather than quartz, while some success has been demonstrated using rapid thermal oxidation [65]. A reduction in D_{it} has been observed [66] for both 1500 °C and 1600 °C MOS capacitor structures formed via dry oxidation. For a lateral MOSFET structure utilizing a p-type epitaxial layer, a 1500 °C oxidation anneal produced a channel mobility of 40 cm^2/(V·s) [66]. It has also been highlighted [67] that, as well as a higher temperature being important, the final D_{it} values are proportional to the flow rate of oxygen through the tube furnace, and hence minimizing flow rate can improve ultimate channel resistance. SIMS (secondary ion mass spectroscopy) and XPS (x-ray photoelectron spectroscopy) on thermally grown SiC/SiO_2 interfaces up to 1350 °C [68] have shown that those oxides grown in higher temperatures have a thinner transition layer of SiO_x (0 < x < 1), which is suggested as the reason for reduced D_{it} under these conditions.

Work on rapid thermal oxidation ranging from 1200 °C – 1700 °C has shown an optimum process temperature of 1450 °C for obtaining the lowest D_{it} value [65]. 1700 °C is deemed unreasonably high as this is too close to the melting point of SiO_2 at 1710 °C. Cooling after the rapid thermal oxidation process, at any temperature, can result in unintentional oxide growth and shows the importance of gas flow control during but also after oxidation.

3C-SiC does not suffer as much as 4H-SiC in terms of channel mobility degradation due to its narrower bandgap, with fewer traps lying within it. One study [69] has shown it is also less sensitive to growth temperature, with a channel mobility of 70 cm^2/(V·s) achieved at oxidation temperatures of 1200 °C, 1300 °C and 1400 °C. Processing beyond

1400 °C is not possible due to 3C-SiC usually being epitaxially grown on a Si substrate [69].

4.4 Low temperature oxidation

Shen and Pantelides [56] suggested that immobile C di-interstitial defects $(C_i)_2$ are formed in SiC as a result of thermal oxidation. As already mentioned, defects of this kind may be responsible for the poor channel mobility observed in SiC MOSFETs. It can therefore be proposed to grow a thin SiO_2 layer at low temperature as a route to decrease the density of these C related defects. This approach requires the additional deposition of a gate dielectric over the thin SiO_2 layer to reduce the gate leakage current.

Hatayama *et al.* [25] measured peak mobilities as high as 300 cm^2/(V·s) in SiC MOSFETs utilising a 0.7 nm thick SiO_2 gate dielectric grown at low temperature (600 °C) on SiC, followed by a deposition of Al_2O_3 dielectric. They concluded that an interfacial oxide layer with thickness above 2 nm degrades the interface and channel mobility.

A field effect mobility of 125 cm^2/(V·s) and a subthreshold slope (S) of 130 mV/dec were obtained by Arith *et al.* [43] in enhancement mode 4H-SiC MOSFETs with a channel length of 2 µm. This combination of high mobility, D_{it} levels in the range from 6×10^{11} - 5×10^{10} 1/(cm^2·eV) and low S is strong evidence of a good control of charged defects in the channel region [70]. S is the inverse gradient of the transfer characteristic, $\log(I_D)$ versus V_{GS}. Given the exponential dependence of I_D on V_{DS} in the subthreshold regime of MOSFET operation, S is given by:

$$S = n\frac{kT}{q}\ln(10) \tag{18}$$

where k is the Boltzmann's constant, and T is the absolute temperature. For an ideal MOSFET $n = 1$ and $S = 60$ mV/dec. The ideality factor n in Eq. 18 can be written as:

$$n = \frac{C_{ox}+C_{dep}+C_{it}}{C_{ox}} \tag{19}$$

where C_{dep} is depletion capacitance per unit area and C_{it} is the capacitive term per unit area associated with D_{it} [11].

The devices were fabricated using an oxidation at 600 °C for 3 min in dry oxygen at atmospheric pressure [43]. The resulting oxide layer thickness was only 0.7 nm thick, thereby limiting the formation of defects in SiC following oxidation, and a 40 nm thick Al_2O_3 formed by ALD. A high mobility was maintained over a wide gate voltage range, as shown in Fig. 6. The fabricated gate stack can also withstand electric field up to 6.5 MV/cm with a leakage current density of 1×10^{-6} A/cm^2, thus indicating that this gate

Advancing Silicon Carbide Electronics Technology II
Materials Research Foundations **69** (2020) 63-106

Materials Research Forum LLC
https://doi.org/10.21741/9781644900673-2

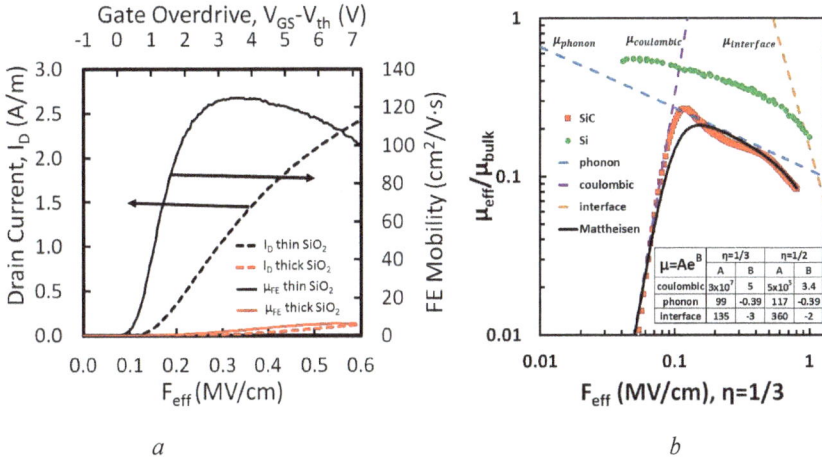

Figure 6. (a) I_D-V_{GS} *transfer curve and field effect mobility by utilizing low temperature oxidation techniques. Reprinted, with permission, from [43] © 2018 IEEE. (b) effective mobility peaks at 265 cm²/(V·s) and device performance is up to 50% of that observed in Si MOSFETs as shown in a plot of mobility versus effective electric field. Reprinted, with permission, from [71]. © 2019 IEEE.*

oxide stack is robust. This work was extended by Urresti *et al.* [71] who demonstrated peak effective mobility of 265 cm²/(V·s) and device performance was shown to be up to 50% of that observed in Si MOSFETs, when compared by normalised universal mobility versus effective electric field as shown in Fig. 6b. The temperature dependence of field effect mobility revealed that coulombic scattering has been sufficiently reduced to make phonon scattering the dominant mechanism controlling carrier transport, further indicating that the use of a thin (0.7 nm) SiO_2 layer in the gate stack can control defects related to C that remain after oxidation. While these results are promising, at the time of writing there is no peer reviewed evidence whether the low temperature grown thin oxides suffer from V_t instability or how stable they are in voltage or temperature stressing.

In addition, Kim *et al.* [72] have grown SiO_2 layers for MOS capacitors using direct plasma-assisted oxidation at room temperature. They claimed that the concentration of silicon oxycarbides (SiO_xC_y) is substantially reduced compared to thermally grown SiO_2 because the oxidation reaction mechanisms are different, and this leads to a reduced $D_{it} \sim 10^{11}$ 1/(cm²·eV).

4.5 Post oxidation annealing

Performing POA in a particular gas ambient at temperatures of around 900 – 1400 °C after oxide formation is another method used to enhance oxide/4H-SiC interface quality and thus to improve electron mobility. One of the most promising POA methods is nitridation, which is employed in nitrogen rich gases such as nitric oxide (NO) [73, 74], nitrous oxide (N_2O) [75, 76] or ammonia (NH_3) [77, 78]. Chung et al. [73] demonstrated an improvement in electron mobility from single digits in MOSFETs using the as-grown oxide as a gate dielectric to 37 cm^2/(V·s) in MOSFETs with the gate dielectric formed by thermal oxidation followed by POA in NO at 1175 °C for 2 hours [74]. During the nitridation process, nitrogen passivates interface traps by forming strong bonds with Si dangling bonds. In addition, the residual clustered C produced during thermal oxidation is also effectively removed [79, 80]. For example, a POA in NO at 1250 °C for 70 min [81], has been shown to reduce interface trap densities down to below 1×10^{12} 1/(cm^2·eV). A correlation between N concentration and electron mobility was reported by Rozen et al. [82, 83], who showed that interface trap density decreased and peak mobility increased as a function of nitridation time. A minimum trap density was achieved after 4 hours in NO at 1175 °C, resulting in a mobility of 45 cm^2/(V·s).

POA in N_2O is widely used as an alternative solution because NO gas is highly toxic [75]. Jamet et al. [84] revealed that POA with either NO or N_2O produced an almost identical effect at the interface, due to the fact that N_2O decomposes into NO gas at a temperature of around 1200 °C [8, 85].

Another effective technique to reduce interface traps and enhance electron mobility is to perform POA in $POCl_3$ [86]. Peak mobility as high as 89 cm^2/(V·s) was achieved in MOSFETs with thermally grown gate oxide annealed in $POCl_3$ at 1000 °C. This mobility level was further improved to 101 cm^2/(V·s) with multistep $POCl_3$ annealing [87]. The conversion of thermally grown SiO_2 into phosphosilicate glass (PSG) during $POCl_3$ annealing can be attributed to the suppression of interface traps and an improvement of channel mobility [7]. Jiao et al. [88] reported that the percentage of phosphorus uptake at the PSG/SiC interface within the channel region depends on the $POCl_3$ annealing temperature. The lowest value of D_{it} was achieved after POA in $POCl_3$ at 900 °C, the lowest temperature tested. However, PSG gate oxide has a polarization effect and thus a gate instability issue arises with higher phosphorus uptake at the interface [88]. One of the main reasons for this is the presence of oxide traps in the PSG and near the interface traps [24].

POA in boron gas is another alternative to passivate interface traps and improve electron mobility in the 4H-SiC MOSFET. By a similar mechanism to that with PSG, thermally grown SiO_2 is converted to borosilicate glass (BSG) by a two-step annealing process. A mobility of 102 $cm^2/(V \cdot s)$ has been obtained with a low D_{it} value of 9.0×10^{11} $1/(cm^2 \cdot eV)$ [89]. Boron atoms were uniformly distributed in the oxide and effectively passivated the active interface traps. The improvement in electron mobility has been suggested to be due to stress relaxation in the SiO_2 structure [89]. Recently, Cabello *et al.* [90] reported peak electron mobility as high as 160 $cm^2/(V \cdot s)$ with a nitrided gate oxide followed by boron annealing. Relatively good V_t control was observed under positive and negative bias stress instability testing at room temperature with a boron annealed gate oxide [90].

4.6 Thermal oxidation conclusion

SiC benefits from being the only compound semiconductor with a stable thermally-grown oxide (SiO_2), but the residual carbon from the oxidation process significantly impacts on thermally-grown oxides. Some carbon escapes as CO and CO_2, but some remains as graphitic carbon clusters at the interface and contributes significantly to D_{it} and mobility degradation. Oxidation at higher temperature and with low oxygen flow rate can mitigate the formation of carbon clusters, promoting the formation of CO and CO_2, but is limited by the melting point of quartz chambers and the SiC itself. High-quality thin interfacial layers can be formed with low temperature oxidation, producing a foundation structure for a deposited dielectric. Post-oxidation annealing (POA) offers further opportunities to passivate residual carbon: phosphorus-based and boron-based processes have demonstrated the highest channel mobility, but nitrogen-based processes offer good channel mobility and excellent stability and reliability. The preferred POA process is nitridation, using NO or N_2O.

5. Other methods to improve channel mobility

5.1 Sodium enhanced oxidation

Sodium enhanced oxidation is a method to improve channel mobility where sodium (Na) is present in the furnace during thermal oxidation [91]. A mobility up to 170 $cm^2/(V \cdot s)$ was reported [92], the result of Na ions increasing the oxidation rate and reducing the formation of interface traps [93]. However, mobile Na ions can diffuse into the gate oxide and cause instability in threshold voltage, and this behaviour has been well recognised in Si technology [70]. Lichtenwalner *et al.* [94] reported the use of group I alkali elements Rb and Cs and group II alkaline earth elements Ca, Sr and Ba as interface passivation materials prior to thermal oxidation of the MOSFET channel region. Those MOSFETs with gate stacks consisting of an ultrathin layer of Ba (~ 0.6 - 0.8 nm) and 30 nm of deposited SiO_2 produced a mobility of 85 $cm^2/(V \cdot s)$. This technique also results in a reduction in D_{it} close to the conduction band compared to thermally grown gate oxide and NO treated gate oxide. Unlike Na-contaminated gate oxide, which contains a large number of mobile ions, the Ba interlayer gate stack demonstrated a consistent threshold voltage with a slight hysteresis of 0.8 V during positive bias temperature stress (BTS) measurements. This suggests that the Ba atoms are less mobile, and hence become strongly bonded and effectively passivate the interface traps.

5.2   Counter doped channel regions

Counter doped MOSFETs have been investigated by implanting group-V elements into the channel region of n-channel MOSFETs. This technique was first introduced by Ueno *et al.* [95] who implanted nitrogen ions into the SiC channel region prior to the formation of the gate oxide. Channel mobility increased and the threshold voltage decreased with increasing nitrogen dose [95]. Instead of direct implantation, interface doping in the channel region can also be formed by POA. Fiorenza *et al.* [96] observed the heavily doped nitrogen and phosphorus atoms at the channel after POA in N_2O and $POCl_3$ using a Scanning Capacitance Microscopy (SCM) probe [96]. The SCM probe also detected an electrically active region underneath the deposited SiO_2.

Counterdoping the channel region of a MOSFET with antimony (Sb) prior to thermal oxidation and NO annealing is a process that resulted in a mobility of up to 110 $cm^2/(V \cdot s)$ [97]. Zheng *et al.* [98] later reported that by replacing the standard NO POA after the Sb treatment with a POA in O_2 at 950 °C for 30 min with a B_2O_3 planar diffusion source, a mobility of 180 $cm^2/(V \cdot s)$ was achieved. However, the borosilicate glass (BSG) gate exhibited a high threshold voltage hysteresis of up to 8 V at 1.5 MV/cm stress at 150 °C during the BTS measurement [98]. The poor BTS performance was suggested to be due to the large concentration of oxide traps, which is a characteristic of BSG gate oxide [99].

5.3   Alternative SiC crystal faces

The Si face (0001) of 4*H*-SiC, though the most commonly used and most well characterized, has a slow oxidation rate and a low channel mobility. Pictured in Chapter 1 of this volume in Fig. 3 are alternative crystal planes, which have been shown to have lower interface state densities, and higher mobility than the Si face. These include the a-plane (11$\bar{2}$0), which is commonly used for vertical channels in the formation of trench MOSFETs, and C face (000$\bar{1}$), which is being used in the formation of SiC IGBTs [50].

4*H*-SiC MOSFETs formed on the a-plane exhibited a field effect mobility of 85 $cm^2/(V \cdot s)$ after conventional NO passivation [100]. Using a PSG treatment instead of NO, this reached 125 $cm^2/(V \cdot s)$. Research benchmarking the mobility of the a- and m-planes and the Si face after NO treatment at 1300 °C for 80 min demonstrated that the a-

Advancing Silicon Carbide Electronics Technology II   Materials Research Forum LLC
Materials Research Foundations **69** (2020) 63-106   https://doi.org/10.21741/9781644900673-2

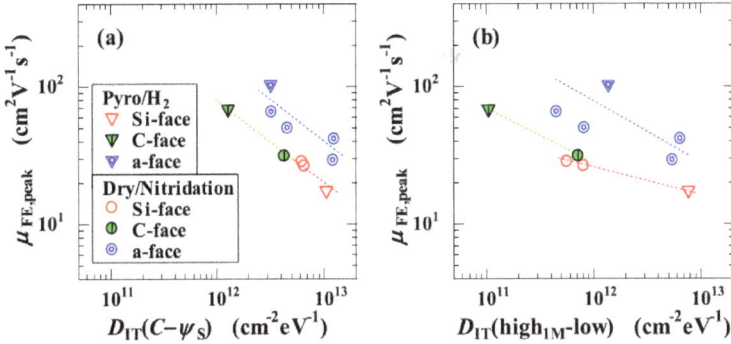

Figure 7. Peak MOSFET field-effect mobility and MOSCAP D_{it} (extracted using (a) the C - ψ_S method and (b) the high-low method) for dry oxidation with nitridation and wet oxidation with H_2 anneal on Si face, C face and a-plane surfaces. Reproduced from [104] with the permission of AIP Publishing.

plane has a peak channel mobility of up to 108 cm^2/(V·s), compared to 46 and 37 cm^2/(V·s) for the C and Si faces, respectively [101]. The authors show that a high channel mobility was observed for the a-plane regardless of nitridation conditions, be it using NO or N$_2$O [101]. However, oxide thickness varies on each crystal face due to different growth rates and hence must be precisely controlled to guarantee a V_t value.

The Si face has shown poor response to wet oxidation and H$_2$ annealing. However, the a-plane and C face have shown low D_{it} and high channel mobility following wet oxidation and H$_2$ anneal. Channel mobility greater than 100 cm^2/(V·s) has been obtained after wet oxidation and H$_2$ POA, with wet oxidation alone showing higher mobility than dry oxidation alone for these faces and H$_2$ POA showing further improvement [102, 103]. On the a-plane and C face, this combination of wet oxidation with H$_2$ POA gives higher channel mobility and lower D_{it} than dry oxidation with nitridation, and gives higher channel mobility than either process on the Si face. A summary of μ_{FE} and D_{it} for these process/plane combinations is shown in Fig. 7 [104].

6.   Surface passivation by dielectrics

The development of high voltage power devices necessitates a passivating layer to dissipate the electric field at the semiconductor surface. SiO$_2$ is the natural material choice for surface passivation, when grown by thermal oxidation of the SiC substrate. As with thermally grown SiO$_2$ for the gate stack, post oxidation anneals improve electrical performance but not to the extent of silicon [105, 106].

The impact of interfacial charge on the design of termination structures for SiC power devices of all types is of importance. Interface charge will lead to increased leakage current when the device is in the blocking state, though this can be reduced using POA treatments [107], as in the MOS interface. In the case of using a bevel-edged termination structure, it has been shown [108] that implanting the surface with argon can create a highly resistive area beneath the passivation region, reducing the field at the surface.

For bipolar devices, charge at the SiC/SiO$_2$ interface increases surface recombination [109, 110], which will affect carrier lifetime and the current gain (β) of bipolar junction transistors (BJTs) [109]. The passivation of SiC BJTs, which will include sidewall passivation of emitter/base mesa structures, is strongly affected by the density of interface traps present at a grown SiC/SiO$_2$ interface. On the sidewall of a BJT emitter/base region, these cause electrons to recombine at this interface, degrading emitter injection efficiency, and hence gain [109, 111, 112]. The impact of passivation oxide thickness on BJT gain has been demonstrated also [113], with a passivating layer of 100-150 nm delivering 60% greater gain ($\beta > 200$) compared with thinner 50 nm passivation. The gain can be traded off in further bipolar design to improve other electrical parameters.

Polyimide is used extensively in the electronics industry for surface passivation [114]. A thick layer of polyimide has been evaluated as the passivation for high voltage Schottky [115] and PiN [116] diodes, and compared to thick deposited SiO$_2$ layers. In the case of the PiN diode, after 40 nm of thermal oxide was formed on the surface of the JTE, a 4 μm polyimide layer was used as a second passivating layer. This resulted in a breakdown voltage that was on average 25% higher than when a 1.8 μm SiO$_2$ layer was used as a second passivating layer. However, the difference in thickness of these layers is substantial, leaving the benefit of polyimide compared to reported thick PECVD SiO$_2$ depositions unclear [117].

7. Summary

The focus of this chapter has been around the impact of dielectrics on the performance of SiC devices and, in particular, 4H-SiC MOSFETs. Fig. 8 shows the current status of (field effect) mobility as a function of D_{it} at 0.2 eV near to the conduction band edge, covered in this review. To date, encouraging improvements have been achieved in increasing channel mobility in 4H-SiC MOSFETs. However, the value of peak channel mobility of 4H-SiC MOSFET is still behind that achieved by silicon MOSFETs. Fig. 8 shows that the field effect mobility is inversely proportional to D_{it}, indicating that interface state traps contribute to the limiting of electron mobility. The correlation in the

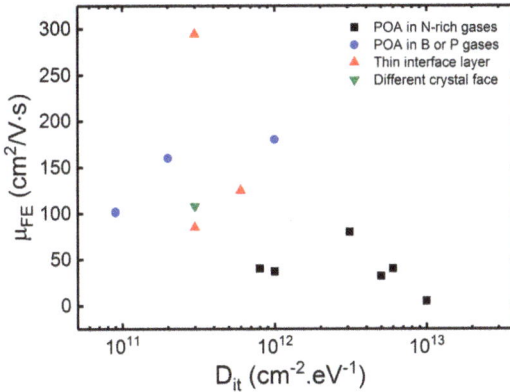

Figure 8.   *Current status of field effect mobility as a function D_{it} at 0.2 eV below the conduction band edge.*

data is somewhat weak, which may result from different device specifications and/or electrical characterization parameters. Nevertheless, the best performing MOSFETs lead to data towards the top left corner of Fig. 8, corresponding with high mobility and low D_{it}. Fig. 8 shows that POA in N-rich gases does not improve mobility as well as POA in B or P gases. However, there are issues with *I-V* hysteresis in devices using B or P gases, resulting on N-rich gases being preferred, despite lower channel mobility. The use of thin interfacial layers offers significant improvements in device performance, using low temperature oxidation followed by dielectric deposition.

Mobilities in excess of 100 cm²/(V·s) have been achieved via a number of different techniques, in particular by the use of high-κ dielectrics, thin thermal oxides or the use of alternative crystal planes. POA and high temperature oxidation has also been shown to improve mobility, although to a lesser extent using N-rich gases.

Despite improvements in channel mobility, new processes have not supplanted thermal oxidation and POA in NO or N_2O, principally because of compromises in leakage current, threshold voltage stability, hysteresis and general reliability. The importance of controlling residual C close to the SiC/SiO_2 interface has been stressed, and factors that mitigate its effect have been discussed.

Acknowledgements

This work was supported by the UK EPSRC (Grant EP/L007010/1 and Grant EP/R00448X/1). F. Arith is grateful for support from the Ministry of Education Malaysia and Faculty of Electronic & Computer Engineering, Universiti Teknikal Malaysia Melaka (UTeM).

References

[1] L. A. Lipkin, J. W. Palmour, "Insulator investigation on SiC for improved reliability," *IEEE Transactions on Electron Devices,* vol. 46, no. 3, pp. 525-532, 1999. https://doi.org/10.1109/16.748872

[2] Y. Hiroshi, K. Tsunenobu, M. Hiroyuki, "Shallow states at SiO_2/4H-SiC interface on ($11\bar{2}0$) and (0001) faces," *Applied Physics Letters,* vol. 81, no. 2, pp. 301-303, 2002. https://doi.org/10.1063/1.1492313

[3] J. M. Knaup, P. Deák, T. Frauenheim, A. Gali, Z. Hajnal, W. J. Choyke, "Theoretical study of the mechanism of dry oxidation of 4H-SiC," *Physical Review B,* vol. 71, no. 23, pp. 235321, 2005.

[4] T. E. Rudenko, I. N. Osiyuk, I. P. Tyagulski, H. Ö. Ólafsson, E. Ö. Sveinbjörnsson, "Interface trap properties of thermally oxidized n-type 4H–SiC and 6H–SiC," *Solid-State Electronics,* vol. 49, no. 4, pp. 545-553, 2005. https://doi.org/10.1016/j.sse.2004.12.006

[5] A. F. Basile, J. Rozen, J. R. Williams, L. C. Feldman, P. M. Mooney, "Capacitance-voltage and deep-level-transient spectroscopy characterization of defects near SiO_2/SiC interfaces," *Journal of Applied Physics,* vol. 109, no. 6, pp. 064514, 2011. https://doi.org/10.1063/1.3552303

[6] S. Dhar, X. D. Chen, P. M. Mooney, J. R. Williams, L. C. Feldman, "Ultrashallow defect states at SiO_2/4H–SiC interfaces," *Applied Physics Letters,* vol. 92, no. 10, pp. 102112, 2008. https://doi.org/10.1063/1.2898502

[7] O. Dai, Y. Hiroshi, H. Tomoaki, F. Takashi, "Removal of near-interface traps at SiO_2/4H–SiC (0001) interfaces by phosphorus incorporation," *Applied Physics Letters,* vol. 96, no. 20, pp. 203508, 2010. https://doi.org/10.1063/1.3432404

[8] T. Kimoto, J. A. Cooper, "Fundamentals of silicon carbide technology: growth, characterization, devices, and applications," John Wiley & Sons, 2014, 538 pages.

[9] K. Matocha, I.-H. Ji, X. Zhang, S. Chowdhury, "SiC Power MOSFETs: Designing for Reliability in Wide-Bandgap Semiconductors." pp. 1-8.

[10] S. Takagi, A. Toriumi, M. Iwase, H. Tango, "On the universality of inversion layer mobility in Si MOSFET's: Part I-effects of substrate impurity concentration,"

IEEE Transactions on Electron Devices, vol. 41, no. 12, pp. 2357-2362, 1994.
https://doi.org/10.1109/16.337449

[11]   D. K. Schroder, *Semiconductor material and device characterization*: John Wiley
& Sons, 2015.

[12]   S. Y. Chou, D. Antoniadis, "Relationship between measured and intrinsic
transconductances of FET's," *IEEE Transactions on Electron Devices,* vol. 34,
no. 2, pp. 448-450, 1987.

[13]   T. Ohashi, Y. Nakabayashi, R. Iijima, "Investigation of the universal mobility of
SiC MOSFETs using wet oxide insulators on carbon face with low interface state
density," *IEEE Transactions on Electron Devices,* vol. 65, no. 7, pp. 2707-2713,
2018.

[14]   F. Roccaforte, P. Fiorenza, F. Giannazzo, "Impact of the Morphological and
Electrical Properties of SiO2/4H-SiC Interfaces on the Behavior of 4H-SiC
MOSFETs," *ECS Journal of Solid State Science and Technology,* vol. 2, no. 8,
pp. N3006-N3011, 2013. https://doi.org/10.1149/2.002308jss

[15]   S. Tanimoto, H. Tanaka, T. Hayashi, Y. Shimoida, M. Hoshi, T. Mihara, "High-
Reliability ONO Gate Dielectric for Power MOSFETs," *Materials Science Forum,*
vol. 483-485, pp. 677-680, 2005.
https://doi.org/10.4028/www.scientific.net/MSF.483-485.677

[16]   S. Tanimoto, "Impact of Dislocations on Gate Oxide in SiC MOS Devices and
High Reliability ONO Dielectrics," *Materials Science Forum,* vol. 527-529, pp.
955-960, 2006. https://doi.org/10.4028/www.scientific.net/MSF.527-529.955

[17]   T. Satoshi, "Highly reliable $SiO_2/SiN/SiO_2$(ONO) gate dielectric on 4H-SiC,"
Electronics and Communications in Japan (Part II: Electronics), vol. 90, no. 5,
pp. 1-10, 2007. https://doi.org/doi:10.1002/ecjb.20329

[18]   S. Tanimoto, T. Suzuki, S. Yamagami, H. Tanaka, T. Hayashi, Y. Hirose, M.
Hoshi, "Negative Field Reliability of ONO Gate Dielectric on 4H-SiC," *Materials
Science Forum,* vol. 600-603, pp. 795-798, 2009.
https://doi.org/10.4028/www.scientific.net/MSF.600-603.795

[19]   K. J. Kuhn, "Reducing Variation in Advanced Logic Technologies: Approaches to
Process and Design for Manufacturability of Nanoscale CMOS." pp. 471-474.

[20]   R. Mahapatra, A. K. Chakraborty, A. B. Horsfall, N. G. Wright, G. Beamson,
K. S. Coleman, "Energy-band alignment of $HfO_2/SiO_2/SiC$ gate dielectric stack,"
Applied Physics Letters, vol. 92, no. 4, 2008. https://doi.org/10.1063/1.2839314

[21] C. M. Tanner, J. Choi, J. P. Chang, "Electronic structure and band alignment at the HfO$_2$/4H-SiC interface," *Journal of Applied Physics,* vol. 101, no. 3, 2007. https://doi.org/10.1063/1.2432402

[22] R. Suri, C. J. Kirkpatrick, D. J. Lichtenwalner, V. Misra, "Energy-band alignment of Al$_2$O$_3$ and HfAlO gate dielectrics deposited by atomic layer deposition on 4H–SiC," *Applied Physics Letters,* vol. 96, no. 4, pp. 042903, 2010. https://doi.org/10.1063/1.3291620

[23] D. J. Lichtenwalner, V. Misra, S. Dhar, S.-H. Ryu, A. Agarwal, "High-mobility enhancement-mode 4H-SiC lateral field-effect transistors utilizing atomic layer deposited Al$_2$O$_3$ gate dielectric," *Applied Physics Letters,* vol. 95, no. 15, pp. 152113, 2009. https://doi.org/10.1063/1.3251076

[24] S. Hino, T. Hatayama, J. Kato, E. Tokumitsu, N. Miura, T. Oomori, "High channel mobility 4H-SiC metal-oxide-semiconductor field-effect transistor with low temperature metal-organic chemical-vapor deposition grown Al2O3 gate insulator," *Applied Physics Letters,* vol. 92, no. 18, pp. 183503, 2008. https://doi.org/10.1063/1.2903103

[25] T. Hatayama, S. Hino, N. Miura, T. Oomori, E. Tokumitsu, "Remarkable Increase in the Channel Mobility of SiC-MOSFETs by Controlling the Interfacial SiO$_2$ Layer Between Al$_2$O$_3$ and SiC," *Electron Devices, IEEE Transactions on,* vol. 55, no. 8, pp. 2041-2045, 2008. https://doi.org/10.1109/TED.2008.926647

[26] X. Yang, B. Lee, V. Misra, "High Mobility 4H-SiC Lateral MOSFETs Using Lanthanum Silicate and Atomic Layer Deposited SiO$_2$," *IEEE Electron Device Letters,* vol. 36, no. 4, pp. 312-314, 2015. https://doi.org/10.1109/LED.2015.2399891

[27] M. O. Aboelfotoh, R. S. Kern, S. Tanaka, R. F. Davis, C. I. Harris, "Electrical characteristics of metal/AlN/n-type 6H-SiC(0001) heterostructures," *Applied Physics Letters,* vol. 69, no. 19, pp. 2873-2875, 1996. https://doi.org/10.1063/1.117347

[28] C. M. Tanner, Y.-C. Perng, C. Frewin, S. E. Saddow, J. P. Chang, "Electrical performance of Al2O3 gate dielectric films deposited by atomic layer deposition on 4H-SiC," *Applied Physics Letters,* vol. 91, no. 20, pp. 203510, 2007. https://doi.org/10.1063/1.2805742

[29] J. Robertson, R. M. Wallace, "High-K materials and metal gates for CMOS applications," *Materials Science and Engineering: R: Reports,* vol. 88, no. Supplement C, pp. 1-41, 2015. https://doi.org/10.1016/j.mser.2014.11.001

[30]   Stanley Wolf and Richard N. Tauber, *Silicon Processing for the VLSI Era*, 2nd ed., USA: Lattice Press, 2000.

[31]   S. Sridevan, V. Misra, P. K. McLarty, B. J. Baliga, J. J. Wortman, "Rapid thermal chemical vapor deposited oxides on N-type 6H-silicon carbide," *IEEE Electron Device Letters,* vol. 16, no. 11, pp. 524-526, 1995.
        https://doi.org/10.1109/55.468288

[32]   S. Sridevan, B. J. Baliga, "Lateral n-channel inversion mode 4H-SiC MOSFETs," *IEEE Electron Device Letters,* vol. 19, no. 7, pp. 228-230, 1998.
        https://doi.org/10.1109/55.701425

[33]   H. Yano, T. Hatayama, Y. Uraoka, T. Fuyuki, "High Temperature NO Annealing of Deposited SiO_2 and SiON Films on N-Type 4H-SiC," *Materials Science Forum,* vol. 483-485, pp. 685-688, 2005.
        https://doi.org/10.4028/www.scientific.net/MSF.483-485.685

[34]   H. K. T. Kimoto, M. Noborio, J. Suda, H. Matsunami, "Improved Dielectric and Interface Properties of 4H-SiC MOS Structures Processed by Oxide Deposition and N2O Annealing," *Materials Science Forum,* vol. 527-529, pp. 987-990, 2006.

[35]   N. Masato, G. Michael, J. B. Anton, P. Dethard, F. Peter, S. Jun, K. Tsunenobu, "Reliability of Nitrided Gate Oxides for N- and P-Type 4H-SiC(0001) Metal–Oxide–Semiconductor Devices," *Japanese Journal of Applied Physics,* vol. 50, no. 9R, pp. 090201, 2011.

[36]   Y. Nanen, H. Yoshioka, M. Noborio, J. Suda, T. Kimoto, "Enhanced Drain Current of 4H-SiC MOSFETs by Adopting a Three-Dimensional Gate Structure," *IEEE Transactions on Electron Devices,* vol. 56, no. 11, pp. 2632-2637, 2009.
        https://doi.org/10.1109/TED.2009.2030437

[37]   R. Esteve, A. Schöner, S. A. Reshanov, C. M. Zetterling, "Comparative Study of Thermal Oxides and Post-Oxidized Deposited Oxides on n-Type Free Standing 3C-SiC," *Materials Science Forum,* vol. 645-648, pp. 829-832, 2010.
        https://doi.org/10.4028/www.scientific.net/MSF.645-648.829

[38]   A. Pérez-Tomás, P. Godignon, N. Mestres, R. Pérez, J. Millán, "A study of the influence of the annealing processes and interfaces with deposited SiO_2 from tetra-ethoxy-silane for reducing the thermal budget in the gate definition of 4H–SiC devices," *Thin Solid Films,* vol. 513, no. 1-2, pp. 248-252, 2006.
        https://doi.org/10.1016/j.tsf.2005.12.308

[39] A. Pérez-Tomás, P. Godignon, J. Camassel, N. Mestres, V. Soulière, "PECVD Deposited TEOS for Field-Effect Mobility Improvement in 4H-SiC MOSFETs on the (0001) and (11-20) Faces," *Materials Science Forum,* vol. 527-529, pp. 1047-1050, 2006. https://doi.org/10.4028/www.scientific.net/MSF.527-529.1047

[40] K. Kawase, S. Noda, T. Nakai, Y. Uehara, "Densification of Chemical Vapor Deposition Silicon Dioxide Film Using Ozone Treatment," *Japanese Journal of Applied Physics,* vol. 48, no. 10, 2009. https://doi.org/10.1143/jjap.48.101401

[41] Y. K. Sharma, A. C. Ahyi, T. Issacs-Smith, M. R. Jennings, S. M. Thomas, P. Mawby, S. Dhar, J. R. Williams, "Stable Phosphorus Passivated SiO$_2$/4H-SiC Interface Using Thin Oxides," *Materials Science Forum,* vol. 806, pp. 139-142, 2015. https://doi.org/10.4028/www.scientific.net/MSF.806.139

[42] R. W. Johnson, A. Hultqvist, S. F. Bent, "A brief review of atomic layer deposition: from fundamentals to applications," *Materials Today,* vol. 17, no. 5, pp. 236-246, 2014. https://doi.org/10.1016/j.mattod.2014.04.026

[43] F. Arith, J. Urresti, K. Vasilevskiy, S. Olsen, N. Wright, A. O'Neill, "Increased Mobility in Enhancement Mode 4H-SiC MOSFET Using a Thin SiO$_2$ / Al$_2$O$_3$ Gate Stack," *IEEE Electron Device Letters,* vol. 39, no. 4, pp. 564-567, 2018. https://doi.org/10.1109/LED.2018.2807620

[44] S. S. Suvanam, M. Usman, D. Martin, M. G. Yazdi, M. Linnarsson, A. Tempez, M. Götelid, A. Hallén, "Improved interface and electrical properties of atomic layer deposited Al$_2$O$_3$/4H-SiC," *Applied Surface Science,* vol. 433, no. Supplement C, pp. 108-115, 2018. https://doi.org/10.1016/j.apsusc.2017.10.006

[45] X. Yang, B. Lee, V. Misra, "Electrical Characteristics of SiO$_2$ Deposited by Atomic Layer Deposition on 4H-SiC After Nitrous Oxide Anneal," *IEEE Transactions on Electron Devices,* vol. 63, no. 7, pp. 2826-2830, 2016. https://doi.org/10.1109/TED.2016.2565665

[46] D. Dutta, D. De, D. Fan, S. Roy, G. Alfieri, M. Camarda, M. Amsler, J. Lehmann, H. Bartolf, S. Goedecker, "Evidence for carbon clusters present near thermal gate oxides affecting the electronic band structure in SiC-MOSFET," *Applied Physics Letters,* vol. 115, no. 10, pp. 101601, 2019.

[47] B. E. Deal, A. S. Grove, "General Relationship for the Thermal Oxidation of Silicon," *Journal of Applied Physics,* vol. 36, no. 12, pp. 3770-3778, 1965. https://doi.org/10.1063/1.1713945

[48]   Y. Song, S. Dhar, L. C. Feldman, G. Chung, J. R. Williams, "Modified Deal Grove model for the thermal oxidation of silicon carbide," *Journal of Applied Physics,* vol. 95, no. 9, pp. 4953-4957, 2004. https://doi.org/10.1063/1.1690097

[49]   Y. Hijikata, S. Yagi, H. Yaguchi, S. Yoshida, "Thermal Oxidation Mechanism of Silicon Carbide," in: *Physics and Technology of Silicon Carbide Devices,* Y. Hijikata, ed., Rijeka: InTech, 2012, p. Ch. 07.

[50]   Y. Yonezawa, T. Mizushima, K. Takenaka, H. Fujisawa, T. Kato, S. Harada, Y. Tanaka, M. Okamoto, M. Sometani, D. Okamoto, "Low V f and highly reliable 16 kV ultrahigh voltage SiC flip-type n-channel implantation and epitaxial IGBT." pp. 6.6. 1-6.6. 4.

[51]   K. C. Chang, N. T. Nuhfer, L. M. Porter, Q. Wahab, "High-carbon concentrations at the silicon dioxide–silicon carbide interface identified by electron energy loss spectroscopy," *Applied Physics Letters,* vol. 77, no. 14, pp. 2186-2188, 2000. https://doi.org/10.1063/1.1314293

[52]   T. Zheleva, A. Lelis, G. Duscher, F. Liu, I. Levin, M. Das, "Transition layers at the SiO2/SiC interface," *Applied Physics Letters,* vol. 93, no. 2, 2008. https://doi.org/10.1063/1.2949081

[53]   W. Lu, L. C. Feldman, Y. Song, S. Dhar, W. E. Collins, W. C. Mitchel, J. R. Williams, "Graphitic features on SiC surface following oxidation and etching using surface enhanced Raman spectroscopy," *Applied Physics Letters,* vol. 85, no. 16, pp. 3495-3497, 2004. https://doi.org/10.1063/1.1804610

[54]   V. N. Brudnyi, A. V. Kosobutsky, "Electronic properties of SiC polytypes: Charge neutrality level and interfacial barrier heights," *Superlattices and Microstructures,* vol. 111, pp. 499-505, 2017. https://doi.org/10.1016/j.spmi.2017.07.003

[55]   T. Hiyoshi, T. Kimoto, "Elimination of the Major Deep Levels in n- and p-Type 4H-SiC by Two-Step Thermal Treatment," *Applied Physics Express,* vol. 2, no. 9, 2009. https://doi.org/10.1143/apex.2.091101

[56]   X. Shen, S. T. Pantelides, "Identification of a major cause of endemically poor mobilities in SiC/SiO2 structures," *Applied Physics Letters,* vol. 98, no. 5, 2011. https://doi.org/10.1063/1.3553786

[57]   S. Wang, M. Di Ventra, S. G. Kim, S. T. Pantelides, "Atomic-Scale Dynamics of the Formation and Dissolution of Carbon Clusters in SiO_2," *Physical Review Letters,* vol. 86, no. 26, pp. 5946-5949, 2001.

[58] H. Yan, R. Jia, X. Tang, Q. Song, Y. Zhang, "Effect of re-oxidation annealing process on the SiO$_2$ /SiC interface characteristics," *Journal of Semiconductors,* vol. 35, no. 6, pp. 066001, 2014.

[59] Z. Q. Zhong, L. D. Zheng, G. J. Zhang, S. Y. Wang, L. P. Dai, Y. L. Gong, "Effect of Ar Annealing Temperature on SiO$_2$/SiC: Carbon-Related Clusters Reduction Causing Interfacial Quality Improvement," *Advanced Materials Research,* vol. 997, pp. 484-487, 2014. https://doi.org/10.4028/www.scientific.net/AMR.997.484

[60] D. Peter, M. K. Jan, H. Tamás, T. Christoph, G. Adam, F. Thomas, "The mechanism of defect creation and passivation at the SiC/SiO$_2$ interface," *Journal of Physics D: Applied Physics,* vol. 40, no. 20, pp. 6242, 2007.

[61] J. Powell, J. Petit, J. Edgar, I. Jenkins, L. Matus, W. Choyke, L. Clemen, M. Yoganathan, J. Yang, P. Pirouz, "Application of oxidation to the structural characterization of SiC epitaxial films," *Applied physics letters,* vol. 59, no. 2, pp. 183-185, 1991.

[62] Y. Nakano, T. Nakamura, A. Kamisawa, H. Takasu, "Investigation of pits formed at oxidation on 4H-SiC." pp. 377-380.

[63] H. Kurimoto, K. Shibata, C. Kimura, H. Aoki, T. Sugino, "Thermal oxidation temperature dependence of 4H-SiC MOS interface," *Applied Surface Science,* vol. 253, no. 5, pp. 2416-2420, 2006. https://doi.org/10.1016/j.apsusc.2006.04.054

[64] H. Naik, T. P. Chow, "4H-SiC MOS Capacitors and MOSFET Fabrication with Gate Oxidation at 1400°C," *Materials Science Forum,* vol. 778-780, pp. 607-610, 2014. https://doi.org/10.4028/www.scientific.net/MSF.778-780.607

[65] T. Hosoi, D. Nagai, M. Sometani, Y. Katsu, H. Takeda, T. Shimura, M. Takei, H. Watanabe, "Ultrahigh-temperature rapid thermal oxidation of 4H-SiC(0001) surfaces and oxidation temperature dependence of SiO2/SiC interface properties," *Applied Physics Letters,* vol. 109, no. 18, 2016. https://doi.org/10.1063/1.4967002

[66] S. M. Thomas, Y. K. Sharma, M. A. Crouch, C. A. Fisher, A. Perez-Tomas, M. R. Jennings, P. A. Mawby, "Enhanced field effect mobility on 4H-SiC by oxidation at 1500 °C," *IEEE Journal of the Electron Devices Society,* vol. 2, no. 5, pp. 114-117, 2014. https://doi.org/10.1109/JEDS.2014.2330737

[67] S. M. Thomas, M. R. Jennings, Y. K. Sharma, C. A. Fisher, P. A. Mawby, "Impact of the Oxidation Temperature on the Interface Trap Density in 4H-SiC MOS Capacitors," *Materials Science Forum,* vol. 778-780, pp. 599-602, 2014. https://doi.org/10.4028/www.scientific.net/MSF.778-780.599

[68] Y. Jia, H. Lv, Q. Song, X. Tang, L. Xiao, L. Wang, G. Tang, Y. Zhang, Y. Zhang, "Influence of oxidation temperature on the interfacial properties of n-type 4H-SiC MOS capacitors," *Applied Surface Science,* vol. 397, pp. 175-182, 2017. https://doi.org/10.1016/j.apsusc.2016.11.142

[69] Y. K. Sharma, F. Li, M. R. Jennings, C. A. Fisher, A. Pérez-Tomás, S. Thomas, D. P. Hamilton, S. A. O. Russell, P. A. Mawby, "High-Temperature (1200–1400°C) Dry Oxidation of 3C-SiC on Silicon," *Journal of Electronic Materials,* vol. 44, no. 11, pp. 4167-4174, 2015. https://doi.org/10.1007/s11664-015-3949-4

[70] S. M. Sze, K. K. Ng, "Physics of semiconductor devices," Wiley-Interscience, 2007, 815 p.

[71] J. Urresti, F. Arith, S. Olsen, N. Wright, A. O'Neill, "Design and Analysis of High Mobility Enhancement-Mode 4H-SiC MOSFETs Using a Thin-SiO_2/Al_2O_3 Gate-Stack," *IEEE Transactions on Electron Devices,* vol. 66, no. 4, pp. 1710-1716, 2019. https://doi.org/10.1109/ted.2019.2901310

[72] D.-K. Kim, Y.-S. Kang, K.-S. Jeong, H.-K. Kang, S. W. Cho, K.-B. Chung, H. Kim, M.-H. Cho, "Effects of spontaneous nitrogen incorporation by a 4H-SiC(0001) surface caused by plasma nitridation," *Journal of Materials Chemistry C,* vol. 3, no. 19, pp. 5078-5088, 2015. https://doi.org/10.1039/C5TC00076A

[73] G. Y. Chung, C. C. Tin, J. R. Williams, K. McDonald, R. K. Chanana, R. A. Weller, S. T. Pantelides, L. C. Feldman, O. W. Holland, M. K. Das, J. W. Palmour, "Improved inversion channel mobility for 4H-SiC MOSFETs following high temperature anneals in nitric oxide," *IEEE Electron Device Letters,* vol. 22, no. 4, pp. 176-178, 2001. https://doi.org/10.1109/55.915604

[74] L. Chao-Yang, J. A. Cooper, T. Tsuji, C. Gilyong, J. R. Williams, K. McDonald, L. C. Feldman, "Effect of process variations and ambient temperature on electron mobility at the SiO_2/4H-SiC interface," *IEEE Transactions on Electron Devices,* vol. 50, no. 7, pp. 1582-1588, 2003. https://doi.org/10.1109/TED.2003.814974

[75] L. A. Lipkin, M. K. Das, J. W. Palmour, "N_2O processing improves the 4H-SiC: SiO_2 interface," *Materials Science Forum,* vol. 389-393, pp. 985-988, 2002. https://doi.org/10.4028/www.scientific.net/MSF.389-393.985

[76] K. Fujihira, Y. Tarui, M. Imaizumi, K.-i. Ohtsuka, T. Takami, T. Shiramizu, K. Kawase, J. Tanimura, T. Ozeki, "Characteristics of 4H–SiC MOS interface annealed in N_2O," *Solid-State Electronics,* vol. 49, no. 6, pp. 896-901, 2005. https://doi.org/10.1016/j.sse.2004.10.016

Materials Research Forum LLC
https://doi.org/10.21741/9781644900673-2

[77]   J. Senzaki, T. Suzuki, A. Shimozato, K. Fukuda, K. Arai, H. Okumura, "Significant Improvement in Reliability of Thermal Oxide on 4H-SiC (0001) Face Using Ammonia Post-Oxidation Annealing," *Materials Science Forum,* vol. 645-648, pp. 685-688, 2010. https://doi.org/10.4028/www.scientific.net/MSF.645-648.685

[78]   N. Soejima, T. Kimura, T. Ishikawa, T. Sugiyama, "Effect of NH$_3$ post-oxidation annealing on flatness of SiO$_2$/SiC interface," *Materials Science Forum,* vol. 740-742, pp. 723-726, 2013. https://doi.org/10.4028/www.scientific.net/MSF.740-742.723

[79]   P. Jamet, S. Dimitrijev, P. Tanner, "Effects of nitridation in gate oxides grown on 4H-SiC," *Journal of Applied Physics,* vol. 90, no. 10, pp. 5058-5063, 2001.

[80]   V. V. Afanas'ev, A. Stesmans, F. Ciobanu, G. Pensl, K. Y. Cheong, S. Dimitrijev, "Mechanisms responsible for improvement of 4H–SiC/SiO2 interface properties by nitridation," *Applied Physics Letters,* vol. 82, no. 4, pp. 568-570, 2003. https://doi.org/10.1063/1.1532103

[81]   H. Yoshioka, T. Nakamura, T. Kimoto, "Generation of very fast states by nitridation of the SiO$_2$/SiC interface," *Journal of Applied Physics,* vol. 112, no. 2, pp. 024520, 2012. https://doi.org/doi:http://dx.doi.org/10.1063/1.4740068

[82]   J. Rozen, A. C. Ahyi, X. Zhu, J. R. Williams, L. C. Feldman, "Scaling Between Channel Mobility and Interface State Density in SiC MOSFETs," *IEEE Transactions on Electron Devices,* vol. 58, no. 11, pp. 3808-3811, 2011. https://doi.org/10.1109/ted.2011.2164800

[83]   J. Rozen, S. Dhar, M. E. Zvanut, J. R. Williams, L. C. Feldman, "Density of interface states, electron traps, and hole traps as a function of the nitrogen density in SiO2 on SiC," *Journal of Applied Physics,* vol. 105, no. 12, 2009. https://doi.org/10.1063/1.3131845

[84]   P. Jamet, S. Dimitrijev, "Physical properties of N2O and NO-nitrided gate oxides grown on 4H SiC," *Applied Physics Letters,* vol. 79, no. 3, pp. 323-325, 2001. https://doi.org/10.1063/1.1385181

[85]   A. Morales-Acevedo, G. Santana, J. Carrillo-López, "Thermal oxidation of silicon in nitrous oxide at high pressures," *Journal of The Electrochemical Society,* vol. 148, no. 10, pp. F200-F202, 2001.

[86]   D. Okamoto, H. Yano, H. Kenji, T. Hatayama, T. Fuyuki, "Improved Inversion Channel Mobility in 4H-SiC MOSFETs on Si Face Utilizing Phosphorus-Doped

Gate Oxide," *Electron Device Letters, IEEE,* vol. 31, no. 7, pp. 710-712, 2010. https://doi.org/10.1109/LED.2010.2047239

[87]   H. Y. Dai Okamotoa, Tomoaki Hatayamac and Takashi Fuyuki, "Development of 4H-SiC MOSFETs with Phosphorus-Doped Gate Oxide," *Materials Science Forum,* vol. 717-720, pp. 733-738, 2012. https://doi.org/10.4028/www.scientific.net/MSF.717-720.733

[88]   C. Jiao, A. C. Ahyi, C. Xu, D. Morisette, L. C. Feldman, S. Dhar, "Phospho-silicate glass gated 4H-SiC metal-oxide-semiconductor devices: Phosphorus concentration dependence," *Journal of Applied Physics,* vol. 119, no. 15, pp. 155705, 2016. https://doi.org/10.1063/1.4947117

[89]   D. Okamoto, M. Sometani, S. Harada, R. Kosugi, Y. Yonezawa, H. Yano, "Improved channel mobility in 4H-SiC MOSFETs by boron passivation," *IEEE Electron Device Letters,* vol. 35, no. 12, pp. 1176-1178, 2014. https://doi.org/10.1109/LED.2014.2362768

[90]   M. Cabello, V. Soler, J. Montserrat, J. Rebollo, J. M. Rafí, P. Godignon, "Impact of boron diffusion on oxynitrided gate oxides in 4H-SiC metal-oxide-semiconductor field-effect transistors," *Applied Physics Letters,* vol. 111, no. 4, pp. 042104, 2017. https://doi.org/10.1063/1.4996365

[91]   F. Allerstam, G. Gudjónsson, E. Ö. Sveinbjörnsson, T. Rödle, R. Jos, "Formation of Deep Traps at the 4H-SiC/SiO$_2$ Interface when Utilizing Sodium Enhanced Oxidation," *Materials Science Forum,* vol. 556-557, pp. 517-520, 2007. https://doi.org/10.4028/www.scientific.net/MSF.556-557.517

[92]   G. Gudjonsson, H. O. Olafsson, F. Allerstam, P. A. Nilsson, E. O. Sveinbjornsson, H. Zirath, T. Rodle, R. Jos, "High field-effect mobility in n-channel Si face 4H-SiC MOSFETs with gate oxide grown on aluminum ion-implanted material," *IEEE Electron Device Letters,* vol. 26, no. 2, pp. 96-98, 2005. https://doi.org/10.1109/LED.2004.841191

[93]   A. F. Basile, A. C. Ahyi, L. C. Feldman, J. R. Williams, P. M. Mooney, "Effects of sodium ions on trapping and transport of electrons at the SiO$_2$/4H-SiC interface," *Journal of Applied Physics,* vol. 115, no. 3, pp. 034502, 2014. https://doi.org/10.1063/1.4861646

[94]   D. J. Lichtenwalner, L. Cheng, S. Dhar, A. Agarwal, J. W. Palmour, "High mobility 4H-SiC (0001) transistors using alkali and alkaline earth interface layers," *Applied Physics Letters,* vol. 105, no. 18, 2014. https://doi.org/10.1063/1.4901259

[95] K. Ueno, T. Oikawa, "Counter-doped MOSFETs of 4H-SiC," *IEEE Electron Device Letters,* vol. 20, no. 12, pp. 624-626, 1999. https://doi.org/10.1109/55.806105

[96] P. Fiorenza, F. Giannazzo, M. Vivona, A. L. Magna, F. Roccaforte, "SiO$_2$/4H-SiC interface doping during post-deposition-annealing of the oxide in N$_2$O or POCl$_3$," *Applied Physics Letters,* vol. 103, no. 15, pp. 153508, 2013. https://doi.org/10.1063/1.4824980

[97] A. Modic, G. Liu, A. C. Ahyi, Y. Zhou, P. Xu, M. C. Hamilton, J. R. Williams, L. C. Feldman, S. Dhar, "High channel mobility 4H-SiC MOSFETs by antimony counter-doping," *IEEE Electron Device Letters,* vol. 35, no. 9, pp. 894-896, 2014. https://doi.org/10.1109/LED.2014.2336592

[98] Y. Zheng, T. Isaacs-Smith, A. C. Ahyi, S. Dhar, "4H-SiC MOSFETs with borosilicate glass gate dielectric and antimony counter-doping," *IEEE Electron Device Letters,* vol. 38, no. 10, pp. 1433-1326, 2017. https://doi.org/10.1109/LED.2017.2743002

[99] W. L. Warren, M. R. Shaneyfelt, D. M. Fleetwood, P. S. Winokur, "Nature of defect centers in B- and P-doped SiO$_2$ thin films," *Applied Physics Letters,* vol. 67, no. 7, pp. 995-997, 1995. https://doi.org/10.1063/1.114970

[100] G. Liu, A. C. Ahyi, Y. Xu, T. Isaacs-Smith, Y. K. Sharma, J. R. Williams, L. C. Feldman, S. Dhar, "Enhanced Inversion Mobility on 4H-SiC (11$\bar{2}$0) Using Phosphorus and Nitrogen Interface Passivation," *IEEE Electron Device Letters,* vol. 34, no. 2, pp. 181-183, 2013. https://doi.org/10.1109/led.2012.2233458

[101] Y. Nanen, M. Kato, J. Suda, T. Kimoto, "Effects of nitridation on 4H-SiC MOSFETs fabricated on various crystal faces," *IEEE Transactions on Electron Devices,* vol. 60, no. 3, pp. 1260-1262, 2013. https://doi.org/10.1109/TED.2012.2236333

[102] K. Fukuda, M. Kato, K. Kojima, J. Senzaki, "Effect of gate oxidation method on electrical properties of metal-oxide-semiconductor field-effect transistors fabricated on 4H-SiC C(000$\bar{1}$) face," *Applied Physics Letters,* vol. 84, no. 12, pp. 2088-2090, 2004. https://doi.org/10.1063/1.1682680

[103] M. Okamoto, Y. Makifuchi, M. Iijima, Y. Sakai, N. Iwamuro, H. Kimura, K. Fukuda, H. Okumura, "Coexistence of small threshold voltage instability and high channel mobility in 4H-SiC (0001) metal–oxide–semiconductor field-effect transistors," *Applied Physics Express,* vol. 5, no. 4, pp. 041302, 2012.

[104] H. Yoshioka, J. Senzaki, A. Shimozato, Y. Tanaka, H. Okumura, "Effects of interface state density on 4H-SiC n-channel field-effect mobility," *Applied Physics Letters,* vol. 104, no. 8, pp. 083516, 2014.

[105] S. Asada, T. Kimoto, J. Suda, "Effect of postoxidation nitridation on forward current-voltage characteristics in 4H-SiC mesa p-n diodes passivated with SiO_2," *IEEE Transactions on Electron Devices,* vol. 64, no. 7, pp. 3016-3018, 2017. https://doi.org/10.1109/TED.2017.2700336

[106] R. Ghandi, B. Buono, M. Domeij, R. Esteve, A. Schoner, J. Han, S. Dimitrijev, S. A. Reshanov, C. M. Zetterling, M. Ostling, "Surface-passivation effects on the performance of 4H-SiC BJTs," *IEEE Transactions on Electron Devices,* vol. 58, no. 1, pp. 259-265, 2011. https://doi.org/10.1109/TED.2010.2082712

[107] T. Kimoto, H. Kawano, M. Noborio, J. Suda, H. Matsunami, "Improved Dielectric and Interface Properties of 4H-SiC MOS Structures Processed by Oxide Deposition and N_2O Annealing," *Materials Science Forum,* vol. 527-529, pp. 987-990, 2006. https://doi.org/10.4028/www.scientific.net/MSF.527-529.987

[108] W. Sung, B. Jayant Baliga, A. Q. Huang, "Area-Efficient Bevel-Edge Termination Techniques for SiC High-Voltage Devices," *IEEE Transactions on Electron Devices,* vol. 63, no. 4, pp. 1630-1636, 2016. https://doi.org/10.1109/ted.2016.2532602

[109] H. Miyake, T. Kimoto, J. Suda, "Improvement of current gain in 4H-SiC BJTs by surface passivation with deposited oxides nitrided in N_2O or NO," *IEEE Electron Device Letters,* vol. 32, no. 3, pp. 285-287, 2011. https://doi.org/10.1109/LED.2010.2101575

[110] T. Okuda, T. Kobayashi, T. Kimoto, J. Suda, "Impact of annealing temperature on surface passivation of SiC epitaxial layers with deposited SiO_2 followed by $POCl_3$ annealing." pp. 233-235.

[111] M. Domeij, H. S. Lee, E. Danielsson, C. M. Zetterling, M. Ostling, A. Schoner, "Geometrical effects in high current gain 1100-V 4H-SiC BJTs," *IEEE Electron Device Letters,* vol. 26, no. 10, pp. 743-745, 2005. https://doi.org/10.1109/led.2005.856010

[112] T. Daranagama, V. Pathirana, F. Udrea, R. McMahon, "Novel 4H-SiC bipolar junction transistor (BJT) with improved current gain." pp. 1-6.

Materials Research Forum LLC
https://doi.org/10.21741/9781644900673-2

[113] L. Lanni, B. G. Malm, M. Östling, C. M. Zetterling, "Influence of passivation oxide thickness and device layout on the current gain of SiC BJTs," *IEEE Electron Device Letters,* vol. 36, no. 1, pp. 11-13, 2015. https://doi.org/10.1109/LED.2014.2372036

[114] S. Zelmat, M.-L. Locatelli, T. Lebey, S. Diaham, "Investigations on high temperature polyimide potentialities for silicon carbide power device passivation," *Microelectronic Engineering,* vol. 83, no. 1, pp. 51-54, 2006. https://doi.org/10.1016/j.mee.2005.10.050

[115] I. H. Kang, M. K. Na, O. Seok, J. H. Moon, H. W. Kim, S. C. Kim, W. Bahng, N. K. Kim, H.-C. Park, C. H. Yang, "Effect of surface passivation on breakdown voltages of 4H-SiC Schottky barrier diodes," *Journal of the Korean Physical Society,* vol. 71, no. 10, pp. 707-710, 2017. https://doi.org/10.3938/jkps.71.707

[116] S. Diaham, M. L. Locatelli, T. Lebey, C. Raynaud, M. Lazar, H. Vang, D. Planson, "Polyimide Passivation Effect on High Voltage 4H-SiC PiN Diode Breakdown Voltage," *Materials Science Forum,* vol. 615-617, pp. 695-698, 2009. https://doi.org/10.4028/www.scientific.net/MSF.615-617.695

[117] A. Siddiqui, H. Elgabra, S. Singh, "The Current Status and the Future Prospects of Surface Passivation in 4H-SiC Transistors," *IEEE Transactions on Device and Materials Reliability,* vol. 16, no. 3, pp. 419-428, 2016. https://doi.org/10.1109/TDMR.2016.2587160

Advancing Silicon Carbide Electronics Technology II
Materials Research Foundations **69** (2020) 107-174

Materials Research Forum LLC
https://doi.org/10.21741/9781644900673-3

CHAPTER 3

Silicon Carbide Doping by Ion Implantation

Philippe Godignon[1], Frank Torregrosa[2], Konstantinos Zekentes[3]*

[1]Centre Nacional de Microelectrònica (CNM), Campus UAB, 08193, Bellaterra, Barcelona (Spain)

[2]IBS (Ion Beam Services), rue G. Imbert Prolongée, ZI Rousset, 13109 Peynier, France

[3]MRG-IESL/FORTH, Vassilika Vouton, PO Box 1385, Heraklion, Greece
and Grenoble INP, IMEP-LAHC, 38000 Grenoble, France

* zekentesk@iesl.forth.gr

Abstract

Ion implantation allows incorporating dopants, or atoms in general, in specific areas of the semiconductor surface. This technique is extensively used in silicon technologies for all kind of devices and circuits integration. An ion implanter is a highly complex machine, with many parameters to set-up. In addition, ion implantation process is always associated with an activation thermal annealing used for the dopants incorporation in the crystal. As we will see in this chapter, the implantation and activation processes in silicon carbide require significantly different parameters than in silicon, and it is today a limiting factor in the development of SiC devices mass volume production.

The chapter starts with a short introduction on ion implantation, which is followed by an overview of the current SiC ion implantation technology. A detailed presentation of all aspects of SiC ion implantation is presented in the remaining parts. More precisely, it is presented the use of different elements as *p*- and *n*-type implanted dopants, the optimum hot implantation conditions for the different elements, the post-implantation annealing, which is still subject of intense studies, the important in SiC channeling effect and the various physical characterization methods of the implanted SiC material. The main part of the present chapter deals with the 4*H*-SiC polytype, as it is the mostly used polytype for device fabrication.

Keywords

Silicon Carbide, Implantation, Dopants Activation, Post-Implantation Annealing, Channeling, Implantation Modeling

Contents

List of used symbols and abbreviations 109

1.   **Introduction** 111

2.   **Ion implantation technique** 112

     2.1   Basics of ion implantation physics 112

     2.2   Basics of ion implantation technology 115

3.   **Specificities of ion implantation in SiC** 119

     3.1   General considerations 119

     3.2   SiC ion-implanted dopants 120

     3.3   Implantation damage 120

     3.4   Hot implantation 121

     3.5   Post-implantation annealing. Activation and diffusion 121

     3.6   SiC devices requirements 123

     3.7   Other SiC implantation reviews 123

4.   *n-Type* **doping** 124

     4.1   *n*-Dopant atoms 124

     4.2   Heating during implantation of *n*-dopants 126

5.   *p*-**Type doping** 127

     5.1   *p*-Type dopants 127

     5.2   *P*-dopant atoms diffusion 127

     5.3   Aluminum doping 128

     5.4   Heating during implantation 129

6.   **Post-implantation annealing** 132

     6.1   Fast thermal annealing 132

     6.2   Very high temperature conventional (CA) and microwave
          annealing (MWA) 133

     6.3   Laser annealing 134

     6.4   Others techniques 135

6.5   Optimization of post-implantation annealing in the case of
Al-implantation 135

6.6   Surface roughness 137

6.8   Capping layer 138

6.9   Electrical activation.. 139

7.   **Crystal quality and electrically active defects** 141

8.   **Channeling and straggling**.. 143

8.1   Channeling in SiC crystal 144

8.2   Lateral/transverse straggling 148

8.3   Box profile 151

9.   **Plasma implantation** 152

10.   **Ion implantation simulation** 153

11.   **Diagnostic techniques of implanted layers**... 155

11.1   Secondary Ion Mass Spectroscopy (SIMS) 156

11.2   Electrical measurements... 156

11.3   Rutherford backscattering spectrometry (RBS)....................................... 156

11.4   Transmission Electron Microscope (TEM).. 157

11.5   Raman spectroscopy... 158

11.6   X-ray diffraction (XRD) 158

11.7   Cross-section imaging techniques.. 159

12.   **Implant services suppliers** 161

13.   **Implant tools for SiC**... 162

Conclusions and challenges 163

Acknowledgements 165

References... 165

List of used symbols and abbreviations

AFM: Atomic Force Microscopy;

AMU: Atomic Mass Unit;

AUTH: Aristotle University of Thessaloniki;

CA: Conventional Annealing;

CNM: Centro Nacional de Microelectrónica;

DLTS: Deep Level Transient Spectroscopy;

FET: Field Effect Transistor;

FORTH: Foundation for Research & Technology – Hellas;

JBS: Junction Barrier Schottky;

JFET: Junction Field Effect Transistor;

JTE: Junction Termination Extension;

MOSFET: Metal Oxide Semiconductor Field Effect Transistor;

MWA: MicroWave Annealing;

PECVD: Plasma Enhanced Chemical Vapour Deposition;

PIA: Post Implantation Annealing;

PIII: Plasma Immersion Ion Implantation;

PLAD: PLAsma Doping;

PSII: Plasma Source Ion Implantation;

RBS: Rutherford Backscattering Spectrometry;

RF: Radio Frequency;

RIE: Reactive Ion Etching;

RT: Room Temperature;

RTA: Rapid Thermal Annealing;

SCM: Scanning Capacitance Microscopy;

SEM: Scanning Electron Microscopy;

SIMS: Secondary Ion Mass Spectroscopy;

SRIM   Stopping and Range of Ions in Matter;

SSRM: Scanning Spreading Resistance Microscopy;

TCAD: Technology Computer Assisted Design;

TEM: Transmission Electron Microscopy;

TLM: Transmission Line Method;

TRIM: Transport of Ions in Matter;

C_{max}:   maximum(peak) concentration of implanted ions;

$D(t)$:   implanted dose in at/cm² after an implantation duration t;

ΔR_p: straggling (the standard deviation of the implanted concentration Gaussian);

R_p: implanted ion projected range;

R_C: contact resistance;

R_{sh}:   sheet resistance;

S: current-collection surface.

1. Introduction

Ion implantation is a common and key technique used in microelectronic technologies for semiconductor processing. Ion implantation allows incorporating dopants, or atoms in general, in specific areas of the semiconductor surface. This process is usually superficial, as the atoms will not penetrate much more than ~1 µm inside the semiconductor layer. An ion implanter is a highly complex machine, with many parameters to set-up. However, the two main parameters that will define the quantity of incorporated atoms and their depths are the implantation energy and the implantation dose. In former standard silicon technology, the typical energy and dose ranges of implanters were 20 keV-180 keV and $1 \cdot 10^{12}$ cm⁻² to $1 \cdot 10^{16}$ cm⁻². In modern sub-micron Si technology, lower energies are required, in the range of 0.2 keV to 20 keV. The atoms used for Si doping are usually boron, arsenic and phosphorus. As we will see in this chapter, the implantation process in Silicon Carbide requires different parameters (larger energy range with an extension to > 300 keV, hot implantation, use of aluminum instead of boron for p-type doping, higher post-implantation annealing (PIA) temperatures) than in silicon. Key issues with SiC implantation, still under investigation, are high resistivity of p-type doped by ion implantation, the precise control of doping profile in implanted regions, the formation of deep junctions or buried layers (>1 µm) and the reduction of point defects and extended defects in order to further optimize SiC power device operation. These issues limit the use of ion-implantation in the fabrication of SiC devices.

The present chapter aims giving all necessary information to a SiC device developer for employing ion implantation in the fabrication of a targeted device. Ion implantation specialists will find further analysis in the hereby-reported bibliography. The introductory Sections 2 and 3 are dedicated to the ion implantation technique and its application to SiC respectively. Note that the Section 3 gives a short and general description of the current state of the art in terms of SiC implantation while next Sections deal in detail with all aspects of it. Reading Sections 3 and 14 (conclusions) allows a fast overview of the state-of-the art in SiC implantation.

The characteristics (dopant species, resulting electrical characteristics of implanted layers, implantation depth, heating during implantation) of SiC n- and p-type doping by ion implantation are reported in Sections 4 and 5 respectively. Section 6 is dedicated to the relevant aspect of dopant activation annealing and the different techniques available for efficient annealing process. In Section 7, the crystal quality, and defects formation and identification are tackled. The next Section is dealing with a significant issue in SiC implantation, the channeling and straggling effects, which are more pronounced in SiC than in Si crystals. In Section 9, we present a novel implantation technique for low defect's surface doping. Finally, in the last Sections we address the SiC implantation simulation and characterization aspects as well as some practical aspects such as implantation facilities and equipment.

$4H$-SiC is mostly considered and in the following we refer to it when the polytype is not mentioned.

2. Ion implantation technique

2.1 Basics of ion implantation physics

Ion implantation is a technique, which consists in giving enough kinetic energy to specific atoms so that they can penetrate in depth in a material target (as a bullet in a wall). The easiest way to give enough kinetic energy to atoms is to charge them (creating positive or negative ions) and to use electrical fields to accelerate them.

After the collision with the target surface, the accelerated ion will lose its energy by two phenomena (Fig. 1):
- Nuclear collisions with target atoms, which can be considered as elastic collisions and easily simulated by usual mechanic's law.
- Coulombic interaction between the electrons cloud of both the accelerated specie and the target atoms, which can be considered as a viscous phenomenon.

Nuclear collisions create defects: vacancy and recoil (also called *Frenkel pair defect*). The defects density and their in-depth localization depend on the mass of incident ions (heavy ions create lots of defects on all their trajectory, light ions create few defects, localized near the final ion position), their energy and the target atoms bonding strength.

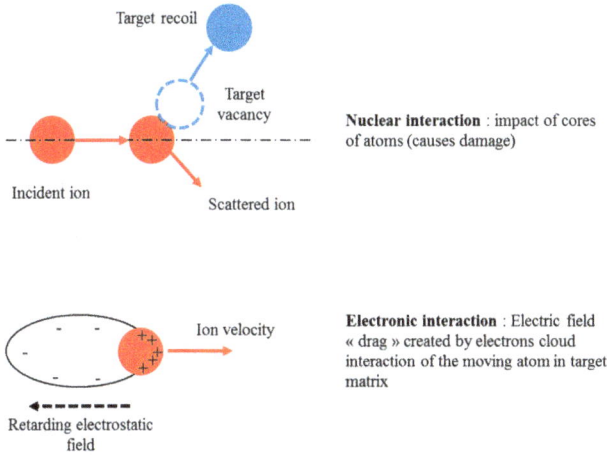

Figure 1.   Schematic of the two main incoming ion energy loss mechanisms: nuclear interaction and electronic interaction.

Defect density also depends on dose rate (overlapping of concomitant collision cascades creating more defects) and temperature of the substrate (cryogenic implantation "freeze the implant defects" while high temperature implantation allows defects "self-recovery").

As illustrated in Fig. 2, the penetration of an energetic ion in a target material creates a cascade of collisions. When collisions occur near the surface, some target atoms can be ejected (sputtering phenomena). As ion implantation is a ballistic phenomenon involving billions of ions, a Gaussian law can roughly represent the final concentration profile of the implanted ions in the crystal target.

The implantation profile, $C(x)$, which is the ions concentration at a given depth x, can be described by using only 2 parameters:

- the mean implantation depth, called *projected range* (R_p)
- the standard deviation of the Gaussian, called *straggling* (ΔR_p)

$$C(x) = C_{max} \cdot exp\left(-\frac{1}{2}\left(\frac{x - R_p}{\Delta R_p}\right)^2\right)$$

$$\tag{1}$$

where C_{max} is the maximum(peak) concentration of implanted ions. If Dose is the total implanted dose (or retained dose), integrating Eq. 1:

Advancing Silicon Carbide Electronics Technology II   Materials Research Forum LLC
Materials Research Foundations **69** (2020) 107-174   https://doi.org/10.21741/9781644900673-3

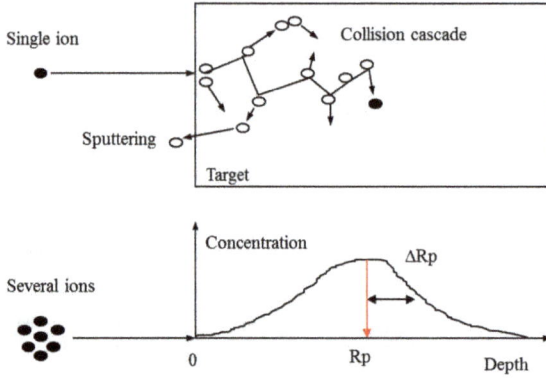

Figure 2. *Schematic of implantation collision mechanisms and resulting doping profile.*

$$Dose = \int_0^\infty C(x) \cdot dx \qquad (2)$$

gives an expression for the maximum/peak concentration C_{max}:

$$C_{max} = \frac{Dose}{\Delta R_p \cdot \sqrt{2\pi}} \qquad (3)$$

Due to sputtering phenomena, for high doses of heavy ions, the retained dose can be lower than the initial dose accelerated in the equipment (or machine dose) and some saturation effect can occur.

Currently, there are more accurate analytical models, which can represent profile dissymmetry and adjust top profile flatness [1].

As most of the implanted substrates or layers in microelectronics (including SiC) are not amorphous, the orientation of the implant beam versus the crystal orientation of the target will impact the number of collisions, and therefore, the implantation profile. If the beam is parallel to a crystal direction, fewer collisions will occur, leading to a deeper profile and less local concentration of the implanted species. This phenomenon is called *channeling*. The lighter the ions and the lower the dose, the more the channeling effect will be important (heavy ions at high doses create amorphous layers, thus breaking the crystal channels). Depending on the implant beam angle dispersion across the wafer, channeling can be a source of inhomogeneity, especially on large substrates (> 150 mm)

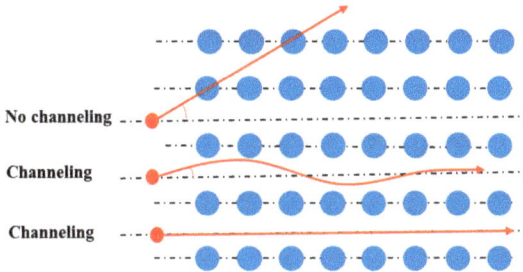

Figure 3. Schematic illustration of the channeling effect (discussed in detail in the following Section 8).

as the incident beam angle at the center of the wafer is not the same than the one at the edge. Nevertheless, modern implanters (ribbon and spot beam systems) have collimating lenses providing the incident beam angle within 0.1 ° across the whole treated surface. As we will see in Section 8, the effect of channeling is very important in 4H-SiC or 6H-SiC. To eliminate channeling, a solution is to orientate the wafer to avoid having the crystal channels parallel to the beam and more precisely the wafer is first tilted (tilt angle) and then rotated on its axis (twist angle). Nevertheless, the channeling can be beneficial in some cases as it results in deeper dopant penetration and lower defect generation [2].

2.2   Basics of ion implantation technology

Russel Ohl from Bell Labs (USA) has built the first ion implanter for semiconductor doping in 1950. In 1957 William Shockley, from the same lab, deposited the first patent on using ion implantation to introduce dopants in semiconductor. Current implantation tools are composed of 3 main parts (Fig. 4):

– The "Terminal" with a gas box (containing gas cylinders and gas lines to provide a stable flow of selected gas precursor to the ion source), an ion source (to create a plasma containing the ions of interest), an extraction electrode (to extract the ions from the source and give them an initial kinetic energy), a mass spectrometer (applying a orthogonal magnetic field to select the ion mass to be implanted through a resolving aperture) and an acceleration column to provide to the selected ions their final kinetic energy.

– The "Beam Line" with electrostatic or magnetic focusing and scanning optics. Depending on the implanters, the scanning of the beam on the wafer can be done either using electrostatic scanning (as represented in Fig. 4) or magnetic scanning (ribbon beam) associated to mechanical scanning. On large substrates (> 200 mm in diameter)

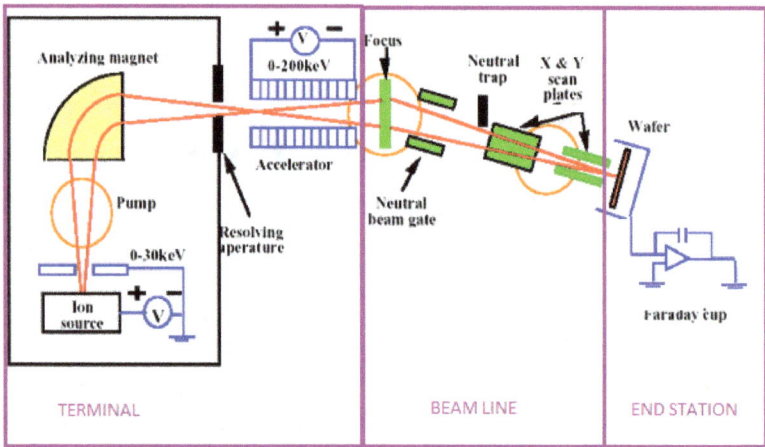

Figure 4.   Schematic of an ion implantation tool.

a parallel scanning is often used to insure a uniform implant angle across the wafer. This feature is not mandatory on the 6 inches SiC wafers used today in production (cost consideration versus need).

– The "End Station" with the substrate holder (cooled or heated) which can be angularly oriented toward the beam (tilt and twist angles), the dose measuring system (also called *the dose integrator* with its Faraday cup) and in some cases additional mechanical scanning motions (translations, or rotations of the substrates) and electrons flood gun to avoid charging effects. Some implanters (as those employed for SiC implantation) also include a heating stage in the substrate holder for high temperature implantation. End station is connected to loading/unloading vacuum robots.

Each of these parts has its own pumping system using turbo-pumps for the ion source and turbo and/or cryogenic pumps for the beam line and in the end-station. To avoid collisions between accelerated ions and residual gas molecules, resulting in possible neutralization (affecting thus, the dose measurement accuracy), ion energy loss and ion beam divergence, the vacuum in the beam line and in the end station must be at least in the 10^{-7} Torr range.

The size of an ion implanter is typically 4 to 6 meters long by 2 to 4 meters wide by 2.5 to about 3 meters high (depending on type, and brand).

Materials Research Forum LLC
https://doi.org/10.21741/9781644900673-3

Even if implantation tools look complex, implantation process is easily controlled by 3 electrical parameters:
- The current in the mass spectrometer, to adjust the required magnetic field and select the desired atomic mass unit to be implanted.
- The voltage applied on the extraction electrode and the acceleration column to give the required kinetic energy and adjust the penetration depth of the implanted ions in the substrate. For instance, 30 kV applied on the extraction electrode and 70 kV applied on the acceleration column will give a total kinetic energy of 100 keV to a single charge ion.
- A precise implantation current measurement at the substrate level (directly on the chuck or on cups surrounding the wafers) to measure the quantity of electrical charges, thus number of ions, reaching the wafer. This allows an accurate real time measurement of the implanted dose using the following equation:

$$D(t) = \frac{1}{q \cdot S} \int_0^t i(t)dt \qquad (4a)$$

where $D(t)$ is the implanted dose in at/cm², q is the charge of the ions (for single charge ions $q = 1.6 \cdot 10^{-19}$ C), S is the current-collection surface in cm² (precisely defined by an aperture), $i(t)$ is the measured implant current in Amperes. If $i(t)$ is constant, then

$$D(t) = \frac{i \cdot t}{q \cdot S} \qquad (4b)$$

Note that the chuck (or current measurement cups) is always combined with an electrostatic Faraday cup (Fig. 4) to avoid secondary electrons, generated by collision of energetic ions with the target material, falsifying the measurement.

Depending on the application, single charge ions (for example B^+), molecular ions (for example BF_2^+) or multiple charged ions (for example B^{2+}) can be implanted.

For multi-charged ions, the kinetic energy will be the acceleration voltage multiplied by the number of charges. For instance, if Al^{3+} is implanted with 100 kV acceleration voltage, the kinetic energy of Al atoms will be $3 \times 100 = 300$ keV. But as only one Al atom is implanted every 3 elementary charges, the programmed machine dose (number of elementary charges per cm²) should be 3 times the required Al dose. Thus, a machine process with $Al^{3+}/3 \cdot 10^{15}$ at/cm²/100 kV corresponds to a physical implantation of Al at 300 keV and a dose of $1 \cdot 10^{15}$ at/cm². Indeed, multiple charged ions are used to access to high energy implantation using medium current implanters with energy ranges often limited to 200 to 500 keV. It is why for SiC p-type doping application, Al^{2+} or Al^{3+} ions are often used, as energies higher than the acceleration voltage capability of the implanter (typical for Si applications) are needed.

In the case of molecular ions, all the atoms of the molecule will be implanted and after the collision with the substrate each atom will share the initial molecule energy in proportion to its atomic mass:

$$E(atom\ A) = E(molecular\ ion) \cdot \frac{M(atom\ A)}{M(Molecule)}$$

(5a)

For example if BF_2^+ is accelerated at 100 keV, the energy of Boron will be:

$$E(B) = 100 \cdot \frac{11}{11 + 2 \times 19} \approx 100 \cdot \frac{1}{4.45} \approx 22\ keV$$

(5b)

As implantation current drops down when decreasing extraction and acceleration voltages, especially below 15 keV, molecular ions are often used to access to lower energy ranges, while keeping enough current. For SiC doping application, this can be efficiently used for n-type contact doping using nitrogen N_2^+ and as 2 nitrogen atoms are implanted per molecular ion, the machine-dose-set in the recipe will be 2 times lower than the needed nitrogen dose, allowing to divide by 2 the implant time.

Moreover, there are possible mass interferences to be taken into account when implanting nitrogen or aluminum for SiC applications:

- Due to close mass values of N_2^+ (28 AMU) and Al^+ (27 AMU), atoms interference errors can easily occur, and presence or absence of N or Al peak at AMU= 14 must always been checked.
- Some ion sources contain boro-nitride (BN) insulators. This means that some quantity of Boron will always be present in the plasma source and may react with residual moisture to create BO^+, which has the same mass (11+16=27) as Al^+. The risk to have boron contamination leading in deeper p-type profile is important. It is, thus, mandatory to replace these BN insulators by Al_2O_3 inside the implanter tool.
- For multi-charged implantations the atomic resolution of the mass spectrometer is crucial. For instance, the *apparent mass* (AMU multiplied by charge) of Al^{3+} (27×3 = 81) is very closed from the one of Ar^{2+} (40×2 = 80) and if the mass spectrometer resolution of the implanter is higher than 1 at AMUs around 80, there is a risk to implant Argon instead of Aluminum.
- For the same reason, the presence of beryllium, silicon, fluorinated compounds or CO_2 gas in the source arc chamber can lead to mass interference or peak selection error in

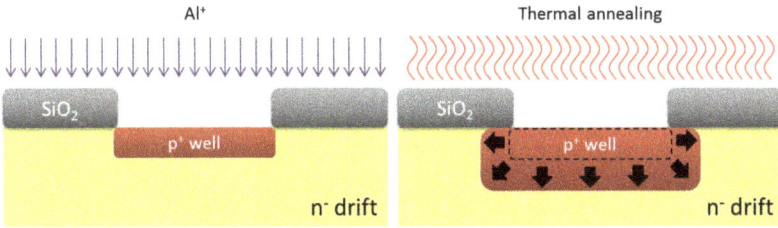

Figure 5.   *Schematic illustration of an implantation and diffusion process.*

the case of aluminum or nitrogen implant. Summarizing, a list of possible mass interferences is given below:

- Be^+ vs. Al^{+++}
- F^{++} vs. Al^{+++}
- Si^+ vs. Al^+
- Si^{++} vs. N^+
- CO^{++} vs. N^+

Therefore, the implantation technique is highly complex and requires a regular and very fine-tuning of the equipment, especially for aluminum, which is probably the most complex atom to implant.

3.   Specificities of ion implantation in SiC

3.1   General considerations

Silicon is a very friendly material for processing compared to other semiconductors like SiC, GaN, Diamond, GaAs. To form a p or n type well in a silicon layer, dopant atoms are implanted at the Si surface and then redistributed by diffusion under a thermal treatment. Depths of tens of microns can be obtained with this process sequence. In Fig. 5, we schematically present the way to form the *p*-well of a Si power MOSFET.

On the contrary, selective doping of SiC is accomplished by multiple ion implantations, in order to obtain a box profile without use of a thermal diffusion step. This is because the diffusion coefficients of aluminum, boron, phosphorous and nitrogen (the most common dopants of SiC) are so low, that doping by thermal diffusion after implantation or from solid or liquid sources, also used for Si, is impractical. This is why, a multi-implantation box profile is typically used in SiC, as described in Section 7.3.

Moreover, the implantation is limited in depth since the ions penetration depth is around 1 nm/keV, because of the higher atom density of SiC compared to Si. Thus, for making a box profile of 0.4 µm depth (typical depth for devices applications like MOSFETs) one would need energies from 20 keV up to 400 keV, which are not available with standard implanters used in the wide majority of Si technologies. Furthermore, for implanting in a depth of 1 µm or more (typical depth for devices applications like JFETs) one needs to employ around 1 MeV of implantation energy.

3.2   SiC ion-implanted dopants

Table 1 summarizes the main properties of the most common dopant ions in 4H-SiC.

Table 1.   Summary of main ion-implanted dopant atoms properties in 4H-SiC

Dopant Ion	Solubility [8] (at·cm^{-3})	Minimum sheet resistance (Ω/□)	Resistivity (mΩ·cm)	Maximum concentration range (cm^{-3})	Activation energy (meV)
Nitrogen	2×10^{20}	290	14.5	$3 \cdot 10^{19}$	50
Phosphorus	1×10^{21}	29	2.3	$2 \cdot 10^{20}$	53
Aluminum	1×10^{21}	800	25	$5. 10^{20}$	190-220

3.3   Implantation damage

The major drawback of ion-implantation-doping technique is the generation of damage as the incoming ions destroy the crystalline structure of the target material. The damage can range from point defects caused by single collision cascades at low implanted dose, to extended (1D and 2D) defects at higher dose and to complete amorphization of the crystal at very high dose. The damage accumulation during ion implantation is approximately proportional to the implant dose until complete amorphization occurs. It is very difficult to obtain good quality high-dose implanted layers of SiC without heating the substrate during implantation or performing annealing between the multi-implantation steps necessary for obtaining the previously mentioned box dopant profile. High-dose implantation at room temperature (RT) leads to complete amorphization of the semiconductor surface layer. Maintaining the original crystalline structure during ion implantation is also delicate because of the complex polytypism and the low stacking-fault formation energy in SiC. Indeed, once the implanted region becomes completely amorphous, as a result of high-dose implantation, lattice recovery to the original polytype is not guaranteed. For instance, PIA of amorphized 4H- or 6H-SiC resulted in most cases in polycrystalline 3C-SiC. This is the main reason why hot implantation of SiC is employed, especially when the implant dose is high.

It is also important to mention that the SiC crystal inflates during the implantation process. A warpage can occur, either during implantation, or after post implantation annealing. This warpage may be problematic for subsequent photolithography steps necessary for completing the device fabrication process.

3.4 Hot implantation

When the sample is heated during implantation, annihilation of defects takes place, the so-called *dynamic annealing*, and a competition exists between the rates of defect generation and annihilation. At low enough fluxes, the annihilation rate can exceed the generation rate and the accumulated defect density may never reach the critical value for amorphization. The substrate temperature plays a crucial role for the dynamic annealing because the annihilation process(es) is thermally activated. Increasing the substrate temperature implies a higher annihilation rate and hence a higher ion flux can be tolerated while still suppressing amorphization.

In the case of hot implantation, the implanted region can retain the original polytype structure and can be easily recovered. Similar results are obtained for all typical dopants ions like nitrogen, phosphorus, aluminum, and boron implantation. Therefore, this is a consistent and unique feature of ion implantation into hexagonal SiC polytypes.

When the implant dose is low, disordering the carbon sub-lattice proceeds at a higher rate than that one for the silicon sub-lattice as a result of the lower threshold-displacement energies for carbon atoms. However, amorphization of the carbon and silicon sub-lattices occurs at almost the same implant dose. The critical implant dose, above which the implanted region becomes amorphous, depends on the implanted species and implantation energy. For instance it is approximately low to mid 10^{15} cm^{-2} for room-temperature nitrogen implantation [3].

3.5 Post-implantation annealing. Activation and diffusion

Post-implant annealing is required to: (1) restore the crystal structure and (2) electrically activate the implanted dopants (shallow acceptors and/or donors). This means that any implanted atom in SiC has to reach a proper lattice site (i.e. a substitutional one) at the end of the annealing process. This atom is an "*activated*" one. Although a substantial part of the implantation-induced damage can be removed by 1200 °C annealing, an annealing at temperatures in excess of 1500 °C should be performed to achieve reasonable electrical activation because of the high bonding strength of the SiC lattice [4]. Thus, high temperature (>1600 °C) PIA is necessary to restore crystal quality and activate implanted atoms. Note that, the annealing temperature cannot be reduced, even if the implantation is conducted at elevated temperatures of 500–1000 °C. One of the main reasons for this is the thermal stability of several deep levels, which are generated by ion implantation and cause compensation of dopants [5]. Typical clean room equipment for silicon annealing reaches 1250 °C only. Again specific and non-standard equipment is needed to perform the SiC implanted dopants activation.

As previously mentioned, the dopant profile exhibits very little diffusion during such high-temperature annealing, retaining the as-implanted profile in the case of N, P and Al implantation. This is expected from the very small diffusion constants and any enhancement of impurity diffusion via implantation-induced defects is negligible. The lack of dopant diffusion in SiC makes it relatively easy to form a shallow junction, but difficult to form a very deep junction. Implanted boron atoms, however, show significant out-diffusion and in-diffusion during the activation annealing [6].

At such elevated PIA temperatures, the SiC crystal surface decomposes due to selective out-diffusion of Si from the SiC lattice. Hence, specific solutions to permit high-temperature annealing of SiC implants while suppressing the out-diffusion of Si from the surface are required (see Section 6). Even higher annealing temperatures result in extended defects, such as dislocation loops, similar to the implanted Si end-of-range defects [7].

The implantation and activation of nitrogen to produce n-type regions has been well understood and optimized in mid-90's, and activated concentrations up to $3 \cdot 10^{19}$ cm^{-3} with a corresponding sheet resistance of typically 300-500 $\Omega/\square$ are obtained routinely [8]. Phosphorus is useful for extending by one decade the doping range (well above a $3 \cdot 10^{19}$ cm^3 donor density, which corresponds to the "solubility"[1] limit for nitrogen [9]) and results in lower sheet resistance for the same implanted depth, as we will detail later.

P-type selective doping has been an area of research during last decades. The two common p-type dopants, aluminum and boron, produce relatively deep acceptor levels but aluminum is generally used because of its smaller ionization energy. An inherent difficulty of Al implantation in comparison to that of nitrogen, is that Al atom is heavier and can dislocate both Si and C atoms while nitrogen dislocate mainly C atoms. Thus, it is more difficult to restore crystal quality and properly activate an Al-implanted layer. Usual values of sheet resistance of high Al dose implanted layers exceed well 1 k$\Omega/\square$.

---

[1] The value mentioned here and in the remaining of the text, is that corresponding to a serious change of electrical and optical properties of the implanted layer and not necessary to the limit of substitutional incorporation and precipitate formation of the involved species.

There is a consensus [8] that an almost 100% activation can be obtained for N and Al and for concentrations up to $3 \cdot 10^{19}$ cm^{-3}, under proper implantation conditions. For P the corresponding concentration increases to $1 \cdot 10^{20}$ cm^{-3}.

3.6 SiC devices requirements

For SiC power devices fabrication, several types of implantation schemes are required. Four of them are mostly used [8]:

1. Low dose (10^{13} cm^{-2}) and deep ($25 \div 400$ keV) aluminum implantation is needed to form Junction Termination Extension (JTE) in most of power devices.

2. Medium dose (10^{14} cm^{-2}) and deep ($25 \div 400$ keV) aluminum implantation for MOSFETs p-well, where the MOS channel is formed, as well as for CMOS integration.

3. High dose (10^{15} cm^{-2}) and thin ($25 \div 150$ keV) aluminum implantation for ohmic contacts to p-type layers, for JBS and pin diodes, bipolar transistors, power MOSFETs, etc.

4. High dose (10^{15} cm^{-2}) and thin ($25 \div 150$ keV) nitrogen or phosphorus implantation for ohmic contacts to n-type layers for most of the power devices or sensors.

There are other more specific needs requiring more complex processing such as:

1. High dose (10^{15} cm^{-2}) and thin ($25 \div 150$ keV) aluminum implantation with high angles ($30° \div 45°$) for side doping of trench JFET gates.

2. High dose (10^{15} cm^{-2}) and deep ($25 \div 400$ keV) aluminum implantation with low damage for buried layers' formation (followed by a re-epitaxy process) used in trench MOSFETs, super-junction FET, JFETs, etc.

3. High dose (10^{15} cm^{-2}) and very deep ($1 \div 2$ MeV) aluminum implantation with low damage for direct buried layers formation.

The above implantations conditions hold also for radiation detectors, biosensors and integrated circuits fabrication.

3.7 Other SiC implantation reviews

A series of comprehensive reviews about ion damage formation and accumulation in $3C$, $4H$ and $6H$-SiC have been published and are suitable for a deep study of the subject [5, 8, 10, 11].

Following the main results and critical issues of SiC ion implantation are reviewed.

Materials Research Forum LLC
https://doi.org/10.21741/9781644900673-3

4.   *n-Type* doping

4.1   *n*-Dopant atoms

Two atoms have been extensively studied for *n*-type doping of SiC, due to their relatively low activation energies: nitrogen and phosphorus (see Table 2) [11].

Table 2.   Ionization energies of the main donor impurities in 6H-SiC and 4H-SiC.

Polytype	Impurity	Ionization energy in hexagonal site (meV)	Ionization energy in cubic site (meV)
6*H*-SiC	Nitrogen	85	140
	Phosphorus	80	110
4*H*-SiC	Nitrogen	50	92
	Phosphorus	53	93

Originally, only nitrogen was used for *n*-type doping, especially for *in-situ* doping during epitaxial growth. However, the free electron concentration in n^+-layers formed by N implantation is saturating at ~$3 \cdot 10^{19}$ cm^{-3} with a corresponding sheet resistance of typically 300-500 $\Omega/\square$ despite high PIA temperatures ($\geq 1600\,°C$) and atomic concentrations in excess of 10^{20} cm^{-3}. Indeed, the lowest resistance of N$^+$-implanted SiC ever reported is 290 $\Omega/\square$, corresponding to a resistivity of 14.5 mΩ·cm (700 °C implanted and 1600 °C annealed, depth: 0.5 µm, dose: $2.7 \cdot 10^{15}$ cm^{-2}) [12]. The saturation of free electron concentration and resistance might be due to the above-mentioned low solubility limit of N in SiC. Schmid *et al.* [13] indicated that the saturation is caused by formation of an electrically neutral defect complex, which is formed during the annealing process and which is responsible for the strong deactivation of implanted N donors.

More recently, phosphorus is used increasingly, especially for MOSFET transistors fabrication, in order to lower the implanted layer resistance. Initial estimates of a low solid solubility for P in SiC have resulted to time-delayed implantation investigation of this specie. However, as it has been shown theoretically [9], the phosphorus ion (P^+) implantation is useful for extending by one decade the doping range i.e. well above $3 \cdot 10^{19}$ cm^{-3} activated (i.e. substitutional) *n*-type dopant density, which is the free electron concentration limit for nitrogen. A lower sheet resistance of 29 $\Omega/\square$, corresponding to a resistivity of 2.3 mΩ·cm (implanted at 800 °C and annealed at 1700 °C; the implantation depth is 0.8 µm and the total implanted dose is $1.4 \cdot 10^{16}$ cm^{-2}) has been achieved for P$^+$-implanted 4*H*-SiC (0001) [13]. So, high-dose P$^+$ implantation at an elevated temperature (higher than 200 °C) followed by annealing at a high temperature above 1600 °C is

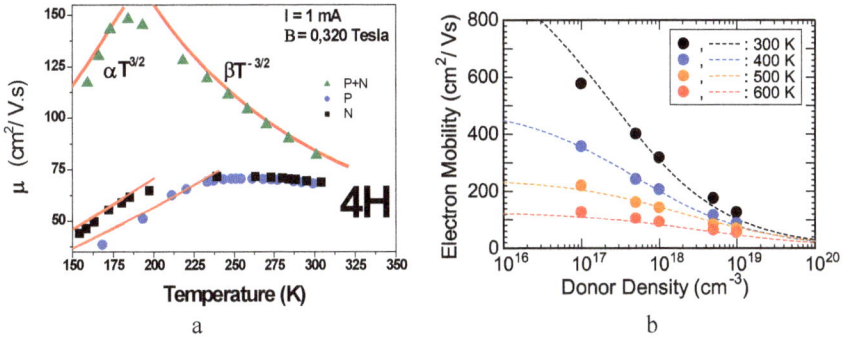

Figure 6.   (a) Comparison of experimental Hall mobility versus temperature for samples implanted with N, with P and co-implanted with N^+P. Reprinted from [15]. (b) Dependence of carrier mobility on doping density for P^+-implanted layers in semi-insulating SiC measured at several temperatures. The present results are plotted using closed circles, while the fitting curves for the data of epitaxial layers are shown by dashed lines. Reprinted from [16] © 2017 the Japan Society of Applied Physics.

effective to reduce sheet resistances to lower values compared to nitrogen implantation and it is generally the dopant of choice for this concentration range.

Ab-initio studies have shown that N dopant atoms substitute C and P atoms substitute Si in the lattice [14]. So, co-implantation of N and P has been considered as an efficient way to reduce sheet resistances, especially in the high doping range [15]. For instance, for a total dose of $7 \cdot 10^{14}$ cm^{-2}, the activation rate measured by Hall effect were 57%, 63% and 86% for P, N and P+N implantation respectively. Corresponding sheet resistances were 940, 830 and 530 $\Omega/\square$. Consequently, the resulting resistivity values (material one and ohmic contact one) were better for the co-implanted P+N samples. The Hall mobility also exhibited a different behavior, as we can infer from Fig. 6. For the P+N implanted sample, the mobility is higher, especially at low temperature. The decrease of the mobility at temperatures below 200 K accounts for scattering from ionized impurities. At high temperature, the mobility is driven by acoustic phonons. Despite the benefits, the drawback of this N and P co-implantation technique is that it is a more complex and costly processing step.

In a recent study [16], performing P implantation in semi-insulating 4H-SiC substrates, the electrical activation ratios was about 88–98% after annealing at 1650 °C and the mobility value (Fig. 6b) were mostly comparable to those of epitaxial layers.

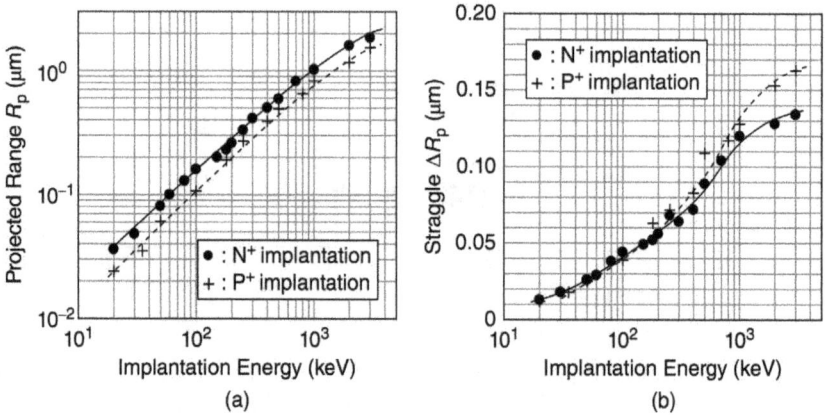

Figure 7. (a) Projected mean range (R_p) and (b) straggle (ΔR_p) versus the implant energy in nitrogen and phosphorus ion implantation into SiC. Reprinted from [8] © 2014 John Wiley & Sons Singapore Pte. Ltd..

Finally, it should be noted that there is no important difference in projected mean range (R_p) and (b) straggle (ΔR_p) versus the implant energy for nitrogen and phosphorus ion implantation into SiC (Fig. 7) [8]. The only one is that P exhibits a 10% lower R_p for all implant energies.

4.2   Heating during implantation of n-dopants

When the implant dose is relatively low ($< 3 \cdot 10^{14}$ cm^{-2}), there are no striking differences in sheet resistance, irrespective of implant species (N$^+$ or P$^+$) or implantation temperature (RT or 500 °C) (Fig. 8) [8].

When the implant dose becomes high ($>10^{15}$ cm^{-2}), the situation is significantly changed. Striking differences are observed between room temperature (RT) implantation and hot implantation, and also between nitrogen implantation and phosphorus implantation (Fig. 8). In the case of room temperature implantation, the sheet resistance exhibits a minimum value at an implant dose of approximately $0.7 \div 1 \cdot 10^{15}$ cm^{-2}, and increases when the implant dose is further increased. In this high-dose region, the lattice damage caused by room temperature implantation is so severe that the implanted region contains a high density of stacking faults and 3C-SiC grains after activation annealing [8]. On the other hand, a continuous decrease in the sheet resistance is observed for hot implantation. As already mentioned, the sheet resistance of the nitrogen-implanted region is almost saturated at 300 Ω/□, which may be limited by the relatively low solubility limit of nitrogen atoms in SiC (Fig. 8). A much lower sheet resistance of 30 ÷ 50 Ω/□ can be obtained by hot implantation of phosphorus owing to its higher solubility limit.

Newer experimental results confirmed these findings exhibiting resistivity values (~1 mΩ·cm) for *P*-implanted samples [17].

5. *p*-Type doping

5.1 *p*-Type dopants

Aluminum is the main acceptor used in SiC epitaxy and ion implantation. Boron has been also investigated but as described below, boron implantation and PIA can cause several unwanted phenomena and is not usually employed for device fabrication in industry.

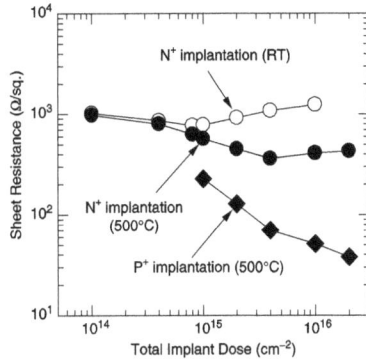

Figure 8. Sheet resistance versus the total implant dose for nitrogen or phosphorus implanted 4H-SiC(0001). Multistep implantation at RT or 500 °C was performed to form a 200 nm-deep box profile. A 30 min PIA at 1700 °C has been also performed. Reprinted from [8]. © 2014 John Wiley & Sons Singapore Pte. Ltd..

The implantation depth depends on the atom properties. The boron being the lightest one, it typically reaches between two and three times the depth of Al atom for a given implantation energy (Fig. 9).

One cannot easily achieve a low sheet resistance of *p*-type layers with any implanted material since both Al and B suffer from relatively deep states in addition to the high associated ionization energy. Indeed, it is necessary to have at the same time, a high dose implantation with a reasonable depth, a good electrical activation, and a high level of holes mobility. This is a very difficult compromise to achieve and usual values of sheet resistance well exceed 1 kΩ/□.

5.2 *P*-dopant atoms diffusion

As it was found for nitrogen or phosphorus ion implantation, implanted aluminum atoms exhibit very little diffusion, even after high-temperature annealing at 1600 ÷ 1700 °C.

Implanted boron atoms, however, show significant out-diffusion and in-diffusion during the activation (post-implantation) annealing [6], especially at high implantation doses. As

a   b

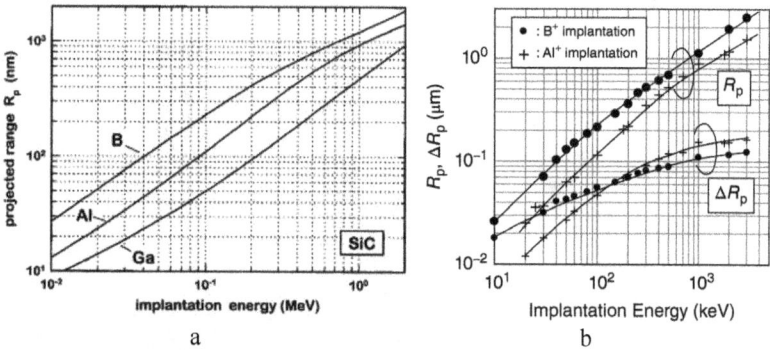

Figure 9.   (a) Calculated projected range of implanted B, Al and Ga as function of
implantation energy in 4H/6H-SiC (from [6]).
(b) Experimental R_p and ΔR_p versus the implant energy for aluminum and boron.
Reprinted from [8] © 2014 John Wiley & Sons Singapore Pte. Ltd..

a result of the out-diffusion, some portion of the implanted boron atoms is lost; the in-diffusion makes the junction depth extremely large compared with the designed depth.

This is illustrated in the SIMS measurement shown in Fig. 10 [18]. After the first annealing at 1500 °C, the boron profile is almost the same as implanted. But after a second annealing at 1700 °C, we can infer a strong distortion of the profile, with a dose loss of 32% for the implanted dose of $2.5 \cdot 10^{13}$ cm^{-2}, and a dose loss of 76% for an initial dose of $2.5 \cdot 10^{15}$ cm^{-2}. It has been accepted that diffusion of boron atoms is enhanced by implantation induced damage via a kick-out mechanism [19]. Boron interstitials created by implantation have a large diffusion constant, and can diffuse already at 1500 °C. Damage-enhanced diffusion is also observed in boron-doped epitaxial layers. Thus, boron is not a good choice for device development.

5.3   Aluminum doping

Aluminum-implanted SiC layers exhibit lower resistance values (Fig. 11) [8] than B-implanted ones. Implanted Al activation was optimized for post-implantation annealing in the range 1600-1700 °C for *4H* [20] and *6H* [21] polytypes, but only low concentrations (< 10^{19} cm^{-3}) were used. This resulted in a layer resistivity always larger than 0.5 Ω·cm (see Ref. [22]). By employing high-dose implantation ($3.0 \cdot 10^{16}$ cm^{-2}) followed by RTA at 1800 °C for 1 min, the sheet resistance could be reduced down to 2.3 kΩ/□ (corresponding resistivity of 46 mΩ·cm) [15].

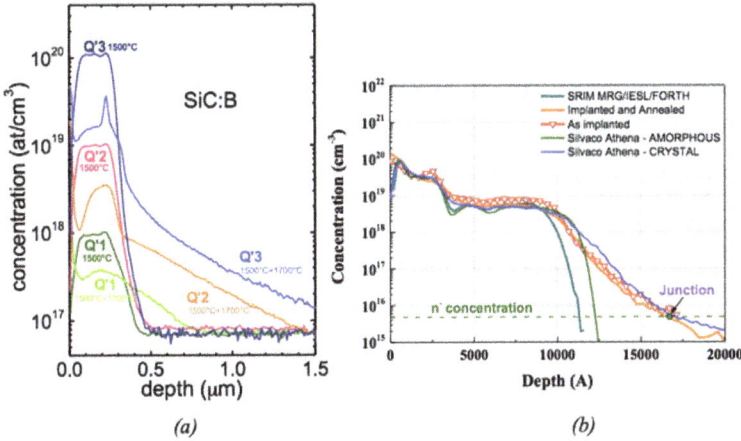

Figure 10. (a) Experimental SIMS profiles of Boron implanted 6H-SiC samples after annealing at 1500 °C and 1700 °C for 3 different implantation doses: Q1 - $2.5 \cdot 10^{13}$, Q2 -$2.5 \cdot 10^{14}$, Q3 - $2.5 \; 10^{15}$ cm^{-2}. Reprinted from [18].
(b) Experimental and simulated SIMS profiles from multistep Al-implanted 4H-SiC at 400 °C with a total dose of $1.58 \cdot 10^{15}$ cm^{-2} Experimental data are from as-implanted samples as well from samples post-implantation annealed at 1750 °C. Courtesy of FORTH.

Use of microwave induced heating (see Section 6) at temperatures higher than 2000 °C resulted sheet resistance down 0.8 kΩ/□ (resistivity of 25 mΩ·cm) while preserving a very low surface roughness [23]. Same order values have been obtained by induction annealing of samples with heavy Al^+ implanted concentrations (> $1 \cdot 10^{20}$ cm^{-3}) at temperatures > 1950 °C resulting in sheet resistance down 1 kΩ/□ (resistivity of 30 mΩ·cm) while preserving a very low surface roughness (< 2 nm) [24].

However, due to large size of Al atom, a high (> 10^{19} cm^{-3}) concentration of Al-dopants induces a lattice strain along the *c*-axis. For very high (> 10^{20} cm^{-3}) concentration, this strain can be relaxed by dislocation formation. Therefore, the optimization of process conditions to obtain a low resistivity by high Al dose implantation with acceptable crystal quality is quite delicate (see discussion below on crystal quality).

5.4 Heating during implantation

During 90's many research groups [25] agreed that the temperature range between 500 and 700 °C is the optimum SiC substrate temperature during Al implantation. The

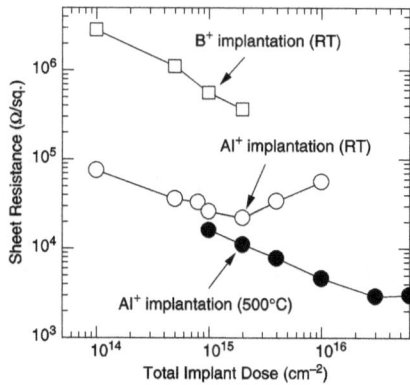

Figure 11.   Sheet resistance versus the total implant dose for aluminium-implanted 4H-SiC(0001) annealed at 1800 °C for 30 min. Reprinted from [8]. © 2014 John Wiley & Sons Singapore Pte. Ltd..

corresponding physical explanation is that the melting temperatures of the Al-Si eutectic (577 °C) and the metallic Al (660 °C) are within this range. Implantation at higher temperatures has to be avoided as it leads to the formation of Al and Si metallic precipitates.

However, the formation of extended defects during implantations at temperatures higher than 500 °C has been demonstrated [26]. For this reason, most groups do not make a hot implantation at temperatures higher than 500 °C, to exclude formation of defects non-recoverable with post-implantation annealing. Nevertheless, a heating above 300 °C during implantation is necessary in the case of Al implantation and for doses higher than 10^{14} cm^{-2}, to avoid significant sample damaging that cannot be easily recovered during annealing while this limit is $5 \cdot 10^{14}$ cm^{-2} for B [27].

Note that the amorphization limit, for room-temperature-implantation, can be at higher doses ($1 \cdot 10^{15}$ cm^{-2} for 100 keV Al$^+$ ions [28]) but well lower doses have to be used in order to be on the safe side.

The above conclusion, about optimum Al implantation temperature in the range of 400 °C to 500 °C, has been corroborated by new results showing the evolution of sheet resistance as a function of implantation temperature (Fig. 12) [29]. In the same study sheet resistance (R_{sh}) and specific contact resistance (R_c) measured by Transmission Line Method (TLM) technique have been compared for hot aluminum implantation and carbon/aluminum co-implantation. Indeed, it has been considered that C$^+$ co-implantation

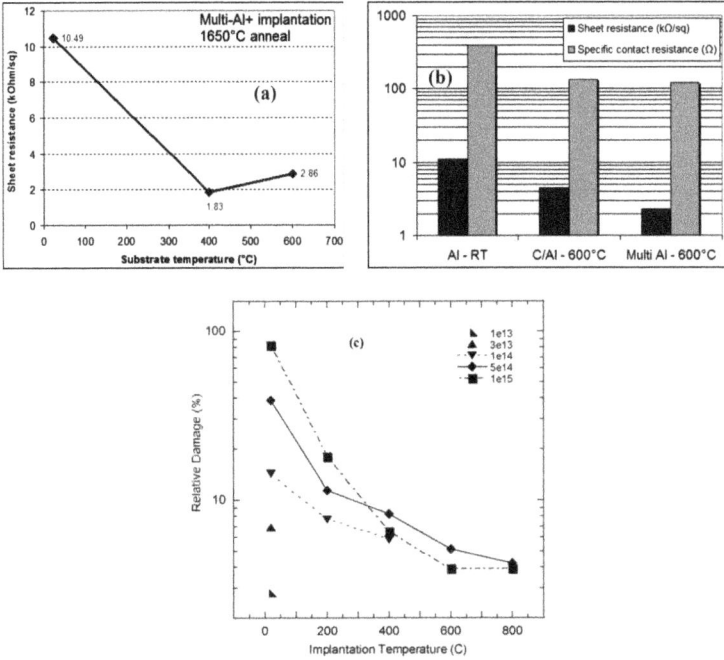

Figure 12.   Characterization of Al-implanted 4H-SiC layers. (a) R_{sh} vs. implantation temperature (b) R_{sh} and R_c measurements for different implantation conditions. (After [29]). (c) Relative damage at different doses as a function of implantation temperature. The relative damage is the degree of amorphization around the projected range, where for 100 is fully amorphous. Reprinted with permission from Elsevier from [28] © 2016.

enhances the probability of Al atoms to occupy Si sublattice sites or reduces the number of implantation induced defects. However, this hypothesis has not been confirmed by experimental data showing that the addition of C^+ ions does not result in better resistivity values (Fig. 12b).

Similar conclusions have been drawn by other groups [28, 30] in terms of optimum implantation temperature values (Fig. 12c). Moreover, for the samples implanted at 600 °C and post-implantation annealed at temperatures above 1650 °C, AFM observations showed holes formation indicating extended structural defects [30].

Thus, the optimum substrate holder temperature, during Al implantation, is in the range between 350 and 500 °C.

6. Post-implantation annealing

Post-implantation annealing temperatures exceeding 1500 °C are required for an efficient electrical activation and for restoring crystal quality. Nitrogen will take carbon sites, while aluminum will reside on silicon lattice sites. Doping with nitrogen is therefore relatively easy, since the formation energy of carbon vacancies is considerably lower than the formation energy of Si vacancies, (5 and 8 eV, respectively) [28]. Full activation of donor doping can then be achieved at temperatures of around 1600 °C, while the acceptor doping needs even higher temperature to reach full activation, particularly for higher dose implantations [8]. To perform the thermal annealing at temperatures exceeding 1500 °C, the typical furnaces used in Si processing cannot be employed and specific equipment has been developed.

The most commonly used method for SiC PIA is high temperature thermal annealing performed in argon ambient at atmospheric pressure or performed in high vacuum.

At these high temperatures of post-implantation annealing, a Si out-diffusion takes place resulting in an important surface roughness (step-bunching). Indeed, the off-axis orientation of the employed wafers induces a high surface energy and Si evaporation contributes to minimize this energy by surface reconstruction, the latter leading to step-bunching. A substantial effort has been devoted to address this issue.

Various approaches have been tested for optimizing SiC post-implantation annealing such as:

1. Thermal annealing performed in Ar ambient at atmospheric pressure or performed in high vacuum.
2. Thermal annealing performed in overpressure of silicon using SiH_4/Ar ambient.
3. Thermal annealing performed with capping layer.
4. Microwave induced heating annealing.
5. Laser annealing.

6.1 Fast thermal annealing

The initial effort, on SiC PIA, has been dedicated to Rapid Thermal Annealing (RTA) systems for various reasons. 4H- and 6H-SiC samples need a very high heating rate in temperature to preserve the polytype from cubic inclusions, which may be generated during the solid phase epitaxy at low temperature. In addition, in many RTA processing systems, the temperature uniformity across the wafer is excellent resulting in a reduction in the thermal gradient that can warp wafers.

RF induction RTA furnaces have been also used in initial investigations, due to advantage of the very high heating rate (> 10 °C/s) [31]. Indeed, a better activation is obtained for high heating (ramp-up) rate. As it is explained in the review paper of R. Nipoti [31], the heating rate influences both the surface roughness and sheet resistance as well as the current-voltage characteristics of SiC diodes formed on ion implanted layers. A better activation has been obtained for a heating rate of 1000 °C/min in comparison to that of 250 °C/min [32]. However, there is no a definitive conclusion on the influence of heating rate on implanted crystal quality.

Very short time annealing using ultra-fast RTA or flash lamp [33] annealing of implanted SiC films resulted in an increase of the activation of the implanted species. Moreover, by using short time annealing techniques, temperatures higher than the Si sublimation one are possible without any other precaution. Indeed, the main advantage of the flash lamp RTA techniques was the surface morphology, as the short processing time did not allow Si evaporation and step-bunching formation. The total process time is not much shorter than in the other cases because the duration of each heating step at 1700 °C can not be higher than 1 min otherwise an equipment damage is highly probable. So, for a total annealing time of 5 mins multiple temperature ramp-ups and downs have to be performed. In addition, this type of equipment can accommodate only one wafer and it is very fragile after a series of high temperature anneals. Furthermore, with the adoption of C-cap and the subsequent SiC surface roughening suppression, the uniqueness of the RTA techniques main advantage does not any more hold. Due to the above reasons, only a few research groups still use this type of apparatus for SiC post-implantation annealing.

Despite the increase in dopant activation in the case of short time annealing, the resistance of the annealed samples is still high probably due to the fact that longer total annealing duration is necessary to annihilate electrically active defects.

6.2 Very high temperature conventional (CA) and microwave annealing (MWA)

Recently, it has been shown that low sheet resistance values combined with satisfactory carrier mobility values can be obtained in the case of Al^+ (and P^+) ion implantation at annealing temperatures exceeding 1800 °C [23, 34] by employing RF induction conventional annealing (CA) or microwave (1 GHz) induced heating. SiC is known as an excellent absorber of microwaves. Hence, SiC MWA provides ultra-fast ramp rates (>1000 °C/s) for annealing temperatures up to 2100 °C. Practically, the SiC sample is placed in microwave transparent surroundings, and the microwaves are exclusively absorbed by the SiC sample, allowing fast ramp heating and cooling of the SiC without heating the surrounding environment and thus, limiting thermal inertial mass during the cooling phase.

Record sheet resistance and hole density values have been obtained both with conventional (1800 °C [35]) and microwave annealing at 2000 and 2100 °C [23] in the case of heavily doped Al-implanted layers [36]. Fig. 13a and 13b show hole densities and sheet resistance/resistivity in Al-implanted $4H$-SiC samples after MWA for 30 s at 2000 °C depending on implanted dose. Data for samples after CA are also shown for comparison.

Moreover, for Al doping concentrations higher than $3 \cdot 10^{20}$ cm^{-3}, the p-type resistivity and hole mobility (Fig. 13c) shows a weak temperature dependence associated to the formation of an impurity band conduction around room temperature, which guarantee p-type SiC materials of almost stable transport features in a wide large temperature range around RT [36, 37].

Longer annealing (45 min instead of 5 min) lowers further the implanted layer resistance and increases the hole mobility [38]. However, other results show that long annealing (15 min in comparison to 5 min) at high temperatures (1800 °C) result in lower crystallinity [39]. Note also the unusual hole mobility dependence on temperature for extremely high Al concentrations (Fig. 13c), which is probably due to conduction through intra-band impurity states [36].

On the other hand, recent results [16] have shown that PIA of Al$^+$ implanted layers at 1650 °C is enough to obtain high impurity activation and electrical characteristics similar to that of epitaxial layers (Fig. 13d) and extremely high temperatures of annealing are not necessary.

Finally, according to [40] the cooling rate plays an important role in the formation of carbon vacancies if the temperature of annealing exceeds 1800 °C. C-vacancies have been identified as the origin for so-called $Z_{1/2}$ centers, the dominant deep-level defects in n-type $4H$-SiC. One of the effects of this defect is reduction of the carrier lifetime. A slow cooling rate (0.25 °C/s) is necessary to reduce C-vacancies formation. Indeed, an increased effort is dedicated currently to determine the optimum trade-off in post-implantation annealing temperature and cooling rate to maximize the dopant activation and keep low the concentration of electrical active defects.

6.3 Laser annealing

Laser annealing is now used for back-side ohmic contact formation in $4H$-SiC devices. However, this technique can be also used, with higher power, to anneal semiconductors crystals. It has been also applied as an alternative to the classical thermal annealing processes for activation of ion-implanted dopants in SiC [41, 42]. To reach the required crystal surface temperatures, pulsed-excimer laser beams must be used. The key aspects are the pulse duration (nanoseconds) and the scanning rate to limit energy absorption into the near-surface region, thus, maintaining the substrate at lower temperature. In the work of C. Dutto *et al.* [41], Al implant annealing was performed by single-shot laser with pulse of 200 ns. When the implanted dose was higher than the amorphization limit ($> 1 \times 10^{15}$ cm^{-2}), the laser annealing did not allow the recovery of the crystalline structure. A columnar structure has been formed after recrystallization [42]. For doses lower than the amorphization limit, laser processing was efficient to suppress the induced-implanted structural defects. As for standard thermal annealing, the Al distribution profile was not modified after laser annealing (no diffusion was observed).

6.4   Others techniques

Other techniques are arising for PIA such as the Thermal Plasma Jet Annealing [43]. This method, based on an arc Ar plasma steam allows precise control of heating and cooling phases. By employing this method, a phosphorus box profile with a maximum carrier concentration of 2×10^{20} cm^{-3} has been obtained from a total dose of 10^{16} cm^{-2} annealed at 1650 °C during less than 20 seconds [44]. The activation rate was 40% in this case.

6.5   Optimization of post-implantation annealing in the case of Al-implantation

Despite the above-mentioned inherent advantages of rapid heating rate thermal annealing, many groups (especially the industrial ones) use specific resistive heating furnaces (see for instance [45]) allowing heating rates in the range 20-150 °C/min (max 250 °C/min)^2 and similar dopant activation. The main advantage of these furnaces is the temperature uniformity over large substrates, as well as the possibility to process simultaneously several wafers, a feature not possible with RF induction furnaces and microwave annealing. Indeed, inhomogeneous temperature in the heated wafer support is a usual drawback of the RF/microwave heated techniques and a temperature gradient of 50 °C along 3-5 cm radius is very often observed.

---

^{2} Note that this heating rate value is exceptional for resistive-heating equipment since the usual set-ups offer a heating rate of around 15 °C/min. According to Centrotherm's engineers, their furnaces use special high-power supplies to be able to furnish the full power in short time.

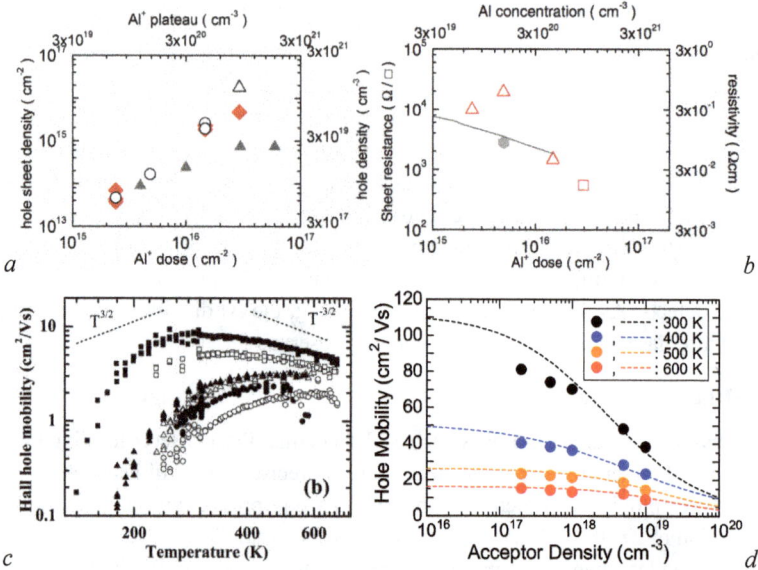

Figure 13. *(a) Hole density and (b) sheet resistance/resistivity versus implanted Al dose in 4H-SiC samples after MWA for 30 s at 2000 °C (circles in (a) and triangles in (b)) and 2100 °C (triangles in (a) and squares in (b)). Open and close symbols in (a) correspond to data obtained with $r_H=1$ and $r_H=0.77$, respectively. Data for samples after CA are shown for comparison by diamonds and hexagons. The R_{sh} curve for completely activated implanted Al and a mobility equal to that one in epitaxial p-type 4H-SiC is also shown for comparison in (b). Reprinted from [23] © IOP Publishing. All rights reserved. (c) Hole mobility in Al implanted 4H-SiC samples depending on temperature. Full symbols - after CA at 1950 °C for 5 min, open symbols after MWA 2000 °C for 30 s. The implanted Al concentrations are $1.5 \cdot 10^{20}$ cm^{-3} (squares), $3 \cdot 10^{20}$ cm^{-3} (triangles), and $5 \cdot 10^{20}$ cm^{-3} (circles). Reprinted from [37] with the permission of AIP Publishing. (d) Hole mobility in Al$^+$ implanted layers on a semi-insulating 4H-SiC substrate measured at several temperatures (circles) and fitting curves for the data of epitaxial layers (dashed lines). Reprinted from [16] © 2017 the Japan Society of Applied Physics.*

Post-implantation annealing of Al$^+$ implanted layers [46] in these industrial furnaces showed that: (i) high ramp rate lowers surface roughness, (ii) low ramp rates shows better activation, (iii) shorter process time ends in better surface roughness and lower activation and (iv) slower cooling annihilates partially carbon vacancies. Typical heating and cooling rates are 100 °C/min and 30 °C/min, respectively. For post-implantation

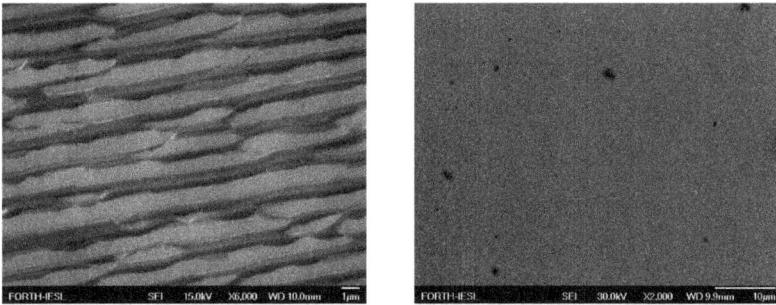

Figure 14.   Scanning Electron Microscopy (SEM) images from two samples implanted simultaneously and post-implantation annealed without a C-cap (left) and with C-cap (right). Courtesy of FORTH.

annealing temperatures higher than 1700 °C and extremely high concentrations ($> 1 \cdot 10^{20}$ cm^{-3}), the typical duration of the annealing is 30 minutes. This annealing duration is necessary in order to have saturation in the activation of the dopants. However, according to [46], for temperatures above 1750 °C, an annealing longer than 15 minutes do not change significantly the activation rates and may affect the surface roughness. Finally, the sheet resistance value is constantly reduced up to annealing temperature of 1950 °C and the activation increases all the way up to 2050 °C but at the same time the mobility decreases. Nevertheless, the variation of activated acceptor concentration is more important than that of holes mobility and thus, optimizing the former is more important for obtaining low sheet resistance value. Somewhere between 1850 °C and 1950 °C the surface roughness seems to increase stronger. So, an annealing temperature around 1750-1800 °C is the best compromise of all the above factors.

6.6   Surface roughness

The surface roughness will depend on the implantations conditions (implanted atom, implantation temperature, doses) but mainly will be fixed by the post implantation annealing parameters. In general, using a thermal annealing process for temperatures in excess of 1400 °C has proven to be problematic due to the out diffusion of Si from the SiC surface, which can cause step bunching [47] (see Fig. 14).

It was also noticed that the surface roughness increased with the annealing time and temperature [47]. Spatial thermal variations as well as a high-temperature annealing can induce SiC "etching" if appropriate environments and configurations rich in silicon and carbon are not provided. As it was mentioned previously, silicon volatilizes at

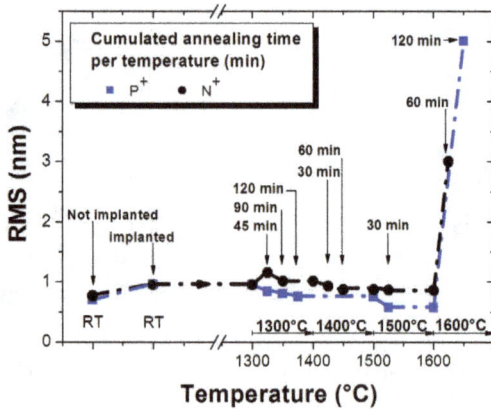

Figure 15.   RMS roughness measured with AFM versus annealing temperature and time for phosphorus and nitrogen implanted samples. Reprinted from [48].

temperatures above 1300 °C at 1 atmosphere resulting in an important step bunching of the SiC surface. The impact of implantation and annealing temperature on RMS roughness is shown for phosphorus and nitrogen implantation in Fig. 15 [48]. We clearly see a significant increase of the roughness after 1600 °C annealing. In addition, above this temperature, the roughness increases with the annealing time.

6.8   Capping layer

To reduce Si sublimation and related roughness, Jones *et al.* proposed to use an AlN capping layer during activation of *n*-type dopants up to a temperature of 1600 °C [49]. However, at temperature higher than 1600 °C, AlN starts to degrade and is not efficient anymore. In addition, AlN can be selectively etched with KOH etch but this post treatment may damage the SiC surface, especially at the defect edges.

Then, as an alternative, various groups developed a capping process based on a graphite-capping layer [50, 51]. Carbon cap (C-cap) is the most widely used method currently. The C-cap is formed in most cases by heating a standard photoresist (like the AZ 5214E) at temperatures around 800 °C. Higher temperatures are sometimes used in industrial environments as it results in minimum furnace and vent line contamination. For the same reason (reduced contamination) some groups use sputtered carbon instead of photoresist. After the PIA, the C-cap is removed by various methods employing oxygen (oxygen RIE, dry oxidation at 800 ÷ 900 °C or combination of these techniques by using a plasma asher with a heating at 800 °C in the presence of oxygen). According to the authors of [50] the

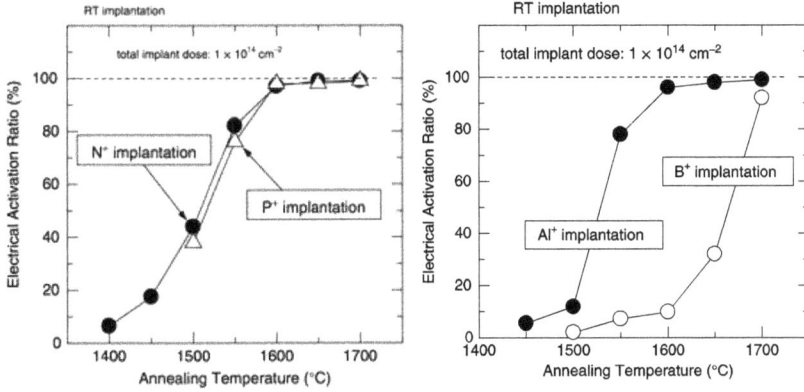

Figure 16. Annealing-temperature dependence of the electrical activation ratio in nitrogen or phosphorus (left) as well as aluminum- or boron-implanted SiC (right). The implantation was performed at RT (total implant dose: $1 \cdot 10^{14}$ cm^{-2} for forming a 400 nm deep box profile by multistep ion implantation). Reprinted from [8] © 2014 John Wiley & Sons Singapore Pte. Ltd..

thermal oxidation gives the best results in terms of surface preservation. The use of C-cap is also beneficial for avoiding dopants out-diffusion during post-implantation annealing which is important in the case of B and, to a lesser extent, Al-implantation.

6.9 Electrical activation

Ion implanted SiC can be activated efficiently by furnace annealing at 1500 °C or more and for around 30 minutes. Higher activation is obtained as the post-implantation annealing temperature increases [52] (see Figs. 16 - 18). Note that raising the annealing temperature by as little as 50 °C increases drastically the electrical activation (e.g. compare ≈10% at 1500 °C vs. ≈78% at 1550 °C for Al). According to [53], annealing for 30 min at temperatures between 1500 °C and 1700 °C results in a complete activation of Al acceptors for an Al concentration of $5 \cdot 10^{18}$ cm^{-3}. In contrast, for an Al concentration $5 \cdot 10^{19}$ cm^{-3} even higher annealing temperatures are required. A decrease of compensation with increasing annealing temperature is found [54]. So, the electrical activation process becomes more efficient with the increase of the PIA temperatures. It can be also inferred that implanted B activation needs higher temperatures than that of Al.

Note that the annealing temperature cannot be reduced very much, even if the implantation is performed at elevated temperatures (300 ÷ 800 °C).

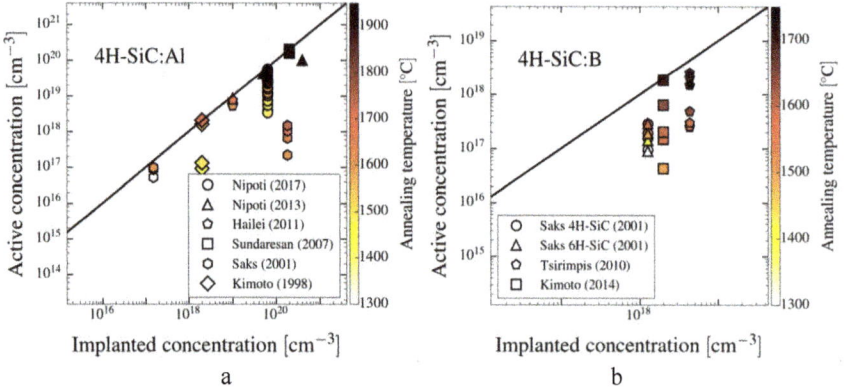

Figure 17. *(a) Al and (b) B acceptor concentrations as a function of implanted concentration for various annealing temperatures taken from various experimental studies. Lines represent 100% activation. Reprinted from [55].*

An analysis of the reported in the bibliography results in terms of the activation rate for the four main dopants (Al, B, N, P) has been performed in [55]. Figs. 17 and 18 show a summary of this work.

The above analysis, considers the activation state in equilibrium (steady-state activation) supposing enough annealing time, resulting thus, in the maximum possible electrical activation for the given implanted dopant concentration and annealing temperature. The reported in various studies time-dependent (i.e., transient) behavior has been also analyzed, thoroughly in the case of Al and P and less in the case of B and N [56, 57]. Fig. 19 summarizes reported results of post-implantation annealing transient behavior for Al and P. The time constants for dopant activation are in the order of several hours for annealing temperatures lower than 1400 °C and are significantly decreased for elevated annealing temperatures. A similar graph is reported for B in [57]. However, there is only one study [15] for the activation kinetics of N.

Empirical models for the transient behavior of the four main dopants (P, N, Al, B) have been proposed in [55, 56, 57] and can be useful for determining optimum post-implantation annealing.

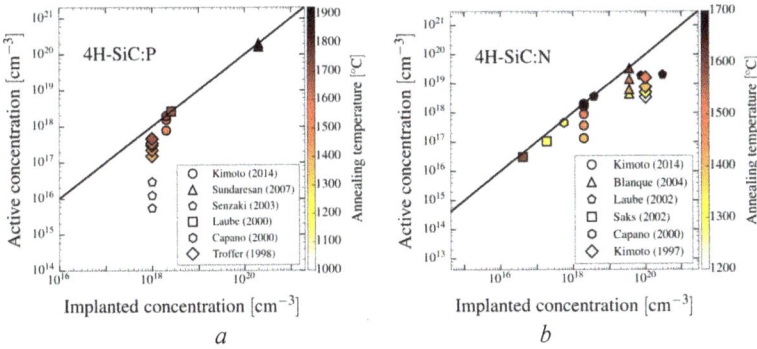

Figure 18. (a) P and (b) N donor concentrations as a function of implanted concentration for various annealing temperatures taken from many experimental studies. Reprinted from [55].

7. Crystal quality and electrically active defects

As it has been already mentioned it is very difficult and almost impossible to completely restore the crystal quality of high-dose implanted SiC layers. Usually, point defects are obvious in Transmission Electron Microscopy (TEM) micrographs especially for Al and N concentrations above $8 \cdot 10^{18}$ cm^{-3} and $2 \cdot 10^{19}$ cm^{-3}, respectively [58, 59]. At higher concentrations (above $3 \cdot 10^{19}$ cm^{-3} and $1 \cdot 10^{20}$ cm^{-3}, for Al and N respectively) these point defects aggregate and form precipitates (Fig. 20). The TEM studies showed also that local strain is present in the implanted layers.

Furthermore, some studies [60] identified extended defects as small as 2 nm for high Al ion flux (5.9×10^{12} Al$^+$ cm^{-2}s^{-1}) and low (180 °C) implantation temperature.

The point defects (vacancies and interstitials) as well as extended defects in the crystalline structure can be electrically active. For instance, the poor electrical activation of N$^+$ ions in 6*H*-SiC was correlated to the formation of a high density of precipitates in the implanted layer when increasing the doping concentration [58].

Post-implantation annealing is sufficient to substantially reduce the point defect concentration although many point defects can still be found in the high temperature annealed samples, for instance by electrical methods such as deep level transient spectroscopy (DLTS) [28]. The concentration of these defects may be in the order of 10^{15} cm^{-3} and they may decrease charge carrier lifetime, lead to increased leakage current

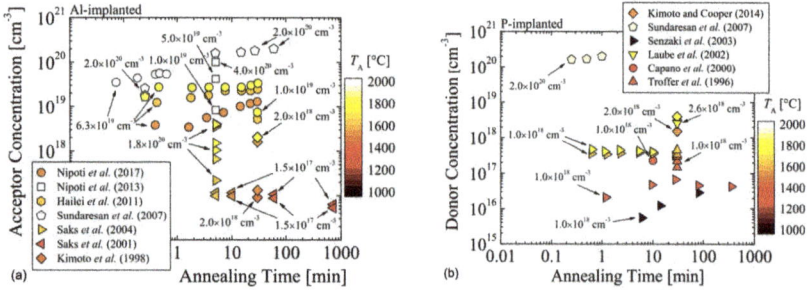

Figure 19. (a) Al acceptor and (b) P donor concentrations as a function of annealing time for various total concentrations and annealing temperatures taken from various experimental studies. The text and arrows within the figures refer to the total implanted concentration of the corresponding data sets. Reprinted from [56] with the permission of AIP Publishing.

under reverse bias, reduced injection efficiency and increased voltage drop under forward biasing [28].

In the case of SiC, two main electrically active levels, $Z_{1/2}$ and $EH_{6/7}$, arise as a result of point defects creation at ion implantation. An almost linear relationship of the $Z_{1/2}$ level (originating from the lifetime-killer carbon vacancy (V_C)) concentration with the implantation dose has been demonstrated in [5].

Apparently, $Z_{1/2}$ level is present for all implanted species (see Fig. 21) [61]. As the $Z_{1/2}$ center and many other deep levels are thermally stable, high-temperature (> 1400 °C) annealing is required to reduce the defect density.

Moreover, recent results [62, 40] showed that an increase of post-implantation annealing temperature is beneficial in terms of crystalline quality but on the other hand, increases the concentration in the bulk of the $Z_{1/2}$ and V_C. A slow cooling reduces the density of V_C and cooling rates lower than 15 °C/min have to be employed. Therefore, there is a need of careful and delicate optimization of the post-implantation annealing temperature to lower to the minimum possible these defects concentration.

The above discussion shows clearly why *pn* junctions formed by ion implantation exhibit higher on-resistance and faster switching speed (shorter carrier lifetime) than epitaxial *pn* junctions [61].

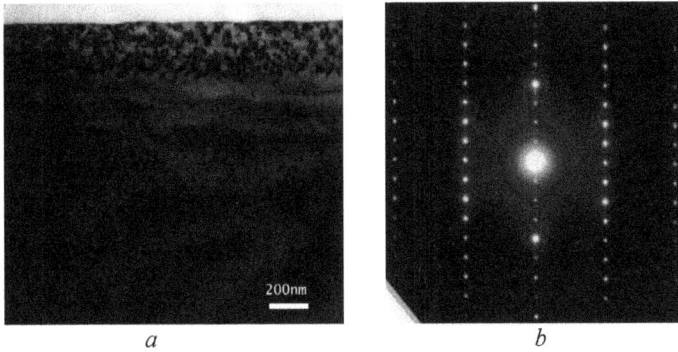

Figure 20. Cross-section TEM image (a) and corresponding electron diffraction pattern (b) from 4H-SiC sample multistep Al-implanted at 400 °C with a total dose of $1.58 \cdot 10^{15}$ cm^{-2} and post-implantation annealed at 1750 °C. Courtesy of Prof N. Frangis from AUTH-Greece.

8. Channeling and straggling

Channeling is most often an unwanted effect and usually "random" implantations are performed. The implantation process is said to be "random" when the direction of the ion beam with respect to the crystal planes is such that the implanted ions experience the same amount of energy loss and collisions as they would in a material with the same chemical composition but of amorphous structure. In spite of any attempt to have a random implantation of a crystal, there is always a nonzero probability that some energetic ions are scattered along major axial or planar directions, giving origin to channeled trajectories.

The channeling effect depends on the crystal structure and on the implanted atom size. If the implantation is performed along low index crystallographic directions, ion channeling occurs resulting in deeper dopant penetration and lower defect generation [2, 47]. The smaller the ions and the lower the dose, the more the channeling effect will be important.

For high dose, the damage build-up has to be considered since it causes de-channeling. Moreover, thermal vibrations have to be taken into account for correct modeling of channeled ions behavior. For low dose implantations or in the first stages of high dose implants, thermal vibrations are the main de-channeling physical mechanism and thus, they influence the final distribution of implanted ions.

As stated in the introductory Section 2, the channeling can present specific advantages like reaching large depths and reducing at the same time the number of displaced atoms. So, the channeling effect could be used to generate deep junction for high voltage applications with relatively low defects at the junction.

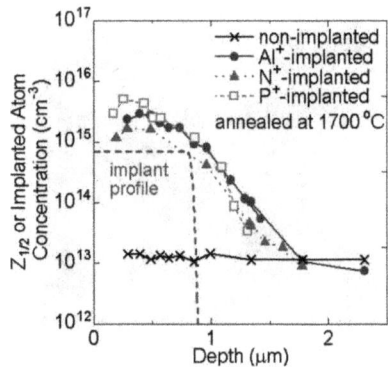

8.1 Channeling in SiC crystal

For many years, channeling was not considered as an issue in the case of 8° off-axis 4*H*-SiC wafers with the miscut toward <11$\bar{2}$0> and an implantation process with the

Figure 21. Depth profiles of the $Z_{1/2}$ center observed in nonimplanted (filled circles) and Al$^+$-implanted n-type 4H-SiC after Ar annealing at 1700 °C for 30 min (filled triangles) and after the annealing followed by oxidation at 1150 °C (squares and inverse triangles). Reprinted from [61] with the permission of AIP Publishing.

ion beam normal to the wafer surface was assumed to be practically equivalent to a tilted (random) implantation performed on an on-axis substrate [63].

However, many studies implying SIMS measurements, especially on Al implanted SiC samples revealed that a channeling tail was systematically observed (Fig. 22) [63, 64]. According to [65], the channeling phenomenon is even stronger in SiC than in Si due the difference in crystal structure and the higher atomic density of SiC. Indeed, the higher (×2) atomic density of SiC compared to Si increases the ion's "guiding" efficiency through the crystal channels. In other words, the channels' walls are denser, and consequently more efficient to guide the implanted ions. In addition, the hexagonal structure implies that the channel planes are every 30°, while they appear every 45° in a cubic crystal.

In what it concerns implantation in 4*H*-SiC, an important channeling has been observed for implantation along [0001], [11$\bar{2}$3] and [11$\bar{2}$0] directions with relatively low doses [63].

Materials Research Forum LLC

https://doi.org/10.21741/9781644900673-3

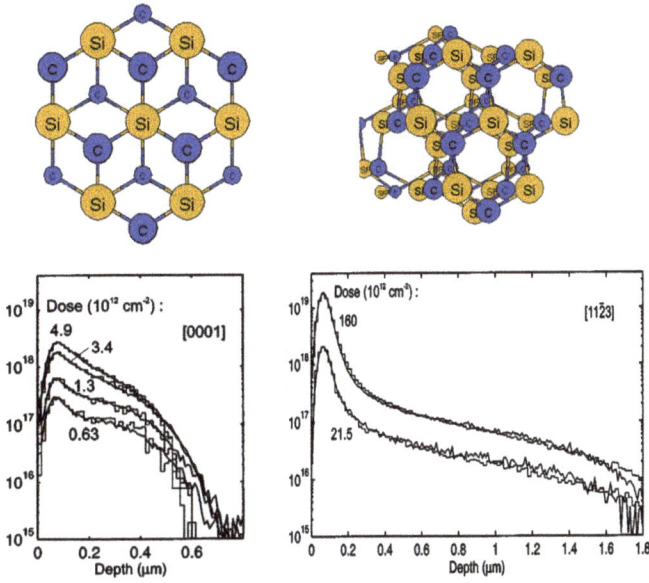

Figure 22. Channeling effect in 4H-SiC for Al implantation. Top: atomic arrangement for [0001] (left) and [112̄3] (right) direction. From [64]. Bottom: comparison of experimental (SIMS) and Monte Carlo simulated profiles for different doses of Al implanted at 60 keV, RT in 4H-SiC (0001) with an off-axis miscut of about 8 ° toward [112̄0]. The left (right) graphs are for implantation along [0001] ([112̄3]) crystallographic direction. Reprinted from [63] with the permission of AIP Publishing.

Two implantation parameters are fundamental in the control of channeling: the tilt and the twist (or rotation angle). These two angles define the direction of the beam with respect to the wafer. In the best case, the tilt angle is mentioned in papers related with implantation in SiC. However, and more specifically on the off-axis wafers, the twist or rotation angle (usually defined with respect to the wafers major flat along a (1̄100) plane) should be also given as it can have a significant impact on the implanted profile. In [66], a specific simulator has been used to make a mapping of the preferential channels. An example is shown in Fig. 23, where a 3D mapping of the mean penetration depth of a boron implant at 100 keV in 6H-SiC is shown. We can see than that a factor up to 5 in the implantation depth can be observed only by 10 ° wafer rotation.

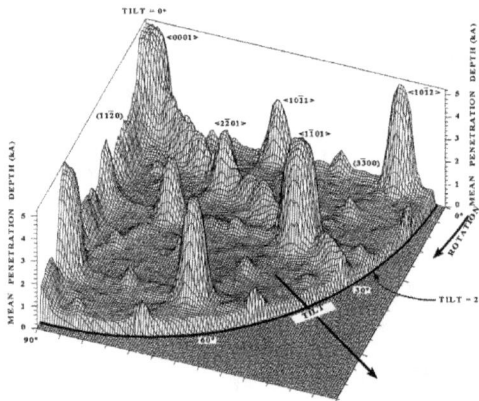

Figure 23.   Mean penetration depth, R_p, versus implantation parameters (tilt and rotation) to identify the channeling for Boron in 6H-SiC. The tilt is varied between 0 ° and 30 ° and the rotation between 0 ° and 90 °. This mapping is then symmetrical over 360 °. Reprinted from [66] © (1999), with permission from Elsevier.

Currently, commercial software is used for performing this type of modeling work as it shown in Fig. 24.

From the above analysis, it is obvious that to eliminate channeling, a solution is to orientate the wafer to avoid having the crystal channels parallel to the beam and choosing optimum twist and tilt angles performs this. In 4H-SiC for instance, a tilt angle of 7 ° and a twist angle with respect to the major flat of 22 ° are commonly used to minimize channeling. These values are a compromise between optimum values for minimum channeling and practical considerations in terms of mask shadowing resulting in important lateral device asymmetry. As some 4H-SiC wafers are provided with a 4° cut-off, this angle must also be taken into account and the tilt adjusted consequently. Nevertheless, even by taking all these precautions to have a "random" implantation a fraction of the ions will be channeled to larger depths than that of amorphous material.

Implantation energy has also an influence on the channeling effect. For instance, near random implantation conditions are reached at relatively low tilt angle 2 ° instead of the above mentioned 7 °, in the case of high-energy (>1 MeV) implantation (see Fig. 25 [67]). Note, however, that there is no a systematic experimental study of the implantation energy influence on channeling effect in the case of SiC.

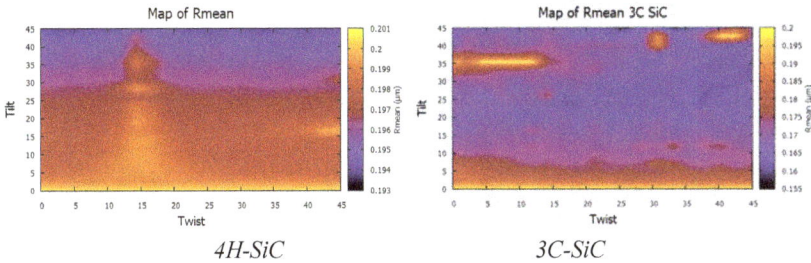

Figure 24.   Map of mean implant depth in 4H and 3C-SiC as a function of tilt and twist angles (calculated by using MC module of Silvaco software). Al$^+$ multi implant: (180 keV; 1.12·10^{15} cm^{-2})   +   (110 keV; 7·10^{14} cm^{-2})   +   (60 keV; 5·10^{14} cm^{-2})   + (30 keV; 2.7·10^{14} cm^{-2}). Courtesy of IBS.

Fig. 22 [63] and 26 [67] demonstrate also the impact of the dose on the channeled profile. Note that these results have been obtained for room temperature implantation. For Al doses from $2 \cdot 10^{12}$ to $4 \cdot 10^{14}$ cm^{-2} [67] an identical implanted-ion distribution is obtained when normalized with the dose. Similar results have been shown in [63], testing up to $1 \cdot 10^{13}$ cm^{-2} in this case. For higher doses a dynamical de-channeling effect due to implantation-induced defects is observed.

The de-channeling is indicated by the saturation of the channeling tail and by the appearance or the increase of concentration peaks at low depth corresponding to the "Gaussian" peaks of a random implantation. The above-mentioned de-channeling threshold doses are for implantations along [0001] and can be different for implantation in other orientations [63]. Obviously, there is necessity of controlling this dose related phenomenon in order to avoid it, or to take benefit from it to form deep, low defect Al-doped layers into 6*H* or 4*H*-SiC. Furthermore, these threshold doses for de-channeling are even higher for boron implantation due to lower implantation damage generated in this case.

Nevertheless, it is widely accepted that hot implantation above 350 °C is high enough to prevent the increase of point defects concentration up to a value which reduces channeling (de-channeling) in comparison with the one occurring in a perfect lattice.

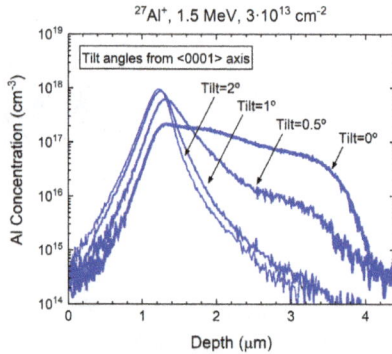

Figure 25.   SIMS profiles of 1.5 MeV room-temperature Al$^+$ implanted 6H-SiC misoriented 3.5 ° for various implantation tilt angles from 0 ° to 2 °. The tilt is measured from the <0001> axis. Each implantation has been normalized to a dose of 3×10^{13} cm^{-2}. Reprinted from [67] with the permission of AIP Publishing.

The effect of 4H-SiC substrate temperature on the channeling in the case of 100 keV energy Al and B implantation along [000$\bar{1}$] has been studied very recently [68, 69]. At room temperature, the penetration depth of channeled ions is 6 times larger than that one of non-channeled ions. Increasing the implantation temperature up to 600 °C reduces the channeling effect, but still the penetration depth of Al atoms is 4 times larger than non-channeled implant. This is not the case for B implantation where there is no relevant difference in channeling between room and 600 °C implantation due to larger channels and less interaction with the SiC lattice. The reduction of Al penetration depth when temperature is increased is mainly caused by an increase of thermal vibrations. These results are also showing that the dynamic annealing of defects during implantation resulting in an increased channeling is not so important as the counter acting de-channeling from thermal vibrations.

8.2   Lateral/transverse straggling

Lateral/transverse straggling phenomena has been observed in both silicon and silicon carbide implantation processes. These phenomena are of high technological relevance as they determine the under mask penetration of implanted ions.

The lateral spread of ions under mask edges is mainly due to substantial deflections of implanted ions with energy in the keV range, but this spread may be largely influenced by the channeling phenomenon due to the possibility of capture of random implanted ions

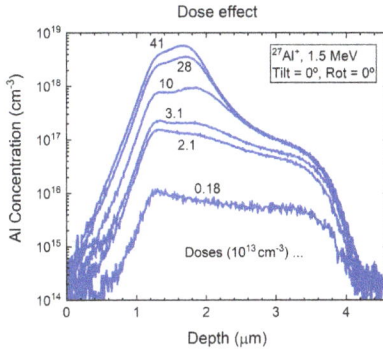

*Figure 26.   SIMS profiles of 1.5 MeV Al implanted at RT in 6H-SiC misoriented 3.5 °
for various implantation doses from 1.8·10¹² to 4.1·10¹⁴ cm⁻², showing the effect of the
implantation dose on the depth profile. An intermediate peak grows up, due to
accumulation of defects between 1 and 4 µm. Implanted concentration saturation of the
deep channeling part of the profile can be observed. Reprinted from [67] with the
permission of AIP Publishing.*

by the crystal channels. The lateral straggling depends on various factors such as the
orientation of the incident beam with respect to crystallographic axes, the mask edge
shape, the ion specie, the implantation energy as well as the dose. For instance, for an
implantation in a non-channeling direction, the lateral scatter/straggling is present from
the surface. This is obvious in Fig. 27 where it is distinguished the channeling along
specific orientations in addition to the Gaussian straggling. The sharp protuberances
along axial or planar channels away from the <0001> axis of 6H-SiC can lead to
unexpected behaviors of implanted junctions (high local electric fields, short channel
effect in lateral implanted MOS devices) and reproducibility problems.

For an implantation in a channeling direction, the ions are trapped in channels and only
scatter out of the channels when their energy drops. This might translate into a lower
lateral scatter as to reach a certain depth, lower energies would be required and much of
the energy will be lost passing down the channel before the ions scatter from the channel
to lateral directions.

Note also that it is possible to shadow an implantation profile with the mask as this is a
3D structure and the beam incident angle is not normal to the surface. So, depending on
the aspect ratio of the mask, some of the direct line of sight to the wafer can be obscured
by the mask layer effectively closing up the implant window.

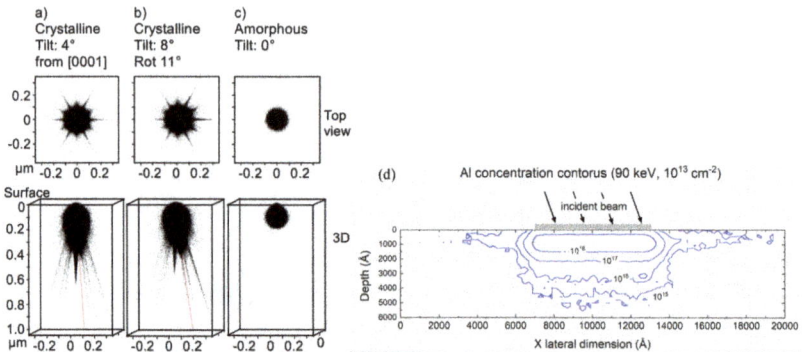

Figure 27. *3D simulations displaying the probability to find an implanted ion at a certain position in a 4H-SiC lattice after ion implantation. The MC-BCA code SIIMPL has been used in the simulations. 200 keV 51V+-ions and an impact area of 1 × 1 nm^2 has been employed. In (a) and (b) the crystal structure of 4H-SiC is used while in (c), an amorphous SiC target is employed with simulation parameters according to SRIM. Different angles for the incidence beam has been utilized as indicated in top of each figure. From [69] © IOP Publishing. All rights reserved. (d) Simulated 2D contours in the XZ plane (X being the [1120] and Z the depth along [0001]) of Al doping level for a 90 keV/1·10^{13} cm^{-2} Al ions implantation of 6H-SiC (tilt angle of 7 ° with respect to the Z-axis and a rotation of 15 ° with respect to the X-axis). Reprinted from [72] with permission from Elsevier © 1999.*

The lateral straggling has been simulated by various models for Al implantation of 4*H* [70, 71] and 6*H*-SiC [72]. According to [70, 72] there is an important lateral straggling along <11$\bar{2}$0> [70] and <1$\bar{1}$00> [72] directions. On the other hand, the simulated undermask iso-concentration contours were the same for the masking edge orientations along [11$\bar{2}$0] or [1$\bar{1}$00] directions according to [71]. Based on these publications, a rough estimation can be done for the case of multistep implantation with a total Al implanted dose exceeding $1·10^{15}$ cm^{-2}: the undermask extension along <11$\bar{2}$0> and <1$\bar{1}$00> directions of a concentration equal to 10^{-3} of the implanted ion concentrations peak value will be of the order of vertical R_p corresponding to the highest energy implantation step.

Finally, the simulations exhibited an asymmetry in the lateral straggling for the two edges of the mask due to the wafer miscut and to the combined substrate tilting/mask shadowing. The latter is more important than the former.

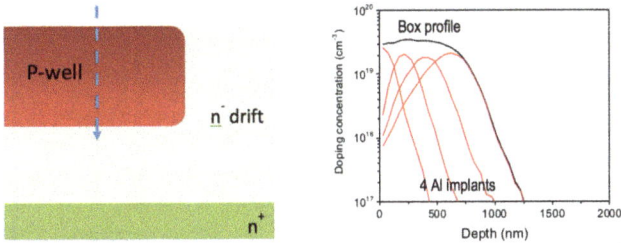

Figure 28.   Illustration of the box doping profile concept to form p-wells with multiple implantations. Courtesy of CNM.

8.3   Box profile

A box dopant profile with a constant concentration over a given depth is often needed, for example, for the formation of MOFETs p and n type well. To obtain such box profile, several implantation steps done at various energies must be successively performed since the diffusion of impurities in SiC is not realistic, as it is illustrated in Fig. 28.

An empirical "rule-of-thumb" is to not exceed a dose of $2 \cdot 10^{14}$ cm^{-2} per each implantation step and a total dose of $5 \cdot 10^{15}$ cm^{-2} to have minimum damages of the crystal and a better recovery following the post-implantation annealing.

In addition, as simulations shown, for obtaining box profiles a higher dose is needed at higher energies, as the channeling and the *straggling* (ΔR_p) are stronger at higher energy values. So, for keeping a "flat" profile, the dose should increase as we implant at higher energies.

In the case of multiple implantations, as used for box profile, the simulations also shown that the order of the implantation may have an impact on the defects formations especially for room or lower than 400°C implantations. Depending on the energies and doses combination, it is sometime more interesting to start with the highest energy, and implant with decreasing energies order. If the order of implantation steps is from lower to higher energies, it will result in deeper implantation with a simultaneous higher damage, as the ions of the next implantation will "push" that of the previous one deeper.

Typically, the only deviation from the targeted "flat" profile is done for the first layer nearby the surface where a high dose is required for ohmic contact formation. In this case an oxide or even an Al-layer in the case of Al implantation should be deposited on the surface to have the peak of the Gaussian of the first step very close to the surface and thus, having the maximum concentration nearby the surface. Very often the thickness of the above mentioned screen oxide is a compromise between the "scattering" of the incoming ions and the need of approaching the Gaussian peak near the surface.

9. Plasma implantation

Plasma Immersion Ion Implantation (PIII), also called Plasma doping

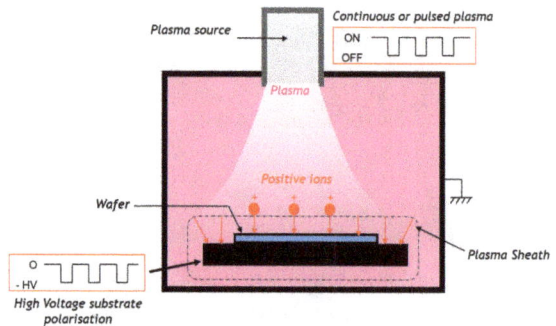

Figure 29.   Schematic of a Plasma Immersion Ion Implanter.

(PLAD) or Plasma Source Ion Implantation (PSII) is an alternative ion implantation technique. Briefly it consists of creating plasma containing the ions of interest above a negatively polarized wafer to accelerate all the positive ions of the plasma toward the wafer to be implanted. To avoid some side effects related to plasma/surface interactions (plasma sheath extension, charging), the acceleration voltage is either DC pulsed or sometime RF (see Fig. 29). This technique allows implantation at low energy, from 30 eV to typically 10/20 keV (even if tools up to 100 keV were built for metallurgical applications) while keeping a high implant current allowing small implant time for high doses (10 to 100× faster than a beam line implanter).

Under certain process conditions PIII also allows 3D conformal implantation (implant of trenches sidewalls, vias, FinFET) [73]. PIII is a multi-energetic process (meaning that ions are not all implanted with the same energy) and the resulting implantation profiles generally show a maximum concentration at the surface. Compared with beam line ion implantation, the structure of the tool is much simple (looks like a PECVD or RIE tool) allowing very cost effective solution for high dose & low energy application. The drawback is that the technique is limited in energy (most of commercial tools are limited to 10 or 20 keV) and there is no mass separation, so, all ions present in the plasma are implanted and precise dose measurement using current measurement as in a beam line is not possible (multi-species, implantation of neutral due to collision in the plasma sheath, plasma electrons and displacement current phenomena).

Most of the PIII technique development was focused on the fabrication of Ultra Shallow Junctions (USJ) for the doping of Source/ Drain extension (SDE) of *p*-channel MOSFET to avoid short channel effects [74].

Very few papers were published on the use of PIII on SiC; nevertheless we can highlight 3 possible applications:

- *Fabrication of shallow junctions*: L. Ottaviani and his coworkers [75, 76] employed this technique in the development of SiC UV detectors. Boron (using B_2H_6 plasma) has been implanted in *n*-type 4*H*-SiC to obtain junctions depth less than 30 nm.

- *Optimization of contact resistance*: As medium current beam line implanters generally used for SiC doping are limited in low energy range, surface concentration is always lower than mean concentration of flat profiles engineered by multi implant steps. Adding an additional low energy PIII implantation allows the surface concentration increase for better ohmic contact (Fig. 30).

- *Material surface modification applications*:
  - Preparation for SiC epitaxial growth: as PIII allows implanting very high doses, it can be used to create compounds. For example, implanting high dose of carbon in Silicon allows the synthesis of a SiC layer on top of a Si substrate (an additional thermal treatment is needed to obtain crystallites of SiC). This has been demonstrated by IBS and NOVASIC [77] and used as a preparation treatment to optimize epitaxial growth of 3*C*-SiC on Si.
  - Other research works are ongoing to use PIII in SiC for non-doping applications, specially to optimize gate stack of SiC MOSFET and improve carrier mobility (see, for example, F. Torregrosa, et al. (2018). Advantages and challenges of Plasma Immersion Ion Implantation for Power devices manufacturing on Si, SiC and GaN using PULSION® tool. IIT-22, IEEE Conf. proceedings, 33-37).

10. Ion implantation simulation

The main 1D implantation simulation programs in Silicon are Marlow [78] and SRIM (Stopping and Range of Ions in Matter)/TRIM (Transport of Ions in Matter) [79]. It is usually admitted that TRIM gives only a rough picture since it is valid for amorphous material. Hopefully, there are many programs, developed and used by individual groups, for simulating SiC implantation profiles, which are more accurate as they take correctly into account the channeling effects. These are I^2SiC developed by CNM/CSIC [65, 80] (Fig. 31), SiiMPL developed by Fraunhofer and used by KTH [81], ICESCREM and KING-IV developed by G. Lulli in Bologna. The latter two are free available. The I^2SiC simulator is based on the previously mentioned MARLOW and TRIM codes. The above simulators are very similar in the ion path and treatment of the lattice. There are some differences in the treatment of damage, thermal vibrations, etc., but on the whole they are analogous.

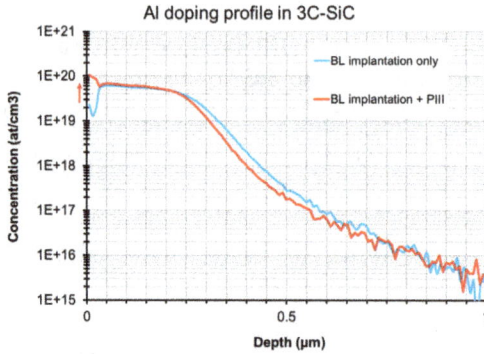

Figure 30.   SIMS profile of an Al implantation in 3C-SiC (Multi-implant steps to have a flat profile) and surface concentration increase using an additional PIII implantation of Aluminum using IBS PULSION tool. (Courtesy of IBS).

As it has been mentioned in Section 2, the first basic mechanisms involved in slowing down a moving ion in condensed matter is *nuclear stopping*. It is modeled by classical two-body scattering within the binary collision approximation (BCA). The calculation of the energy loss after a collision between the incident ion and a target atom requires the evaluation of a scattering integral. For computational efficiency, this is done using approximate formulas, together with an adequate universal potential proposed by Ziegler *et al.* [82]. This approach is valid for a wide range of implanted impurities. The second mechanism of ion-energy loss along its path is by excitation of the electron population of the target material. In the energy range of interest for implantation, the ion mainly transfers its energy to the valence electron gas of the crystal. In this case, the "electronic"-energy losses can be calculated within the framework of the Brandt and Kitagawa effective charge theory [83] and the Echenique's [84] approach for stopping power calculation.

The channeling effect arises naturally as a consequence of the crystal structure and valence electron distribution. The effects due to the existence of a native oxide, of dynamic amorphization and of thermal vibrations of the lattice atoms, affecting the channeling behavior of the ions, must be included in the simulation. Thus, in order to properly model this channeling effect, one has to take into account two major aspects of the crystal structure:

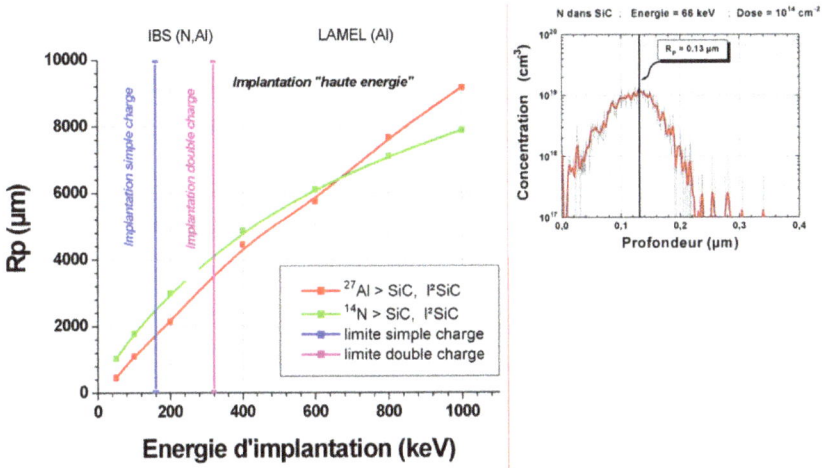

Figure 31. R_p *as a function of N or Al implantation energy according to I^2SiC simulator. Reprinted from [65].*

(i) thermal vibrations of the lattice atoms, which contribute to the dechanneling of the ions. This is done by displacing a target atom under collision around its static ideal position in the lattice. The displacement is calculated in a standard way, assuming a Gaussian distribution along each axis with a standard deviation according to the Debye theory and an empirical Debye temperature for the crystal.

(ii) dynamical process of damage formation during ion implantation, which leads to the amorphization of the SiC crystal.

The chemical doping profiles generated by these simulators can be included in 2D numerical electrical simulator for devices design. However, the activation rate must be properly defined to get reliable electrical modeling results.

Note however, that the new version of TCAD simulators such as that of SILVACO © reproduce very well the implantation profile when the crystalline material is considered as it was shown in Fig. 10b above.

11. Diagnostic techniques of implanted layers

Several analysis techniques are used for the physical (structural, chemical, electrical, etc.) characterization of the implanted SiC layers. All of them are more or less also employed in silicon implantation. The ones that are mainly used in the Si-device-related-industry are SIMS, sheet resistance measurements and thermal wave (TW) probe. TW probe is non-destructive and it is widely used in the industry to prepare implant references for calibration. The technique is still used only for silicon because of commercial TW instruments use lasers with a wavelength not absorbed by SiC and the hot implantation, employed in the case of SiC, would complicate the analysis of the recorded signal.

For more detailed analysis, semiconductor material characterization techniques such us the Rutherford backscattering spectrometry (RBS), the Transmission Electron Microscope (TEM), the Raman spectroscopy, the X-ray diffraction (XRD) are also employed.

A comprehensive discussion of the diagnostic techniques of SiC implanted layers is given in [10] and here only the main points are reported in addition to some new results.

11.1 Secondary Ion Mass Spectroscopy (SIMS)

SIMS is probably the most common characterization technique of implanted layers. It allows implanted impurities depth profiling and the total fluence of implanted ions can be obtained by integration of the SIMS depth profile.

According to the experimental setup there are different detection limits of SIMS for various n- and p-type doping species in SiC and typical values are 10^{13} cm^{-3} for B, 10^{17} cm^{-3} for N, and 10^{14} cm^{-3} for Al. Typical SIMS profiles taken from SiC implanted layers have been already shown in Fig. 7.

11.2 Electrical measurements

The sheet resistance (R_{sh}) of implanted and annealed samples is commonly measured through TLM measurements. However, Van der Pauw–Hall measurements allow simultaneous measurements of both sheet resistance and carrier mobility. There is an excellent presentation on electrical measurements of SiC implanted material in [11] and the interested reader can get all necessary information on measurements and their analysis.

11.3 Rutherford backscattering spectrometry (RBS)

RBS channeling measurements can be employed to evaluate crystal quality. A comparison (Fig. 32) of the RBS spectra after the post-implantation annealing with the corresponding ones of non implanted crystals can give a rough estimation of the crystal quality restore.

Nevertheless, the lattice damage is monitored by the integrated scattering yield in the damaged region of a channeling spectrum divided by the integrated yield of a random

Materials Research Forum LLC
https://doi.org/10.21741/9781644900673-3

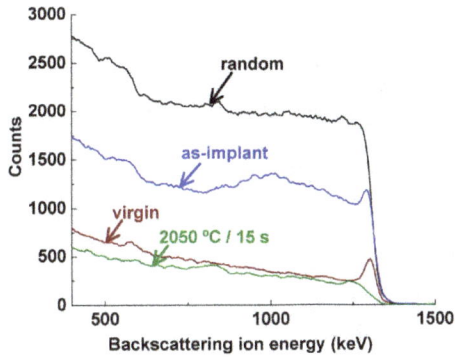

Figure 32.   Aligned (i.e. parallel to the c axis) RBS-C spectra for a virgin 4H-SiC sample, an Al$^+$ as-implanted sample, and a 2050 °C/15 s microwave-annealed sample. The spectrum for a randomly aligned SiC sample is also shown for reference. Reprinted from [52] with the permission of AIP Publishing.

spectrum (χ_{min}). In the case of a high dose implantation at room temperature, the χ_{min} value from the implanted region reaches almost 100%, indicating complete amorphization by implantation induced damage, when the implant dose is higher than $(3\div4)\times10^{15}$ cm^{-2} (*critical implant dose for amorphization*).

On the other hand, the lower damage limit that can be monitored by this technique for room temperature implantations is at a dose around 1×10^{13} cm^{-2} for 100 keV Al ions [28].

11.4   Transmission Electron Microscope (TEM)

The use of the TEM technique in the high-resolution mode allows to identify amorphous and/or polycrystalline layers, extended defects and point defects in high number.

For high dose implantation, point defects are observed. Dark bands on the maxima of implanted specie concentration, probably due to strain from point defects or their clusters, are also observed [85]. Indeed, dark contrast areas (spots or bands) are considered as indicators of local strain areas. In a newer report from the same group [61], the dark areas have been attributed to extrinsic stacking faults (extra planes) in {0001} basal planes.

A detailed study in the case of room temperature/high dose P implantation [86] showed that predominantly prismatic basal plane dislocations are formed. In addition some large defects showing a shear in the basal plane have been observed. Two possible candidates are proposed, namely pure Shockley partials and/or some 0001 sheared interstitial loops bounded by Shockley partials

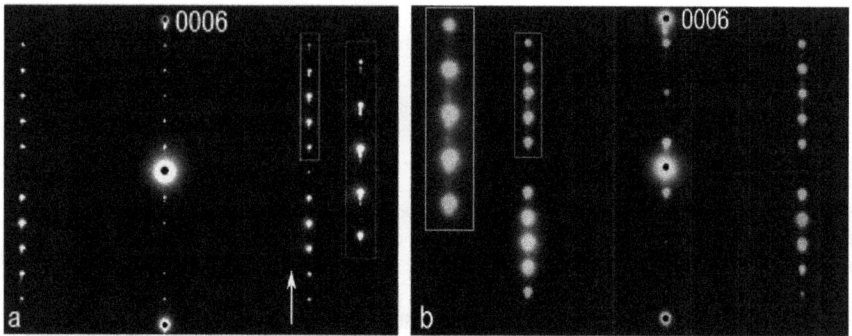

Figure 33.   (a) TEM (11$\bar{2}$0) electron diffraction patterns from N-implanted 6H-SiC wafers (cross-section specimens). In panel (a) satellite spots along the c-axis are present. The inset shows in larger magnification the upper part of the right row, revealing the existence of satellite spots around the basic ones. In panel (b) extra rows parallel to c-axis are observed. Reprinted from [877].

An interesting feature needing further investigation is the appearance of new periodicities (see Fig. 33) for the as-nitrogen-implanted samples quite probably originating from the interstitials [87].

11.5   Raman spectroscopy

Raman spectroscopy is another method, which can be used to study the crystal recovery after implantation and annealing [88].

The implantation damage is related to the change in intensity of the LO-mode/band:

$$Dam=(I_0-I)/I_0=\Delta I/I_0 \qquad\qquad (6)$$

In this expression, I_0 and I are the Raman intensities of LO modes in the non implanted and implanted parts of sample, respectively.

As we can see in Fig. 34, several Raman peaks appear after the implantation process, typically indicating the presence of Si-Si and C-C bonds. The objective, as shown in this figure, is to recover the virgin sample profile after the annealing process.

11.6   X-ray diffraction (XRD)

X-ray diffraction (XRD) is a method of characterization of crystal quality. XRD measurements have been used [89, 90] for SiC implanted layers.

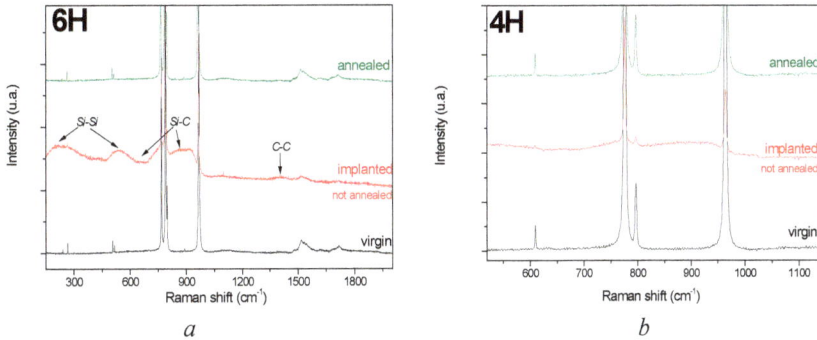

Figure 34. Left hand side figure: Raman spectra of 6H-SiC samples taken before implantation (denoted as virgin), after Al implantation (dose 7·10^14 cm^−2) and after high temperature annealing. Right hand side figure: similar spectra for a nitrogen implantation (dose 7·10^14 cm^−2) in 4H-SiC samples. Reprinted from [18].

It was shown by high resolution XRD that the nitrogen implantation in 4*H*- and 6*H*-SiC crystals introduces new periodicities as indicated by the appearance of extra peaks [87, 91]. The number of the additional peaks depends on the number of the implantation steps (Fig. 35). These periodicities disappear following the high temperature post-implantation annealing and thus, it has been concluded that they originate from the implanted nitrogen atoms. Indeed, each implantation step creates a disordered layer, for which the degree of disorder and the location from the sample surface depend on implantation dose and energy respectively. Furthermore, it is generally accepted that the implanted nitrogen ions are mainly located in interstitials sites. Quite probably they form a new lattice with a longer periodicity than the SiC lattice since their density is lower than that of the matrix material. Following the post-implantation annealing the nitrogen atoms occupy substitutional sites and the periodicity of the SiC matrix material is only observed.

11.7 Cross-section imaging techniques

Various methods have been proposed for the cross-section dopant profiling of implanted SiC such Scanning Electron Microscopy (SEM) employing a through-the-lens (ExB) detector [92], Scanning Spreading Resistance Microscopy (SSRM) [93], Scanning Capacitance Microscopy (SCM) [94] or a combination of them [95].

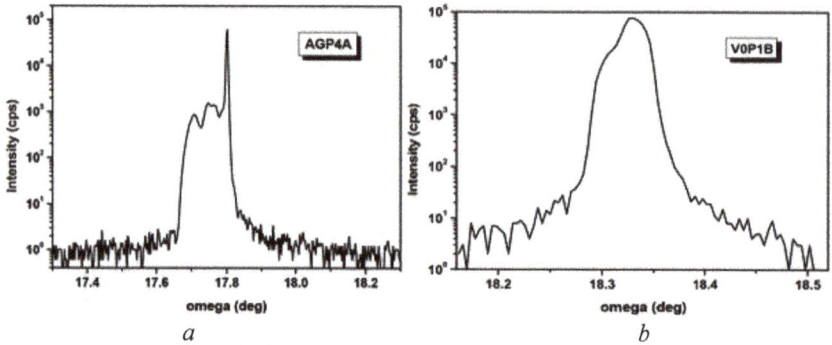

Figure 35.   ω-scan XRD (0004) graphs measured under the same conditions for N-implantation in 4H-SiC (left) and 6H-SiC (right) samples. Both samples have been implanted with identical 5 successive implantation steps although in the case of that on the left (a), has been implanted with 2 additional implantation steps at higher energies. Reprinted from [87].

In a recent study [96] (Fig. 36), it has been shown that the SEM intensity profile is sufficient not only for determining the electrical junction position but in addition, it follows the doping profile as determined by the SIMS. Therefore, it can be used as a much faster and simpler method for an initial evaluation of the doping distribution with a high accuracy in what it concerns doping spatial extension.

SCM provides information on the doping of semiconductor materials, using an atomic

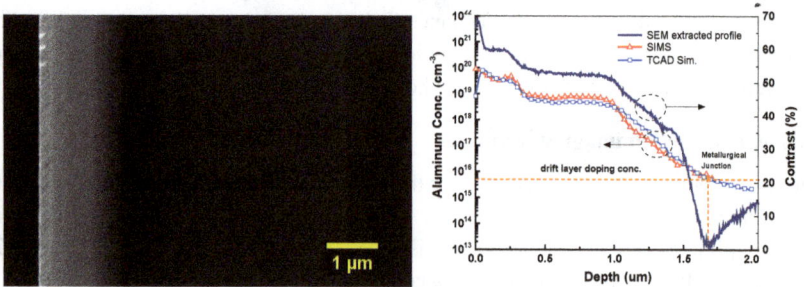

Figure 36.   SEM observation of bare 4H-SiC multilayer structure's cross-section. Left: Typical grey scale SEM photo. Right: The SEM contrast profile, the SIMS one and that TCAD simulated are shown. Modified figure from [96].

Materials Research Forum LLC
https://doi.org/10.21741/9781644900673-3

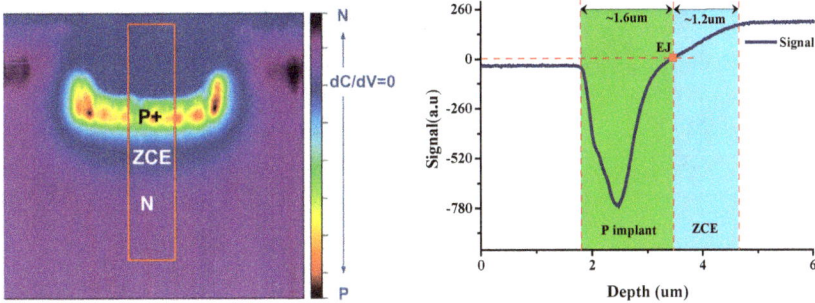

Figure 37. Cross-section SCM image of JFET's gate area (left). Quadrature signal versus depth (right). ZCE is the depletion region formed around the p-gate. Reprinted from [96].

force microscope. In the case of SCM observations, the distinction between n and p type doping is realized using signal's change of sign (for quadrature dC/dV signal, negative signal for p, positive signal for n and 0 corresponds to junction delineation/position). Fig. 37 is a typical SCM image measured on processed samples where the *p*-layer has been formed by Al implantation.

SSRM is based on Atomic Force Microscopy and it is the extension of the Spreading Resistance Probe (SRP) to micro and nano-scale. SSRM measurements on a SiC multilayer pn structure has been reported in [97] and application of the method in the case of implanted SiC layers has been reported in [93, 95, 96].

12. Implant services suppliers

Table 3 summarizes implant capabilities for SiC application from main ion implantation services suppliers around the world.

Table 3.   Implantation services suppliers

Name, location, *web site*	Status, comment	Available species	Substrate size	Energy range capability	Implant temperature
Ion Beam Services (IBS), France *www.ion-beam-services.com*	Private company (device fabrication provider and tool manufacturer)	More than 65 elements. Standard dopants Al, N for SiC	From 5×5 mm coupons to 300 mm	From 100 eV (Using PIII) to 250 keV (single charge) up to 750 keV (triple charge)	From LN2 cooled to 600 °C (limited to 150 mm)

Name, location, web site	Status, comment	Available species	Substrate size	Energy range capability	Implant temperature
RISE ACREO /Uppsala University, Sweden, www.acreo.se	Research institute, uses implant lab of Uppsala University	Standard dopants Al, N for SiC	From 5×5 mm pieces to 100 mm	From 2 keV to 350 keV (single charge)	From RT to 500 °C
Surrey University, UK, www.surrey.ac.uk/ion-beam-centre	University	Several elements and standard dopant Al and N for SiC	From small pieces to 200 mm (100 mm for high energy)	From 2 keV to 2 MeV (single charge)	Up to 800 °C (1100 °C for 1 cm).
Cutting Edge Ions, USA, www.cuttingedgeions.com	Private company	More than 65 elements. Standard dopants Al, N for SiC	From pieces to 150 mm	Up to 760 keV	From LN2 cooled to 1000 °C
Leonard Kroko, USA, www.krokoimplants.com	Private company	15 available elements. N and Al for SiC	3" diameter or smaller wafers / samples	5 - 190 keV (single charge) up to 380 keV (double charge)	Up to 1000 °C
INNOVION, USA, www.innovioncorp.com	Private company	More than 62 elements. Standard dopants Al, N for SiC	From pieces to 300 mm		
Nissin, Japan, www.nissin-ion.co.jp/en/ios	Private company. Tool manufacturer but can provide implant service	Standard dopants. Standard dopants Al, N for SiC	From pieces to 300 mm	5 kV to 960 keV (multiple charge)	From RT to 500 °C

13. Implant tools for SiC

With the recent development of SiC device market, several tool manufacturers have developed specific ion implanters for this substrate with the capability to implant at high temperature. The main differences between these tools are the mass spectrometer resolution, the energy range, the max temperature, the versatility, the size, the throughput and of course the equipment cost.

Available tools (ranked by introduction date on the market) are:

- IMC® (for R&D Lab), FLEXION® (for production) and PULSION® (PIII) tools from IBS (France)

- IMPHEAT® from Nissin (Japan)

- PURION®-M from Axcelis (USA)

- IH-860DSIC from Ulvac (Japan)

Conclusions and challenges

The main conclusions from the present review are:

- Nitrogen and phosphorous are used for n-type doping. Nitrogen implantation is used for targeted concentrations up to $\sim 3 \cdot 10^{19}$ cm^{-3} with a corresponding sheet resistance (resistivity) of typically 300-500 $\Omega/\square$ (14.5 mΩ·cm). A lower sheet resistance of 29 $\Omega/\square$, corresponding to a resistivity of 2.3 mΩ·cm is reached with phosphorous implantation thanks to higher (more than one decade) activated concentration.

- Aluminum is used for p-type doping. Boron is not anymore considered due to higher ionization energy and its diffusion during post implantation annealing. Typical sheet resistance values of Al-implanted layers are above $1 k\Omega/\square$ (corresponding resistivity of > 30 mΩ·cm). It is not possible to lower these values by very high dose ($> 1 \cdot 10^{16}$ cm^{-2}) implantation due to the induced strain in SiC crystal for high Al concentration ($> 10^{19}$ cm^{-3}).

- Heating during implantation is necessary for doses $> 3 \cdot 10^{14}$ cm^{-2} to avoid high values of resistivity due to partial crystal recovery. Heating at temperatures between 300 °C to 500 °C is employed in this case.

- Post implantation annealing at temperatures higher than 1500 °C is necessary for implanted species activation. An annealing around 1650 °C is sufficient in the case of nitrogen implantation while higher temperatures (above 1700 °C) are necessary for aluminum and phosphorous implantation cases. Post-implantation annealing of implanted with Al SiC layer is a current subject of intense investigation.

- Carbon cap (C-cap) is the most widely used method currently to avoid surface roughness following post-implantation annealing at temperature higher than 1600 °C.

- 100% activation is reached by employing post-implantation conditions. Annealing at higher temperatures is needed as the implantation increases. In the case of Al implantation, the annealing at 1650 °C, 1800 °C and 1950 °C results in a complete activation of Al acceptors for an Al concentration of $2.5 \cdot 10^{18}$ cm^{-3}, $2.5 \cdot 10^{19}$ cm^{-3} and $2.5 \cdot 10^{20}$ cm^{-3} respectively. The respective annealing duration (steady state activation) is $(10 \div 25)$ min, $(15 \div 35)$ min and $(30 \div 45)$ min.

- Cooling rates after completing high temperature PIA should be lower than 15 °/min in order to minimize formation of the $Z_{1/2}$ and V_C centers in the bulk (i.e. below implanted area) of the annealed samples.

- Channeling is important in SiC despite the misorientation of SiC substrates and the latter have to be suitably tilted and twisted during implantation. Indeed, even if one set up the "most random" direction there will always be a little channeling tail in the implanted profile. Channeling in SiC is a current subject of intense studies.

- Undermask penetration of implanted ions has to be considered when planning an implantation for device fabrication. For a high dose ($>1\cdot10^{15}$ cm^{-2}) multistep Al implantation the undermask extension along $<11\bar{2}0>$ and $<1\bar{1}00>$ directions of a concentration equal to 10^{-3} of the implanted-ion-concentrations-peak-value will be of the order of vertical R_p corresponding to the highest energy implantation step.

- Box profiles are obtained by multistep implantation. An empirical "rule-of-thumb" is to not exceed a dose of $2\cdot10^{14}$ cm^{-2} per each implantation step and a total dose of $5\cdot10^{15}$ cm^{-2} to have minimum damages of the crystal and a better recovery following the post-implantation annealing. In addition, for keeping a "flat" profile, the dose should increase as we implant at higher energies.

- The specificities of SiC implantation required the development of dedicated simulators in the late 90's. Nowadays, commercial TCAD simulators like the MC Implant module of SILVACO incorporate the main features of the above simulators and address satisfactorily the corresponding simulation needs.

- Plasma Immersion Ion Implantation (PIII) is a suitable method for forming shallow junctions and more importantly to address drop of the implanted concentration close to the surface. Therefore it is suitable for the formation of ohmic contact layers.

- The usually employed physical characterization methods (SIMS and sheet resistance measurements) for Si implanted layers are also used in the case of SiC. As crystal recovery is not fully optimized in the case of SiC, methods like RBS, TEM, Raman spectroscopy, XRD are also employed. Finally, microscopy methods investigating cross-sections of implanted layers are very promising for a fast crystal quality evaluation of implanted layers as well as for doping profiling purposes.

Present major challenges in ion implantation into SiC (0001), include the reduction of sheet resistances for both n- and p-types, and the precise control of doping profile in implanted regions while that of the suppression of surface roughening during high-temperature annealing seems to be resolved with the use of a carbon cap. To address these challenges, current research includes the study of ion flux (dose rate) and the heating/cooling rates of post-implantation annealing as well as investigation of different post-implantation-annealing methods. The use of alternative implantation species and atom complexes as a solution for the high sheet resistance value is not anymore studied. Another challenge is the formation of deep junctions or buried layers (>1 µm). Consequently, a new generation of implanters capable of reaching high energy (>300 keV) is now appearing on the market.

Acknowledgements

The authors would like to express their gratitude to J. Stoemenos, P. Schmidt and M. Negri for their valuable comments on the manuscript. K. Zekentes would like to express his gratitude to R. Nipoti and A. Hallen for the very informative discussions on SiC ion implantation during the last years.

References

[1]   R.G Wilson, The Pearson IV distribution and its application to ion implanted depth profiles. *Radiat. Eff.*, 46, 141 (1980). https://doi.org/10.1080/00337578008209163

[2]   R. Simonton, D. Kamenista, A. Ray, C. Park, K. Klein, A. Tasch, Channeling control for large tilt angle implantation in Si ⟨100⟩, *Nucl. Instrum. Methods Phys. Res. B*, vol 55, 39 (1991). https://doi.org/10.1016/0168-583X(91)96132-5

[3]   M. G. Grimaldi, L. Calcagno, P. Musumeci, N. Frangis and J. Van Landuyt, Amorphization and defect recombination in ion implanted silicon carbide, *J. Appl. Phys.*, 81, 7181 (1997). https://doi.org/10.1063/1.365317

[4]   T. Tsukamoto, M.Hirai, M.Kusaka, M.Iwami, T.Ozawa, T.Nagamura, T.Nakata, Annealing effect on surfaces of 4H(6H)-SiC(0001) Si face, *Appl. Surface Sci.*, Vol 113–114, pp. 467-471, (1997). https://doi.org/10.1016/S0169-4332(96)00903-8

[5]   A. Hallen, M.S. Janson, A.Yu. Kuznetsov, D. Aberg, M.K. Linnarsson, B.G. Svensson, P.O. Persson, F.H.C. Carlsson, L. Storasta, J.P. Bergman, S.G. Sridhara, Y. Zhang, Ion implantation of silicon carbide, *Nucl. Instr. Meth. Phys. Res. B*, 186, 186–194, (2002). https://doi.org/10.1016/S0168-583X(01)00880-1

[6]   T. Troffer, M. Schadt, T.Frank, H. Itoh, G. Pensl, J. Heindl, H. P. Strunk, M. Maier, Doping of SiC by implantation of boron and aluminum, *Phys. Stat. Solidi A*, 162, 277-298 (1997).
https://doi.org/10.1002/1521-396X(199707)162:1<277::AID-PSSA277>3.0.CO;2-C

[7]   L. S. Robertson and K. S. Jones, Annealing kinetics of {311} defects and dislocation loops in the end-of-range damage region of ion implanted silicon, *J. Appl. Phys.*, 87, 2910 (2000). https://doi.org/10.1063/1.372276

[8]   T. Kimoto, J. A. Cooper, *Fundamentals of silicon carbide technology: growth, characterization, devices and applications*, John Wiley & Sons Singapore Pte. Ltd, (2014). https://doi.org/10.1002/9781118313534

[9]   M. Bockstedte, A. Mattausch, and O. Pankratov, Solubility of nitrogen and phosphorus in 4H-SiC: A theoretical study, *Appl. Phys. Lett.,* Vol. 85(1), pp. 58-60, (2004). https://doi.org/10.1063/1.1769075

[10]   A. Hallen, R. Nipoti, S. E. Saddow, S. Rao and B. G. Svensson, *Advances in Silicon Carbide Processing and Applications,* Eds. S. E. Saddow and A. Agarwal, Artech House, Inc., Norwood Ma, p.109, (2004,).

[11]   A. Schoener, Ion implantation and diffusion in SiC, in *"Process Technology for Silicon Carbide Devices"*, C-M Zetterling (Ed.) INSPEC, London, pp.51-84, (2002). https://doi.org/10.1049/PBEP002E_ch3

[12]   F. Schmid, T. Frank, G. Pensl, Experimental Evidence for an Electrically Neutral (N-Si)-Complex Formed during the Annealing Process of Si+-/N+-Co-Implanted 4H-SiC, *Mater. Sci. Forum,* Vols 483-485 p.641-644, (2005). https://doi.org/10.4028/www.scientific.net/MSF.483-485.641

[13]   F. Schmid, M. Laube, G. Pensl, G. Wagner, M. Maier, Electrical activation of implanted phosphorus ions in [0001]- and [11–20]-oriented 4H-SiC, *J. Appl. Phys.,* Vol. 91p. 9182-9186, (2002). https://doi.org/10.1063/1.1470241

[14]   R. Rurali, E. Hernandez, P. Godignon, R. Rebollo, P. Orderjon, *Mater. Sci. Forum*, 433-436, 649-652 (2003). https://doi.org/10.4028/www.scientific.net/MSF.433-436.649

[15]   S. Blanqué, J. Lyonnet, J. Camassel, R. Perez, P. Terziyska, S. Contreras, P. Godignon, N. Mestres, J. Pascual, Mater. Sci. Forum, vols. 483-485 pp. 645-648 (2005). https://doi.org/10.4028/www.scientific.net/MSF.483-485.645

[16]   H. Fujihara, J. Suda, T. Kimoto, Electrical properties of n- and p-type 4H-SiC formed by ion implantation into high-purity semi-insulating substrates, *Jpn. J. Appl. Phys.,* 56, 070306 (2017). https://doi.org/10.7567/JJAP.56.070306

[17]   R. Nipoti, A. Nath, S. Cristiani, M. Sanmartin, M. V. Rao *Mater. Sci. Forum*, Vols. 679-680 pp. 393-396 (2011) and in M. V. Rao, A. Nath, S. B. Qadri, Y. L Tian, R. Nipoti, *AIP proceedings* CP1321 pp.241-244 (2010). https://doi.org/10.4028/www.scientific.net/MSF.679-680.393

[18]   S. Blanqué, *Optimisation de l'implantation ionique et du recuit thermique pour SiC,* PhD Thesis (2004), University of Montpellier II.

[19]   H Bracht, N.A. Stolwijk, M. Laube, and G. Pensl, (2000). *Appl. Phys. Lett.,* 77, 3188 and M. Bockstedte, A. Mattausch, and O. Pankratov, Different roles of

carbon and silicon interstitials in the interstitial-mediated boron diffusion in SiC. *Phys. Rev.* B, 70, 115203, (2004). https://doi.org/10.1103/PhysRevB.70.115203

[20] Y. Tanaka N. Kobayashi, H. Okumura, R. Suzuki, T. Ohdaira, M. Hasegawa, M. Ogura, S. Yoshida, H. Tanoue, Electrical and Structural Properties of Al and B Implanted 4*H*-SiC, *Mater. Sci. Forum* 338-342 pp. 909-912, (2000). https://doi.org/10.4028/www.scientific.net/MSF.338-342.909

[21] T. Kimoto, A. Itoh, N. Inoue, O. Takemura, T. Yamamoto, T. Nakajima, and H. Matsunami, Conductivity Control of SiC by In-Situ Doping and Ion Implantation, *Mater. Sci. Forum* 264-268, pp. 675-678 (1998). https://doi.org/10.4028/www.scientific.net/MSF.264-268.675

[22] Y. Negoro, T. Kimoto, H. Matsunami, Technological Aspects of Ion Implantation in SiC Device Processes, *Mater. Sci. Forum*, Vols. 483-485 pp. 599-604 (2005). https://doi.org/10.4028/www.scientific.net/MSF.483-485.599

[23] R. Nipoti, A. Nath, M. V. Rao, A. Hallen, A. Carnera, and Y. L. Tian, "Microwave Annealing of Very High Dose Aluminum-Implanted 4*H*-SiC", *Appl. Phys. Express* 4, 111301 (2011). https://doi.org/10.1143/APEX.4.111301

[24] R. Nipoti, A. Hallén, A. Parisini, F. Moscatelli, S. Vantaggio Al+ Implanted 4*H*-SiC: Improved Electrical Activation and Ohmic Contacts, *Mater. Sci. Forum* 740-742 767 (2013). https://doi.org/10.4028/www.scientific.net/MSF.740-742.767

[25] V. Heera, W. Skorupa, J. Stoemenos, B. Pécz, High Dose Implantation in 6*H*-SiC *Mater. Sci. Forum*, 353-357 pp.579-582 (2001). https://doi.org/10.4028/www.scientific.net/MSF.353-356.579

[26] E. Wendler, A. Heft, W. Wesh, Ion-beam induced damage and annealing behaviour in SiC, *Nucl. Instr. Meth. Phys. B*, Vol. 141(1-4), p105-117 (1998). https://doi.org/10.1016/S0168-583X(98)00083-4

[27] S. Seshadri, G. W. Eldridge, and A. K. Agarwal, Comparison of the annealing behavior of high-dose aluminum-, and boron implanted 4H–SiC, *Appl. Phys. Lett.* 72, 2026, doi: 10.1063/1.121681 (1998). https://doi.org/10.1063/1.121681

[28] A. Hallén, M. Linnarsson, Ion implantation technology for silicon carbide, Surf. Coat. Technol., 306 pp. 190–193, (2016). https://doi.org/10.1016/j.surfcoat.2016.05.075

[29] S. Morata, G. Mathieu, F. Torregrosa, G. Boccheciampe, L. Roux, G. Grosset, IMC-200 Series from IBS: Ion Implantation solution for SiC doping, *Book of abstracts, ECSCRM'2012.*

[30] J.F. Michaud, X. Song, J. Biscarrat, F. Cayrel, E. Collard, D. Alquier, Aluminum Implantation in 4*H*-SiC: Physical and Electrical Properties, *Mater. Sci. Forum* 740-742, pp. 581-584, (2012). https://doi.org/10.4028/www.scientific.net/MSF.740-742.581

[31] R. Nipoti, Post-Implantation Annealing of SiC: Relevance of the Heating Rate, *Mater. Sci. Forum,* Vols 556-557 pp. 561-566, (2007). https://doi.org/10.4028/www.scientific.net/MSF.556-557.561

[32] M. Lazar, C. Raynaud, D. Planson, M. L. Locatelli, K. Isoird, L. Ottaviani, J. P. Chante, R. Nipoti, A. Poggi, G. Cardinalli, A Comparative Study of High-Temperature Aluminum Post-Implantation Annealing in 6*H*- and 4*H*-SiC, Non-Uniform Temperature Effects, *Mat. Sci Forum*, Vols 389-393 pp. 827-830, (2002). https://doi.org/10.4028/www.scientific.net/MSF.389-393.827

[33] H. Wirth, D. Panknin, W. Skorupa, Efficient p-type doping of 6*H*-SiC: Flash-lamp annealing after aluminum implantation, *Appl. Phys. Lett.,* vol 74, n°7 pp.979-981, (1999). https://doi.org/10.1063/1.123429

[34] M. V. Rao "Ultra-Fast Microwave Heating for Large Bandgap Semiconductor Processing" in *Advances in Induction and Microwave Heating of Mineral and Organic Materials*, ISBN 978-953-307-522-8 Edited by: Stanisław Grundas, Publisher: InTech, (2011). https://doi.org/10.5772/16036

[35] Y. Negoro, T. Kimoto, H. Matsunami, F. Schmid, and G. Pensl, Electrical activation of high-concentration aluminum implanted in 4*H*-SiC. *J. Appl. Phys.,* 96, 4916, (2004). https://doi.org/10.1063/1.1796518

[36] R. Nipoti, A. Hallén, A. Parisini, F. Moscatelli, S. Vantaggio, Al⁺ implanted 4*H*-SiC: improved electrical activation and ohmic contacts, *Mater. Sci. Forum* 740-742 pp. 767-772 (2013). https://doi.org/10.4028/www.scientific.net/MSF.740-742.767

[37] A. Parisini, M. Gorni, A. Nath, L. Belsito, M. V. Rao, and R. Nipoti, Remarks on the room temperature impurity band conduction in heavily Al+ implanted 4*H*-SiC, *J. Appl. Phys.* 118, 035101 (2015). https://doi.org/10.1063/1.4926751

[38] R. Nipoti, A. Parisini, S. Vantaggio, G. Alfieri, U. Grossner, E. Centurioni, 1950°C Annealing of Al+ Implanted 4*H*-SiC: Sheet Resistance Dependence on the Annealing Time, *Mater. Sci. Forum,* Vol. 858, pp.523-526, (2016). https://doi.org/10.4028/www.scientific.net/MSF.858.523

[39]   Y.D. Tang, H.J. Shen, Z.D. Zhou, X.F. Zhang, Y. Bai, C.Z. Li, Z.Y. Peng,
       Y.Y. Wang, K.A. Liu, X.Y. Liu *Book of abstracts, ICSCRM'2015.*

[40]   H. M. Ayedh, R. Nipoti, A. Hallén and B. G. Svensson, Controlling the Carbon
       Vacancy Concentration in 4*H*-SiC Subjected to High Temperature Treatment,
       Mater. Sci. Forum, vol 858 pp. 414-417, (2016).
       https://doi.org/10.4028/www.scientific.net/MSF.858.414

[41]   C. Dutto, E. Fogarassy, D. Mathiot, D. Muller, P. Kern, D. Ballutaud, Long-pulse
       duration excimer laser annealing of Al+ ion implanted 4*H*-SiC for *pn* junction
       formation, *Appl. Surface Sci.* 208-209 pp. 292-297 (2003).
       https://doi.org/10.1016/S0169-4332(02)01357-0

[42]   C. Boutopoulos, P. Terzis, I. Zergioti, A.G. Kontos, K. Zekentes,
       K. Giannakopoulos, Y.S. Raptis, Laser annealing of Al implanted silicon carbide:
       Structural and optical characterization", *Appl. Surf. Sci.* 253 (2007), 7912–7916.
       https://doi.org/10.1016/j.apsusc.2007.02.070

[43]   K. Maruyma, H. Hanafusa, R. Ashihara, S. Hayashi, H. Murakami, and S.
       Higasahi, High-efficiency impurity activation by precise control of cooling rate
       during atmospheric pressure thermal plasma jet annealing of 4*H*-SiC wafer, *Japan.
       J. Appl. Phys.* 54, 06GC01 (2015). https://doi.org/10.7567/JJAP.54.06GC01

[44]   H. Hanafusa, K. Maruyma, R. Ishimaru, and S. Higasahi, High Efficiency
       Activation of Phosphorus Atoms in 4*H*-SiC by Atmospheric Pressure Thermal
       Plasma Jet Annealing, *Mater. Sci. Forum*, Vol 858, pp. 535-539, (2016).
       https://doi.org/10.4028/www.scientific.net/MSF.858.535

[45]   Information on http://www.centrotherm.world/technologien-
       loesungen/halbleiter/produktionsequipment/cactivator-150.html

[46]   Centrotherm's private communication

[47]   R. Nipoti, E. Albertazzi, M. Bianconi, R. Lotti, G. Lulli, M. Cervera and A.
       Carnera, "Ion implantation induced swelling in 6*H*-SiC", *Appl. Phys. Lett.* 70
       (1997) pp. 3425-3427. https://doi.org/10.1063/1.119191

[48]   S. Blanqué, R.Pérez, P. Godignon, N. Mestres, E. Morvan, A. Kerlain, C. Dua, C.
       Brylinski, M. Zielinski and J. Camassel, Room Temperature Implantation and
       activation Kinetics of nitrogen and Phosphorus in 4*H*-SiC Crystals, *Mater. Sci.
       Forum* 457-460, pp. 893-898 (2004).
       https://doi.org/10.4028/www.scientific.net/MSF.457-460.893

[49]   K. A. Jones, P. B. Shah, K. W. Kirchner, R. T. Lareau, M. C. Wood, M. H. Ervin, R. D. Vispute, R. P. Sharma, T. Venkatesan and O. W. Holland, Annealing ion implanted SiC with an AlN cap, *Mater. Sci. & Eng.* Vol. B61-62, p. 281, (2000). https://doi.org/10.1016/S0921-5107(98)00518-2

[50]   K.V. Vassilevski, N.G. Wright, I.P. Nikitina, A.B. Horsfall, A.G. O'Neill, M.J. Uren, K.P. Hilton, A.G. Masterton, A.J. Hydes and C.M. Johnson, Protection of selectively implanted and patterned silicon carbide surfaces with graphite capping layer during post-implantation annealing, *Semicond. Sci. Technol.* Vol. 20, p.271 (2005). https://doi.org/10.1088/0268-1242/20/3/003

[51]   Y. Negoro, K. Katsumoto. T. Kimoto, H. Matsunami, Flat Surface after High-Temperature Annealing for Phosphorus-Ion Implanted 4*H*-SiC(0001) using Graphite Cap, *Mater. Sci. Forum* Vol. 457-460 p. 933-936 (2004). https://doi.org/10.4028/www.scientific.net/MSF.457-460.933

[52]   S. G. Sundaresan, M. V. Rao, Y.J. Tian, M. C. Ridgway, J. A. Schreifels, J. S. Kopanski, Ultrahigh-temperature microwave annealing of Al+- and P+-implanted 4*H*-SiC, *J. Appl. Phys.* 101, 073708 (2007). https://doi.org/10.1063/1.2717016

[53]   M. Rambach, A. J. Bauer, H. Ryssel, *Silicon Carbide: Current trends in Research and Applications*, Eds P. Friedrichs, T. Kimoto, L. Ley, G. Pensl, 2010, Wiley, p. 181

[54]   M. Lazar, L. Ottaviani, M. L. Locatelli, C. Raynaud, D. Planson, E. Morvan, P. Godignon, W. Skorupa, J. P. Chante, High Electrical Activation of Aluminium and Nitrogen Implanted in 6*H*-SiC at Room Temperature by RF Annealing, *Mater. Sci. Forum*, 353-356, pp.571-574 (2001). https://doi.org/10.4028/www.scientific.net/MSF.353-356.571

[55]   A. Toifl, Modeling and Simulation of Thermal Annealing of Implanted GaN and SiC, B.Sc. Thesis, Technical Un. Vienna (2018).

[56]   V. Šimonka, A. Toifl, A. Hössinger, S. Selberherr, J. Weinbub, Transient model for electrical activation of aluminium and phosphorus-implanted silicon carbide, J. Appl. Phys. 123 (23), 235701, (2018). https://doi.org/10.1063/1.5031185

[57]   V. Simonka, A. Hossinger, J. Weinbub, S. Selberherr, Empirical Model for Electrical Activation of Aluminum-and Boron-Implanted Silicon Carbide, IEEE Trans. Electron Dev. 65, 674-679, (2018). https://doi.org/10.1109/TED.2017.2786086

[58]   D. Goghero, F. Giannazzo, V. Raineri, P. Musumeci and L. Calcagno, Structural and electrical characterization of n+-type ion-implanted 6*H*-SiC, *Eur. Physical J. Appl. Phys,* Vol 27(1-3), pp 239-242, (2004). https://doi.org/10.1051/epjap:2004112

[59]   Y. Furukawa, H. Suzuki, S. Shimizu, N. Ohse, M. Watanabe, K. Fukuda, Distribution of Secondary Defects and Electrical Activation after Annealing of Al-Implanted SiC, *Mat. Sci. Forum*, Vols. 821-823, pp. 407-410, (2015). https://doi.org/10.4028/www.scientific.net/MSF.821-823.407

[60]   C. M. Wang, Y. Zhang, W. J. Weber, W. Jiang and L. E. Thomas, Microstructural Features of Al-Implanted 4*H*-SiC, J. Mater. Res., Vol. 18, pp. 772–779 (2003), and in W. J. Weber, W. Jiang, C. M. Wang, A. Hallén, and G. Possnert, Effects of implantation temperature and ion flux on damage accumulation in Al- implanted 4*H*-SiC, J. Appl. Phys., Vol. 93, No. 4, pp. 1954–1960 (2003). https://doi.org/10.1557/JMR.2003.0107

[61]   K. Kawaharaa, G. Alfieri, T. Kimoto, Detection and depth analyses of deep levels generated by ion implantation in n- and p-type 4*H*-SiC, *J. Appl. Phys.* 106, 013719 (2009). https://doi.org/10.1063/1.3159901

[62]   B. Zippelius, J. Suda, T. Kimoto, High temperature annealing of n-type 4*H*-SiC: Impact on intrinsic defects and carrier lifetime, *J. Appl. Phys.* 111, 033515, (2012). https://doi.org/10.1063/1.3681806

[63]   J. Wong-Leung, M. Janson and B. Svensson, Effect of crystal orientation on the implant profile of 60 keV Al into 4*H*-SiC crystals, *J. Appl. Phys*, Vol.93, p.8914 (2003). https://doi.org/10.1063/1.1569972

[64]   SILVACO^TM website: https://www.silvaco.com

[65]   E. Morvan, Modélisation de l'implantation ionique dans α-SiC et application à la conception de composants de puissance, PhD thesis, INSA Lyon (1999) https://www.theses.fr/1999ISAL0115

[66]   E. Morvan, P. Godignon, J. Montserrat, D. Flores, X. Jorda, M. Vellvehi, Mapping of 6*H*-SiC for implantation control, *Diam. Rel. Mat.* 8, pp. 335–340 (1999). https://doi.org/10.1016/S0925-9635(98)00411-7

[67]   E. Morvan, P. Godignon, M. Vellvehi, A. Hallén, M. Linnarsson, and A. Yu. Kuznetsov, Channeling implantations of Al+ into 6H silicon carbide *Appl. Phys. Lett.* 74, 3990 (1999). https://doi.org/10.1063/1.124246

[68]   A. Hallen, M. Linnarsson, L. Vines, B. Swensson, To be published in Proc. of ECSCRM2018, Birmingham (UK), September 2018. https://doi.org/10.1088/1361-6641/ab4163

[69]   M. K. Linnarsson, A. Hallén and L. Vines, Intentional and unintentional channeling during implantation of 51V ions into 4*H*-SiC, *Semicond. Sci. Technol.* 34 (2019) 115006

[70]   G. Lulli, R. Nipoti, 2D Simulation of under-Mask Penetration in 4*H*-SiC Implanted with Al+ Ions, *Mat. Sci. Forum* Vols. 679-680 pp 421-424, (2011). https://doi.org/10.4028/www.scientific.net/MSF.679-680.421

[71]   K. Mochizuki, N. Yokoyama, Two-Dimensional Modeling of Aluminum-Ion Implantation into 4*H*-SiC, *Mat. Sci. Forum* Vols. 679-680, pp 405-408 (2011). https://doi.org/10.4028/www.scientific.net/MSF.679-680.405

[72]   E. Morvan, N. Mestres, J. Pascual, D. Flores, M. Vellvehi, J. Rebollo, Mat. Sci. Eng. B61–62, 373–377 (1999). https://doi.org/10.1016/S0921-5107(98)00537-6

[73]   F. Torregrosa, Y. Spiegel, J. Duchaine, T. Michel, G. Borvon, L. Roux, Recent developments on PULSION® PIII tool: FinFET 3D doping, High temp implantation, III-V doping, contact and silicide improvement, & 450 mm, *Proc. IWJT 2015* Kyoto. https://doi.org/10.1109/IWJT.2015.7467061

[74]   X.Y. Qian, N. W. Cheung, M. A. Lieberman, S. B. Felch, R. Brennan, and M. I. CurrentPlasma immersion ion implantation of SiF_4 and BF_3 for sub-100 nm Pþ/N junction fabrication, *Appl. Phys. Lett.* 59, 348–350, (1991). https://doi.org/10.1063/1.106392

[75]   L. Ottaviani, S. Biondo, M. Kazan, O. Palais, J. Duchaine, F. Milesi, R. Daineche, B. Courtois and F. Torregrosa: Implantation of Nitrogen Atoms in 4*H*-SiC Epitaxial Layer: a Comparison between Standard and Plasma Immersion Processes, *Adv. Mat. Res.* Vol. 324, pp. 265-268, (2011). https://doi.org/10.4028/www.scientific.net/AMR.324.265

[76]   S. Biondi, M. Lazar, L. Ottaviani, W. Vervish, O. Palais, R. Daineche, D. Planson, J. Duchaine F. Milesi and F. Torregrosa: Electrical characteristics of SiC UV-Photodetector device: from the p-i-n structure behaviour to junction Barrier Schottky structure behaviour , *Mat. Sci. Forum* Vol. 711 pp. 114-117, (2012). https://doi.org/10.4028/www.scientific.net/MSF.711.114

[77]   M. Zielinski, S. Monnoye, H. Mank, F. Torregrosa, G. Grosset, Y. Spiegel, Novel Carbon Treatment to Create an Oriented 3*C*-SiC Seed on Silicon. *Book of*

abstracts, ECSCRM 2018, August 2018.
https://doi.org/10.4028/www.scientific.net/MSF.963.153

[78]   M. Robinson, I. Torrens, Computer simulation of atomic-displacement cascades in solids in the binary-collision approximation *Phys. Rev. B,* 9, p 5008 , (1974). https://doi.org/10.1103/PhysRevB.9.5008

[79]   O. Oen, M. Robinson, *Nucl. Inst. Methods*, 132 p647, (1976). https://doi.org/10.1016/0029-554X(76)90806-5

[80]   E. Morvan, P. Godignon, J. Montserrat, J. Fernadez, J. Millan, J.P. Chante, Montecarlo simulation of ion implantation into SiC-6H single crystal including channeling effect, *Mat. Sci. Eng.* B46, pp. 218-222(1997). https://doi.org/10.1016/S0921-5107(96)01982-4

[81]   M.S. Janson, PhD Thesis, KTH 2003, ISSN0284-0545

[82]   J. F. Ziegler, J. P. Biersack, and U. Littmark, In *The Stopping and Range of Ions in Matter*, volume 1, New York, Pergamon. ISBN 0-08-022053-3, (1985).

[83]   W. Brandt, M. Ktitagawa, Effective stopping-power charges of swift ions in condensed matter, *Physics Rev. B*, 25 (9), p5631, (1982). https://doi.org/10.1103/PhysRevB.25.5631

[84]   P.M, Echenique, R.M. Niemen, J.C. Ashley and RX. Ritchie, Nonlinear stopping power of an electron gas for slow ions, *Phys. Rev. A*, 33, pp. 897 (1986). https://doi.org/10.1103/PhysRevA.33.897

[85]   T. Kimoto, N. Inoue, H. Matsunami, Nitrogen Ion Implantation into α‑ SiC Epitaxial Layers, *Phys.Stat. Sol.(a)* 162, pp. 263-276, (1997). https://doi.org/10.1002/1521-396X(199707)162:1<263::AID-PSSA263>3.0.CO;2-W

[86]   J. Wong-Leung, M. K. Linnarsson, B. Svensson and D. J. H. Cockayne, Ion-implantation-induced extended defect formation in (0001) and (11-20) 4*H*-SiC, *Phys. Rev.* B 71, 165210 2005. https://doi.org/10.1103/PhysRevB.71.165210

[87]   K. Zekentes, K. Tsagaraki, A. Breza and N. Frangis, The formation of new periodicities after N-implantation in 4*H*- and 6*H*- SiC samples, *Mat. Sci. Forum* Vols. 740-742 pp 447-450 (2013). https://doi.org/10.4028/www.scientific.net/MSF.740-742.447

[88]   J. Camassel, S. Blanque, N. Mestres, P. Godignon and J. Pascual, Comparative evaluation of implantation damage produced by N and P ions in 6H-SiC, *Phys. Stat. Sol. A*, 195, p.875- 880 (2003). https://doi.org/10.1002/pssc.200306246

[89] K.B. Mulpuri, S.B. Qadri, J. Grun, C.K. Manka and M.C. Ridgway, Annealing of ion-implanted SiC by laser-pulse-exposure-generated shock-waves, *Solid-State Electronics* 50, pp.1035–1040, (2006). https://doi.org/10.1016/j.sse.2006.04.019

[90] Z.C. Feng, S.C. Lien, J.H. Zhao, X.W. Sun and W. Lu, Structural and Optical Studies on Ion-implanted 6H–SiC Thin Films, *Thin Solid Films* 516, pp.5217–5222 (2008). https://doi.org/10.1016/j.tsf.2007.07.094

[91] K. Zekentes, K. Tsagaraki, M. Androulidaki, M. Kayambaki, A. Stavrinidis, H. Peyre and J. Camassel, Room temperature physical characterization of implanted 4*H*- and 6*H*-SiC, *Mat. Sci. Forum. 717-720* pp 589-592 (2012). https://doi.org/10.4028/www.scientific.net/MSF.717-720.589

[92] M. Buzzo, M. Ciappa, J. Millan, P. Godignon, W. Fichtner, Microelectronic Engineering 84 pp. 413–418 (2007). https://doi.org/10.1016/j.mee.2006.10.055

[93] R. Elpelt, B. Zippelius, S. Doering, U. Winkler, Employing Scanning Spreading Resistance Microscopy (SSRM) for Improving TCAD Simulation Accuracy of Silicon Carbide, *Mater. Sci. Forum*, 897, p295, (2017). https://doi.org/10.4028/www.scientific.net/MSF.897.295

[94] F. Giannazzo, L. Calcagno, F. Roccaforte, P. Musumeci, F. LaVia, V. Raineri, Dopant profile measurements in ion implanted 6H–SiC by scanning capacitance microscopy, *Appl. Surface Sci.* 184(1-4), 183 (2001). https://doi.org/10.1016/S0169-4332(01)00500-1

[95] O. Ishiyama, S. Inazato, Dopant Profiling on 4H Silicon Carbide P+N Junction by Scanning Probe and Secondary Electron Microscopy, *J. Surface Analysis* 14(4), pp. 441-443 (2008).

[96] K. Tsagaraki, M. Nafouti, H. Peyré, K. Vamvoukakis, N. Makris, M. Kayambaki, A. Stavrinidis, G. Konstantinidis, M. Panagopoulou, D. Alquier, K. Zekentes, Cross-section doping topography of 4*H*-SiC VJFETs by various techniques, *Mat. Sci. Forum.* 924, pp. 653-656, (2018). https://doi.org/10.4028/www.scientific.net/MSF.924.653

[97] J. Suda, S. Nakamura, M. Miura, T. Kimoto and H. Matsunami, Scanning Capacitance and Spreading Resistance Microscopy of SiC Multiple-pn-Junction Structure, *Jpn. J. Appl. Phys.*, 41, L40 (2002). https://doi.org/10.1143/JJAP.41.L40

Advancing Silicon Carbide Electronics Technology II
Materials Research Foundations **69** (2020) 175-232

Materials Research Forum LLC
https://doi.org/10.21741/9781644900673-4

CHAPTER 4

Plasma Etching of Silicon Carbide

K. Zekentes[1]*, J. Pezoldt[2], V. Veliadis[3]

[1] MRG-IESL/FORTH, Vassilika Vouton, PO Box 1385, Heraklion,
Greece and Grenoble INP, IMEP-LAHC, 38000 Grenoble, France

[2] FG Nanotechnologie, Institut für Mikro- und Nanoelektronik und Institut für Mikro- und
Nanotechnologien MacroNano®, Postfach 100565, 98684 Ilmenau, Germany

[3] PowerAmerica & North Carolina State University, Electrical & Computer Engineering Dept.
930 Main Campus Drive Suite 200, Raleigh, NC 27606, USA

* zekentesk@iesl.forth.gr

Abstract

Plasma etching is the only microelectronics-industry-compatible way to etch SiC for the device pattern transfer process. After more than twenty years of SiC plasma etching technology development, there are still issues such as (i) the etch-rate dependence on plasma parameters, (ii) the surface roughness, (iii) the microtrenching, (iv) the lack of understanding of the very-deep-etching mechanisms, and (v) the not fully understood process optimization that is, in many aspects, based on an empirical approach. The present review deals with all aspects of SiC plasma etching with an emphasis on the above issues that are not well understood.

Keywords

Dry Etching, Plasma Etching, Ion Etching, Sputtering, Chemical Etching, Plasma Chemistry, Gas Phase Chemistry, Etch Rate, Residue Free Etching, Micromasking, Microtrenching, Sidewall Slope, Mask Selectivity, Aspect Ratio Dependent Etching, Microloading

Contents

List of used abbreviations 177
1. Introduction 177
2. Gas chemistry – etching mechanisms 178
    2.1   SiC etching gas chemistry 178
    2.2   Surface carbon rich layer 180
    2.3   Cl-based chemistry 180

2.4   Results relative to the use of different fluorinated gases.........................181

2.5   Role of additives (N_2, H_2, O_2, Ar, He) in the gas mixture 183

3.   Etch rate..186

3.1   Role of pressure 186

3.2   Role of substrate platen RF power / DC self-bias 188

3.3   Role of ICP RF power (source/coil power) 190

3.4   Role of gas flow...191

3.5   Role of crystal face 191

3.6   Role of doping type...191

3.7   Role of chamber/substrate electrode geometry...................................192

3.8   Role of substrate temperature 192

3.9   Loading effects..194

4.   Morphology of etched surfaces/sidewalls 195

4.1   Micromasking 195

4.2   Micromasking after deep (> 10 μm) etching 198

4.3   Ion-etching induced polishing effect of SiC surfaces...........................200

4.4   Microtrenching effect..200

4.5   Isotropic etching..204

4.6   Sidewall shape 205

4.7   Sloped walls from sloped etch mask...207

4.8   Vertical scratches...208

5.   Mask material (adherence, micromasking, selectivity)....................209

6.   Surface conditioning prior-to / following etching............................211

7.   Carrier of the SiC sample under etching...212

8.   DRIE (Deep RIE) process in SiC: via-hole formation - MEMS..................213

8.1   Continuous etching process 213

8.2   Bosch process...214

9.   Nanopillar/nanowire formation..215

10. Electrical properties after etching...217

11. Main conclusions...220

Acknowledgements 222

References...222

List of used abbreviations

AFM: Atomic Force Microscopy;

BHF: Buffered HF;

BJT: Bipolar Junction Transistor;

CCP: Capacitively Coupled Plasma;

DRIE: Deep Reactive Ion Etching;

ECR: Electron-Cyclotron-Resonance;

ER: Etch Rate;

ICP: Inductive Coupled Plasma;

ITO: Indium Tin Oxide;

JFET: Junction Field Effect Transistor;

MEMS: Micro Electro Mechanical Systems;

MOS: Metal Oxide Semiconductor;

NEMS: Nano Electro Mechanical Systems;

NW: Nano Wire;

OES: Optical Emission Spectroscopy;

RCA: Radio Corporation of America;

RIE: Reactive Ion Etching;

SEM: Scanning Electron Microscopy;

UMOSFET: U-shape metal-oxide-semiconductor field-effect-transistors;

XPS: X-rays Photoelectron Spectroscopy.

1. Introduction

Whereas the large Si-C bonding energy and the wide band gap accompanied with extremely low intrinsic carrier concentration at room temperature makes silicon carbide attractive for applications in harsh environments (high temperature, corrosive, etc.), it complicates its etching. Conventional wet chemical etching is not feasible at practical process temperatures, and can only be performed at high temperature in molten salts, by electrochemistry or UV excitation. Hence, plasma-based dry etching is primarily used in SiC device patterning. For instance, plasma etching is used for trench formation in SiC devices such as BJTs, buried-gate JFETs, UMOSFETs, thyristors, JBS and *p-i-n* diodes, or for via-hole etching in SiC.

Various review papers dedicated to SiC plasma etching have been published [1, 2, 3, 4] in an early stage of the SiC plasma etching.

The present review aims to be a comprehensive guide for any SiC device process developer for SiC dry etching. The first part deals with process chemistry showing why fluorine chemistry has been mainly used and the reasons for choosing SF_6 as main gas, as well as the effects of other gases added in the process and the proposed etching mechanisms. The second part is dedicated to the control of the etch rate through the various plasma parameters. The third part covers a wide area related to the morphology of the etched bottom surface as well as of the etched sidewalls. Hard mask material and especially its selectivity with SiC is the subject of the forth part. The next two short parts deal with SiC surface conditioning prior or after the plasma etching and the choice of appropriate carrier for the SiC wafer under etching. The next part concerns deep etching for via-hole and MEMS applications while the top-down formation of nanowires is the theme of the following one. The electrical properties of the etched surfaces are presented in the last part before the final conclusions.

In the following, the capacitively-coupled plasma reactive ion etching (CCP-RIE) and inductively coupled plasma reactive ion etching (ICP-RIE) will be mentioned as RIE and ICP respectively for simplicity purposes which are also compatible with the evolution of the semiconductor plasma etching and the usually employed vocabulary.

2. Gas chemistry – etching mechanisms

2.1 SiC etching gas chemistry

SiC etching typically proceeds via an ion-enhanced mechanism [1, 5]. The atomic bond is broken by the action of ion bombardment causing bond weakening, defect generation, amorphisation and stoichiometric disturbances in the near surface region [6, 7, 8]. The different types of defects enhances the etching rate and the Si-Si and C-C bond formation allows the partially separated atoms to be removed by chemical reactions, the latter being possible in many cases thanks to the presence of the energetic ion flux. Thus, the SiC etching can be treated as a separate etching of Si and C atoms.

High volatile etch byproducts are critical for fast etching. Therefore, etch byproducts should have relatively high room temperature vapor pressure or equivalently low boiling point. SiF_x and CF_x, which are the SiC etch products of fluorinated plasma chemistries are much more volatile than $SiCl_x$ and CCl_x, the corresponding etch products for Cl-based chemistries, as expected by looking the corresponding boiling points (see Table 1 [5]).

Table 1. Boiling points of Potential Etch Products in Plasma Etching of SiC.
Reprinted by permission from [Springer] J. Electron. Mat. [Copyright] (2001) [5].

Etching Product	Boiling Point [ºC]
$SiCl_4$	57.6
SiF_4	-86
CCl_4	76.8
CF_4	-128
CO_2	-78.5 subl
CO	-191.5

Experimental results with F, Br and Cl based chemistries [5, 9] confirm that F-based chemistries lead to higher rates than Cl and Br. Indeed, SiC ion etching in fluorinated plasmas has been shown to produce high etch rates (usually some 100 nm/min) and a high degree of etching anisotropy allowing for the patterning of sub-micron or even above-micron features. ICP etching in a SF_6/O_2 plasma has yielded etch rates up to 0.97 μm/min [10] and a helicon reactor produced $6H$-SiC etch rates as high as 1.35 μm/min [11]. Therefore, fluorine-based plasma etching of SiC has proven successful and this chemistry is primarily used in the fabrication of SiC devices.

For the F-related chemistry the most common chemical reactions associated with the removal of Si and C atoms are given by [1, 12, 11]:

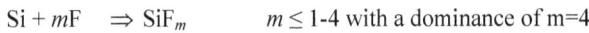

$Si + mF \Rightarrow SiF_m$ $m \leq$ 1-4 with a dominance of m=4

$C + mF \Rightarrow CF_m$ $m \leq$ 1-3

CF and CF_2 are the primary etch products at the etched surface, whereas SiF_2 radicals are mainly produced in the gas phase (probably by electron impact dissociation of SiF_4, the putative major etch product) [11].

According to process conditions the ion energetic flux results in:

1. Damaging or breaking the surface Si-C bond (4.52 eV) and the strong C-C bonding (6.27 eV) thus enhancing the chemical reaction efficiency.

2. Physical removal (sputtering) of non-volatile surface species, which enables the chemical reaction to proceed.

3. Sputtering of SiC surface.

2.2 Surface carbon rich layer

The above mentioned theory, of separate removal of Si and C atoms in SiC etching, is strengthened by experimental results showing that a thin carbon-rich layer (mostly containing CF_x compounds) is formed on the etched surface as reported by several research groups [11, 12, 13]. The thickness of the zone with changed stoichiometry ranges from approximately 1 nm to 15 nm and depends on the surface potential (acceleration voltage) and the mass of the impinging ion [14].

The removal of this layer, which is a rate-limiting step, produces unsaturated CF_x (x=1, 2, 3) radicals. In addition, the relative intensities of these bonds vary with the etch conditions [12]. Therefore, C is not removed sufficiently quickly from the etched surface through the reaction of carbon-fluorine [C-F] or eventually carbon-oxygen [C-O] for O_2 containing gas mixtures, resulting in C-rich etched surface.

The removal of C has been debated in the literature, with some works indicating that the C removal is via physical bombardment [15] while others claim a removal through a reactive chemistry between the fluorine and carbon [12]. In some studies, [2, 10, 12], the addition of oxygen in the gas increases substantially the C removal through chemical reactions. On the contrary in [15], samples of 6*H*-SiC as well as diamond have been etched simultaneously and a weak (SiC case) or no (diamond case) dependence of C removal on the O_2 content has been observed respectively. Thus, the authors have concluded the C is removed by physical reaction (sputtering). It is possible that depending upon the etch chemistry and process parameters (RF power, chamber pressure, electrode area and spacing etc.), any one of the above C-removal mechanisms may actually dominate.

2.3 Cl-based chemistry

As mentioned above, Cl-based chemistry has not been considered as a useful chemistry for SiC etching since it is widely accepted that it results in lower etch rates than the F-chemistry [5] on one hand and uses corrosive Cl-based gases on the other. Nevertheless, there are studies investigating Cl-based chemistry [16, 17, 18].

In one of the first studies, Nieman *et al.* [16] stated that chloride based RIE plasmas have the potential for using non-metallic etch mask materials (e.g. SiO_2) with good selectivity even for high SiC etch rates (100 – 200 nm/min).

Furthermore, the use of Cl-based chemistry is proposed, despite its lower etch rate, because plasma etching induced damage can be annealed out at relatively low temperatures, which is not possible with F-based chemistry [17]. This is possibly due to the much higher reactivity of carbon with fluorine than chlorine forming various kinds of

C–F bonds on the SiC surfaces. These bonds can play an important role in the electrical properties of the etched SiC surface, while Cl radicals create fewer problems of this type.

Note also, in contrast to employing F-containing gas mixtures, the use of Cl_2 or IBr exhibits a threshold level of DC self-bias on the substrate, below which no etching occurs. Obviously, the reaction and removal of etch products needs to be induced by ion-bombardment because of the lower volatility of the heavier halides of silicon and carbon [3] (ion enhanced etching mechanism dominates).

In a recent work [19], Cl-based chemistry has been used to control the bevel of the etched sidewalls quite efficiently since $(BCl_3 + N_2)$ and $(Cl_2 + O_2)$ gas mixtures allowed the control of the etching angle from 40° to 80° and from 7° to 17°, respectively, through the adjustment of the mixture ratio.

The conclusion from the reported studies is that Cl-based chemistries result in lower etch rate values and the only advantage of using them (although not thoroughly studied) is the better control of electrical-induced damages and less microtrenching. Moreover, the toxic character of Cl-based gases was another disadvantage of Cl-chemistry although fluorine-based gases have their own environmental issues.

2.4   Results relative to the use of different fluorinated gases

The plasma etching of SiC polytypes has been widely investigated in fluorinated gases (CHF_3, $CBrF_3$, CF_4, SF_6 and NF_3), usually mixed with oxygen, hydrogen, or argon. A representative summary of these results is given below and in Fig. 1.

The SiC etch rates correlated well with the average bond energies of the feedstock gases, i.e., BF_3 154 kcal/mol, PF_5 126 kcal/mol, CF_4 116 kcal/mol, CHF_3 115 kcal/mol, SF_6 78.3 kcal/mol, and NF_3 66.4 kcal/mol [5, 20, 21, 22]. The lower the bond energy, the more effectively the fluorinate gas dissociation formed atomic fluorine neutrals, which are the active etchant species.

In terms of surface roughness there was no difference between the different gases [5].

There are very few studies on using CF_4, in SiC RIE-etching [2, 15, 21, 22, 23] (mainly on $3C$-SiC etching) and high-plasma-density-etching [3, 24]. Etch rates are clearly lower than usually reported for SF_6 etching. An additional reason for not using CF_4 widely is the high probability of forming polymers at high RF powers due to the presence of carbon in the reacting molecule, which induces C-F polymer formation capable of blocking etching under certain conditions [5, 25].

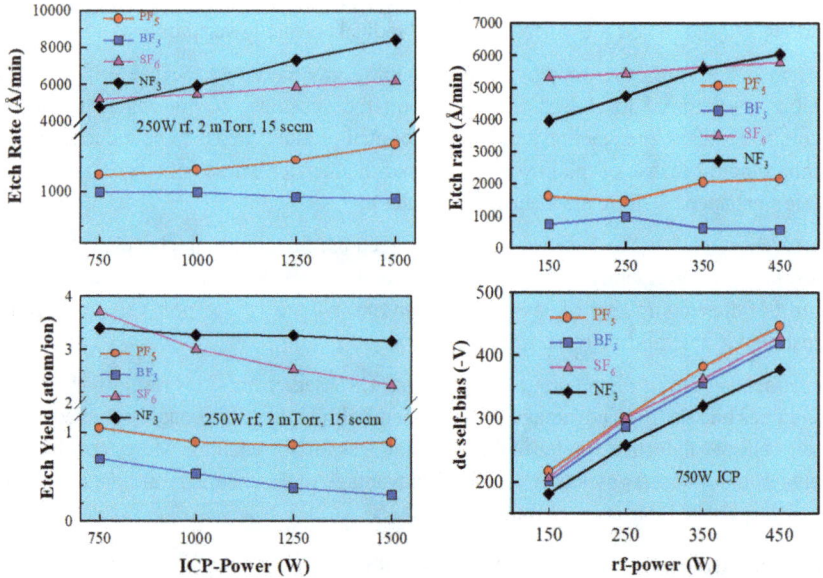

Figure 1.   SiC etching parameter variation for different gas chemistries. Left: Effect ICP source power on the etch rate (top) and yield (bottom). Right: Etch rate (top) and DC self-bias (bottom) dependence on RF electrode power. Reproduced from [20] with the permission of the American Vacuum Society.

High etch rates of 6H- and 4H–SiC have been obtained by using NF_3 in ICP [5], RIE [26, 27] and ECR (Electron Cyclotron Resonance) [28] plasma etching studies. The latter investigation showed that NF_3 gives an etch rate about 4 times higher than SF_6. Although NF_3 results in higher etch rates, it has not been adopted as the standard chemistry for etching SiC probably due to its toxicity and higher price.

The most advantageous of fluorinated gases for SiC etching is SF_6, and a series of studies have been published on this chemistry [23, 29, 30, 31, 32, 33]. Note that SF_6 plasmas are widely used for Si etching in Si semiconductor processing technology as the abundance of [F] in the SF_6 plasma produces a high Si substrate/layers etch rate. This is particularly favourable for applications requiring etching of thick layers such as micromachining and power device patterning. Typical etch rates for RIE and ICP etching of silicon carbide in SF_6 chemistry are 100 – 200 nm/min [31, 32] and 300-600 nm/min [5], respectively. The corresponding reported maximum etch rates are 700 nm/min [33] and 970 nm/min [10].

2.5 Role of additives (N_2, H_2, O_2, Ar, He) in the gas mixture

The following analysis is concentrated on the effect of additives on the etching mechanism and the etch rate. Other consequences such as the etched surface morphology will be detailed later on in Section 4.

Argon

Ar has been a popular additive in the gas mixture in the initial investigations of SiC etching. Indeed, Ar is often used in plasma etching, since it is credited with enhancing the etch rate or facilitating anisotropic profile control. Thus, most experiments use Ar ions for direct surface etching by a physical etching mode (sputtering).

Ar also modifies ion densities in plasma and its effect was well described in a plasma-charge-dynamics work performed in an ICP system [34].

Another reason for employing Ar is for modifying plasma impedance and thus optimizing the efficiency with which power is deposited into the discharge resulting in an increase of the electron and positive ion densities, as well as in the production of radicals. This mechanism has been studied in the case of RIE etching of 6H-SiC in SF_6-based chemistries at the Air Force Research Laboratory [30, 31, 35].

Finally, it has been proposed that the physical sputtering by Ar^+ ions helps to remove non-volatile or low-volatility fluorocarbon or carbon-rich etch products from the etched surface [36].

Many experimental studies [30, 31, 32, 33, 35, 36, 37] showed that the addition of Ar increases the etch rate moderately (~15-20%) despite a net increase of the self-induced DC self-bias at high Ar proportion [32, 33]. The low effect on the etch rate in combination with the high DC self-bias values is the reason for which very few groups currently use Ar for the etching of SiC.

Oxygen

From the very beginning of SiC plasma etching, one of the most commonly used diluents has been O_2. Several reasons have been outlined for adding oxygen in the fluorinated plasmas during the SiC etching:

1. The significantly higher (~4×) etch rates obtained in the plasma etching of Si with O_2 mixtures of fluorinated gases in comparison to the etching with pure fluorinated gas. In this Si-etching case, the higher etch rate of mixtures with O_2 content up to 20% was attributed to the higher density of atomic fluorine through chemical reactions with etching by-products (SiF_x, ...).

2. Oxygen removes carbon from the SiC surface by volatile oxide formation, since the Si atom removal rate is higher than that of C atoms, and a C-rich layer is formed on the RIE-etched SiC surface [2].

3. In complement to the above argument, increasing O_2 into the feed gas results in sufficient [O] chemical absorption on the surface making it more "oxide-like", thus reducing the available Si sites for etching and rendering equal the Si and C removal rates [38].

4. The addition of O_2 to a strongly electronegative attaching gas modifies the electrical characteristics of the discharge enabling more efficient power deposition and resulting in greater ion and radical production [31].

5. Oxygen reacts with SiC-surface-unsaturated-fluoride species generating reactive F atoms, while simultaneously depleting these polymer-forming species [13].

6. The etch selectivity of SiC over Al masks is increased in the presence of O_2 [39].

7. The addition of O_2 in SF_6 gas prevents non-desirable sulphur-based deposition on the chamber walls (by formation of volatile products such as SO_xF_y) [11].

However, the conclusion from numerous studies having employed O_2 for the SiC etching is that the relative increase in SiC etch rate using O_2-rich SF_6 and CF_4 mixtures was not as significant as in the Si case showing a decreased importance of the chemical component in the SiC etching mechanism. The prevalent explanation for this result is that the addition of O_2 in fluorine plasmas provides indeed the pathway for volatilizing C from residual C-F bonds, in the form of CO and CO_2. This increases the etch rate but it also produces SiO_2 on the SiC surface, which can inhibit the etch process [10, 12]. Another possible explanation for the reduced SiC etch rate at high O_2 content is that the reduction of reactive fluorine, and therefore C-F reactions, as the dilution effect cannot be compensated from the increased C-O reactions [25].

An optimum O_2 concentration, in terms of peak etch rate, is obtained from these competing mechanisms, which depends on the employed fluorine gas. Indeed, the peak etch rate occurs for higher O_2% in low fluorine content gases [2]. For instance, a moderate etch rate increase was observed with a peak O_2 content of ~60%, ~40% and ~20% for CHF_3/O_2 [25], CF_4/O_2 [15, 23, 24] and SF_6/O_2 [10, 12, 23, 40, 41], gas mixtures respectively.

Hydrogen

There is a consensus [13, 21, 42] that the addition of H_2 improves etched surface morphology especially in terms of micromasking. This is due to the removal of Al deposited on the SiC surface by AlH_3 formation [14]. The Al deposition on SiC is due to sputtering of chamber walls made by Al. Furthermore, the addition of H_2 decreases the surface-fluoride-atomic-concentration, which is associated with the rate-limiting fluorocarbon film formed on the surface (see above Section 2.2) [13]. According to the same study, H_2 post-etching annealing suppressed further the surface atomic fluoride concentration to less than 3 % [13].

However, there are contradictory results in terms of the H_2 effect on the etch rate, with an increase up to 10% content [13], decrease above 10% content [21] and slight decrease [42]. Moreover, a contamination either of the etched surface (case of CF_4/H_2) or of the RIE chamber (case of the SF_6/H_2) has been observed [42]. In the former case, an organic layer non-removable by any chemical etchant has been deposited on the SiC surface rendering it hydrophobic. In the latter case, a white dust has been deposited on the RIE chamber walls. The authors of [25] agreed with the above observations that the addition of H_2 increases polymer formation not only on the SiC under etching but also on the etch chamber.

H_2 is no longer used for etching SiC primarily due to the above contamination problems, and the employment of other approaches allowing for smooth plasma-etched surfaces without micromasking.

Helium

Similar to Ar, He has been proposed as a diluent for better power transfer [31] producing higher etch rates than Ar-based mixtures. Although its use is not mentioned in scientific publications, industrial foundries are employing it as gas additive in plasma etches for the fabrication of SiC devices.

Nitrogen

Wolf and Helbing [23] have studied the effects of N_2 on the etching of SiC in fluorinated mixtures with oxygen plasmas. The use of N_2 did not demonstrate a decisive advantage to be adopted for SiC etching, and there are no longer studies reporting its use.

3.   Etch rate

3.1   Role of pressure

There are multiple and sometimes-contradictory results in terms of plasma-etching-chamber pressure impact on SiC etch rate, with a noticeable difference between RIE and ICP systems.

Most studies on RIE etching of SiC agree that the increase of the gas pressure induces a corresponding increase of the etch rate up to a certain value and then a decrease. The former is directly related to the atomic fluorine production in the RIE chamber due to a higher collision probability between gas molecules and electrons. The latter is caused by the reduction of the mean free path, i.e. the energy accumulated between two electron – molecule collisions from the electric field decreases reducing the ionisation probability. Additionally, in this high-pressure region surface passivation due to polymer formation, etch-residues and adsorption of etch-products reduce the etching reactions. This result is in agreement with the "bell-shape" trend of the plasma density versus gas pressure dependency whatever the material to be etched.

The rigorous study of Camara *et al.* [32, 33] of the chamber pressure effect, employing Optical Emission Spectroscopy (OES) to monitor the evolution of atomic fluorine concentration, is representative of RIE based studies [24, 43] of the pressure effect on the etch-rate. Very high $4H$-SiC etch rates (up to 700 nm/min) are obtained in SF_6-based processes performed at high chamber pressure [33]. For constant RF power, the etch rate increases smoothly with pressure up to a maximum value, and then decreases sharply beyond that (Fig. 2a). The maximum value of the etch rate (etch rate peak), obtained at a definite RIE chamber pressure, increases almost linearly with the RF power (Fig. 3) [33]. For pressure below 100 mTorr, the etch-rate depends mainly on the DC self-bias value and not on the atomic fluorine content as the latter is almost the same for high enough power (>100 W) showing a purely physical mechanism of etching. However, when the pressure increases above 100 mTorr, the fluorine atom concentration is increasing in agreement with the etch rate increase while the DC self-bias decreases (and so does the energy of impinging ions). Since F atoms are the main chemical agents for SiC etching, increasing the F atom density enhances the etch rate indicating a more "chemical" nature of the etching. According to [35], the direct electron-impact excitation rates increase nonlinearly with pressure and this explains the higher atomic fluorine production. Note that, above the etch rate peak pressure value, a quite abrupt etch rate decrease is observed and various reasons are proposed for this abrupt drop: (i) atomic fluorine production is reduced due to increased collisions and consequent decrease of mean free path/ions lifetime [32, 33], (ii) the DC self-bias decreases below 100 V reducing bond break by the ion energetic flux [32, 33] and, (iii) surface chemistry offsets elevated concentrations of F atoms [35]. The latter can include active species desorption although not explicitly mentioned in [35]. Note that in [35] the above abrupt drop of the etch rate is accompanied by a significant sulphur deposition contamination.

The above net etch rate increase with chamber pressure, obtained in RIE systems, has not been observed in high density plasma reactors [3, 4, 11, 28, 44, 45, 46, 47], where etch rate variation with pressure is not as pronounced as in RIE systems and in some cases the reverse behaviour has been exhibited. It is worth to notice that the typical pressure range in RIE systems is between 20 and 300 mTorr while

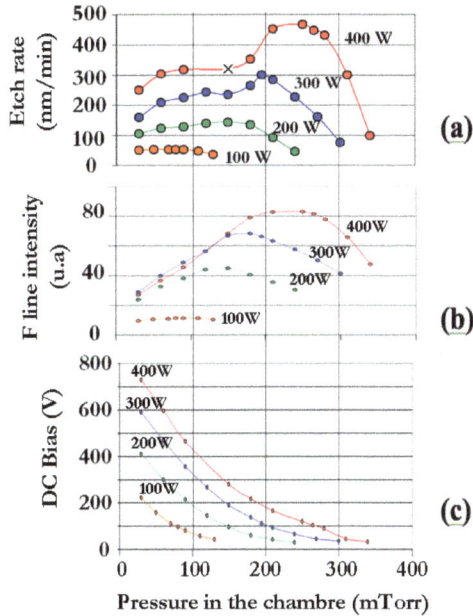

Figure 2. Etch rate (a), fluorine peak intensity (b) RIE RF DC self-bias (c), dependence on chamber pressure for various RIE RF power values. Reprinted from [33].

in ICP systems between 4 and 30 mTorr. Moreover, gas dissociation to atomic fluorine occurs at very low pressures in high-density-plasma reactors, so there is no need for high pressure to produce elevated levels of reactive atomic F as in RIE.

For instance, in a magnetically-enhanced ICP study using SF_6 [44], the SiC etch rates decreased linearly with the increase of operating pressure similarly to the change of F radical density for 1500 W of RF source power [44]. However, when the magnet was turned off, the etch rate remained constant with pressure. In [4], it has been also reported a linear decrease of the SiC etch rate with pressure for an 800 W coil power. On the contrary, the same group reported that the etch rate increases with increasing pressure in a later study [45]. The study in [48] agrees with this conclusion. In [25], the etch rate increased with gas pressure up to a saturation value and the same was observed in [46]. Biscarrat et al. [47] reported a small decrease of the etch rate with SF_6 pressure for a coil power of 600 W, while the etch rate increased slightly for 1200 W and steadily for 1800

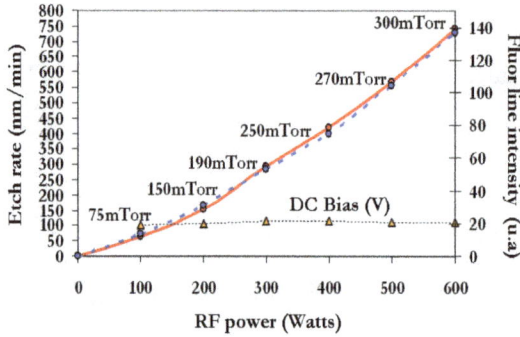

Figure 3.   Maximum etch-rate (blue circles) and the corresponding, to this maximum, values of DC bias (yellow triangles), fluorescence intensity (red circles) and chamber pressure (blue circles) versus RIE RF power. The graph has been constructed by using the data of Fig. 2 above. Reprinted from [33].

W of coil power. Similar results have been reported in [41] in which the etch rate increases for a power of 800 W while it remains stable or slightly decreases for a 500 W coil power.

The above seemingly complex and often contradictory trends could be related to the relatively low dependence of the etch rate on the chamber pressure. Thus, the specific high-density-plasma-process conditions influencing the production of reactive species and positive ion flux can result in a different trend in the various studies. For instance, the temperature of the SiC surface under etching can explain the above results as at high source power levels the thermal budget in the chamber heats substantially the surface of the SiC under etch if a good thermal contact with the electrode is not ensured. Therefore, it is not possible to make the comparison between the various high-density-plasma-based-SiC-etching studies and extracting a single unanimous behaviour as sample self-heating during the process could complicate the whole situation.

3.2   Role of substrate platen RF power / DC self-bias

Generally, it is accepted that for physical or ion-energetic plasma etching the etch rate follows a square root dependence with the substrate energy. However this dependence is not often obvious for various reasons.

First of all, note that in most SiC plasma etching studies the influence of each parameter has been studied separately by keeping the other parameters constant. Thus, most of the substrate-RF-power-variation related studies were conducted at constant pressure by varying the RF power. In such experimental configuration, a common experimental finding was that by increasing the substrate-holder RF power, an almost linear increase of the induced DC self-bias (in both RIE and ICP systems) as well as of the etch rate have been observed [27, 23]. It was thus, widely accepted that the SiC etch rate depends on the DC self-bias and therefore, it was assumed that to obtain high etch rates, a physical bombardment-based process has to be performed, which is possible by increasing RF power and self-induced DC self-bias.

The above assumption does not take into account that in RIE systems, the increase of the RF power results in an increase of both the atomic fluorine density and the DC self-bias value and a distinction of the contribution of each on the etch rate cannot be done. This point, which requires a systematic investigation to clarify the role of DC self-bias and atomic fluorine, has been addressed by Camara *et al.* [32, 33]. The main results of that study (Fig. 3) were: (i) for high enough pressures, the etch rate is primarily related to the intensity of atomic fluorine and not on the value of DC self-bias, (ii) the etch rate increases with DC self-bias value when varying the RF power at constant pressure, (iii) when the chamber pressure increases at constant RF power, there is an increase of the etch rate with a simultaneous decrease of the DC self-bias (Fig. 2) and, (iv) at each RF power value, the maximum etch rate has been obtained at the same DC self-bias value of around 100 V showing that a high DC self-bias is not a prerequisite to get high SiC etch rates. A possible explanation for the 100 V DC self-bias value at maximum etch rate is that at that DC-bias there is the optimum compromise between ion-induced chemical-etching-processes activation and chemical-active species surface desorption. The conclusion was that at high pressures the ion-energetic etching mode is dominant. On the other hand, for low enough RIE chamber pressures (< 100 mTorr), the intensity of atomic fluorine was the same for the different values of RF power and the etch rate was directly related to the DC self-bias as expected for a dominant sputtering-based etching [32, 33].

In most ICP studies using SF_6 or CF_4/O_2 mixtures and at constant gas pressure, the etch-rate increases linearly [44, 47, 24] or super-linearly [3, 41] with substrate DC self-bias, the latter induced by an increase of the RF electrode power [4]. In the case of [41, 49] this increase holds up to 200 V (350 V for [39]) and then a saturation is observed. According to [39] this saturation of the etch rate with the self bias above a certain value, is an indication that Si–C bond breaking is no longer the limiting step. For high-enough fluorine production the etch rate-limiting factor is a surface-based reaction most probably the C-F reaction to produce CF_2 volatiles. DC self-bias seems to not have an influence on the C-F reaction but rather on Si-C bond breaking [46].

An enhanced ion bombardment seems a plausible explanation for this etch-rate increase with RF platen power by taking into account that in ICP systems: (i) the atomic fluorine density is mainly controlled by RF source/coil power, (ii) the RF bias power does not have an appreciable impact on the ion and electron densities [34] and (iii) that the pressure in ICP systems is usually quite low (< 30 mTorr).

An etch rate of 20 nm/min [24] and ~10 nm/min [41] has also been reported even when there is no ICP substrate bias. According to [3], this indicates that SiC is etched chemically by F radicals and does not require the assistance of energetic ions. However, the highly anisotropic etching character of SiC rules out this explanation. A more plausible explanation is that the plasma potential, present even for zero value of self bias, and the presence of the F radicals are enough for the etching of SiC through an ion-energetic etching mechanism widely accepted for the SiC plasma etching[1].

In summary, a steadily increase of the etch rate with RF power has been observed for most studies of SiC RIE etching regardless of experimental configuration. This was attributed to a higher fluorine production controlling the etch rate. An increase of the etch rate with the RF platen power or equivalently with the self-induced bias has been observed for most studies of SiC ICP etching due to an increased sputtering (i.e. physical removal) mechanism of etching.

3.3   Role of ICP RF power (source/coil power)

There is no unanimity about the effect of the RF source power on the etch rate. In many studies there is a clear steady etch rate increase [24, 41, 44, 47, 49]. However, for other studies this increase is not significant (Fig. 1) [5, 48, 50] indicating that the density of ion radicals created by RF plasma remains approximately constant as ICP power varies or there is a saturation in terms of necessary radicals for the etch-rate limiting processes (bonds breaking, deposited surface layer removal, etc.).

Note that the DC self-bias (V_{DC}) decreases with the increase of the source power for constant RF electrode power. Here one considers that $P_{substrate} = J_{i+} \times V_{DC}$ and that the positive ion flux (J_{i+}) is proportional to the source power [47]. This is an approximation as the coupling between source and electrode power is not taken into account.

---

[1] Let's remind that the ion energy is the sum of the plasma potential and the DC self-bias.

3.4 Role of gas flow

The effect of the gas flow has not been extensively studied but most studies [3, 46, 48, 51], that have investigated this dependency, agree for a slight etch-rate increase with the gas flow.

The lack of other studies on the effect of gas flow indicates that its effect is not important given that high enough gas flows are used to replenish reactants and remove volatile etch products.

3.5 Role of crystal face

In principle, differences in the etch rates are expected between distinct crystal faces exposed to plasma. These differences are due more to the dissimilar dangling-bond densities and the corresponding reactivity of the crystal faces than to the different crystal structures. For example, each atom on cubic (001) face has two dangling bonds, whereas only one dangling bond exists on (111) face or similarly to the (0001) face of hexagonal SiC.

Wolf and Helbig have investigated the effect of Si and C-face on the $6H$-SiC RIE etching [23] by employing CF_4 and SF_6 chemistries. For both chemistries and in the absence of O_2 the silicon face is etched faster than the carbon face by a factor of 1.2. With the addition of O_2 the situation initially reverses, and the carbon face is being etched faster. Further increase of oxygen led to an etch ratio of unity for both crystal planes.

An interesting study showing the effects of polytype and crystal face on the etch rate is that of J. Choi *et al.* [52]. Their purpose was the formation of high aspect SiC pillars and after a long etching the initially circular shape turned to a faceted one. The facets were different for the different SiC polytypes under etching and have been also dependent on the misorientation of the SiC wafer. According to the authors, the sufficiently long etching process allows the appearance of the crystal planes with the lowest etch rates according to the etched polytype and crystal orientation.

As all power SiC devices are fabricated on the Si-face there was no need for an extensive research on this aspect.

3.6 Role of doping type

According to [53], under ICP conditions there is no measurable effect in etching rates between n^+ and p^+ SiC indicating that Fermi level effects play no role in the etching mechanism. Similar conclusion was reported in [4] while in [1] it is stated that the etch rate in RIE systems increases when the n-type doping increases.

A comparison of n-type and semi-insulating (S.I.) SiC substrate etching has been performed by Okamoto [54] with the purpose of optimizing via-hole formation. The etch rates for S.I. SiC were clearly lower (by more than 20%) than those for n-SiC. This different behaviour has been attributed to the differences between the substrates in the wafer heating. The thermal conductivity in n-SiC is lower due to the higher free-carrier absorption. Thus, the wafer temperature during etching became higher in the case of n^+ SiC than in S.I. SiC and as a consequence the etch rate was enhanced in n-SiC.

In summary, there is no difference between the various conductivity SiC substrates given that there is no substrate-self-heating during the etching.

3.7 Role of chamber/substrate electrode geometry

Changing the spacing between electrodes in plasma etchers can lead to a change of the plasma properties.

Indeed, a decrease of the electrodes' distance in an asymmetric parallel plate RIE seems to decrease the etch rate quite probably due to a lesser degree of plasma-molecules ionization [29].

In [50], the ICP electrode distance from the source coil has been increased (from 90 mm to 170 mm) with the purpose of increasing further the self-bias (V_{DC}) once the other parameters (mainly platen power and source power) reached the optimum values for a high V_{DC} value. The resulted V_{DC} increase induced an increase of the etch rate.

On the contrary, a high etch-rate and uniform etching regime is obtained by minimizing the distance between the substrate holder and the bottom of the source tube in a helicon chamber according to [11]. This behaviour has been attributed to the fact that in the diffusion regime ($P = 6$ mTorr), the positive ion density decreases when the distance from the source increases. The situation deteriorates at higher pressure since the diffusion length varies with $1/P$. Similar results have been obtained for ECR-reactor experiments [3], as the rate of etching was increasing by a factor of three as the distance decreased from 8 cm to 1 cm.

In summary, decreasing the plasma length by moving the substrate electrode results in an etch rate decrease in the case of RIE and ICP systems and etch rate increase in the case of ECR/helicon systems. The reason for this etch-rate difference is not known.

3.8 Role of substrate temperature

The effect of the surface-temperature variation of the wafer under etching has not been studied extensively despite the fact that substantial heating can happen due to the plasma action and to the thermal isolation of the wafer from the electrode platen. Indeed, ICP

Advancing Silicon Carbide Electronics Technology II   Materials Research Forum LLC
Materials Research Foundations **69** (2020) 175-232   https://doi.org/10.21741/9781644900673-4

involves high levels of RF power (source plus platen), and anything exposed to the plasma, including the sample under etching, will heat up as well. Without an electrode cooled with backside helium gas flow, a substrate can heat up to 300 °C without much effort. This was demonstrated in [55] where SiC etch rates of 2.92 and 1.62 μm/min were measured for simply mounted and In-bonded on Ni carrier samples respectively, under the same ICP etching conditions. On the contrary, RIE (parallel plate configuration) does not often need thermal mounting as the total RF power involved is

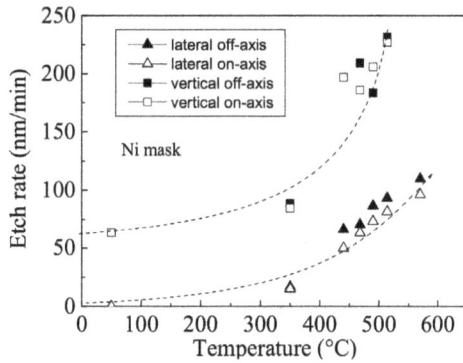

Figure 4.   Vertical and lateral etch rate versus temperature for on- and 8.5° off-axis Si-face 4H-SiC covered with a Ni mask (40 sccm SF_6, 100 W and 0.2 mbar). Reprinted from [56].

much lower than in the ICP case. However, a problem of Al-mask adherence for etching at powers higher than 200 W, due to the substrate heating, has been observed in SiC RIE etching [33].

A systematic study of the temperature effect on the etch rate of on- and off-axis 4H-SiC substrates has been performed in a specially designed RIE reactor [56] (Fig. 4) as well as ECR reactors [57]. The temperature increase of the substrate results in a more isotropic etching with an increase of both vertical and lateral etch rates (Fig. 4). The temperature dependence of the etch rate can be described using the following expression:

$$ER(T) = AT^{1/2} \exp(-E_{etch}/kT)$$

with ER as the etch rate, A as a constant, E_{etch} as the activation energy of the etch rate, k as the Boltzmann constant and T the substrate temperature. The activation energy of the etch rate was determined to be 0.23-0.24 eV. This value corresponds to the evaporation enthalpy of SiF_4 being 0.27 eV indicating a reaction-product-desorption limitation of the etch rate at elevated temperatures.

Thus, high temperature seems to increase the etch rate in fluorinated gas-based etch in contrast to what has been observed in Cl-based etch [18] in one of the very few studies on SiC cryogenic etching. Note that at temperatures below 150 °C the change of the adsorption behaviour and the increased defect formation as well as sputtering rate are dominant.

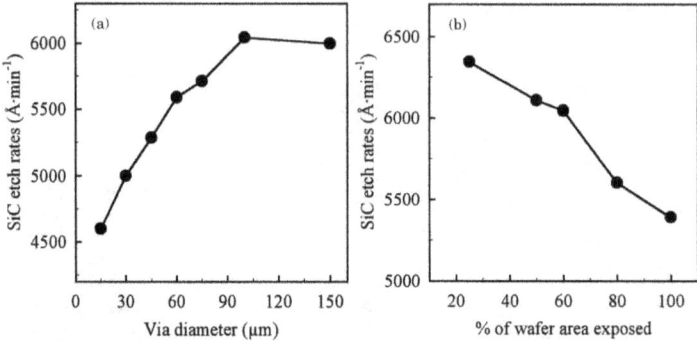

Figure 5. *Dependence of SiC etch rate on (a) via diameter for SF$_6$ discharges (~60% of the wafer was exposed) at 500 W source power and (b) on the percentage of the wafer area exposed for a constant via diameter of 100 µm. Other ICP conditions: 5 mTorr process pressure and 250 W of chuck power. Reprinted by permission from Springer Nature Customer Service Centre GmbH: J. Electron. Mat. [5]. Copyright 2001.*

In summary, heating the SiC substrate above 300°C during fluorine etching results in an important etch rate increase as well as a less anisotropic etching. Thus, an efficient heat dissipation (thermal bonding and high thermal conductivity carrier) configuration is necessary for ICP etching involving high RF power.

3.9   Loading effects

An important loading effect has been observed in via-hole etching optimization experiments with an ICP system [5] even for 5×5 mm² small samples (Fig. 5). The etch rate falls off with both decreasing via diameter (µ-loading) and increasing area of exposed SiC (macro-loading). According to the authors, this is a very clear demonstration of aspect ratio dependent etching, which results from the difficulty of getting reactants into (and etch products out of) a deepening hole as etching proceeds.

The same behaviour has been observed for deep (> 200 µm) etching in a magnetically enhanced, inductively coupled plasma reactive ion etcher (ME-ICP-RIE) with the µ-loading effect becoming important for mask openings smaller than 100 µm [58].

In [59], a mask opening limit of 20 µm (instead of the above 100 µm) has been proposed for the decrease of the etch-rate in the case of stripes. The time evolution of the etch rate as the etching proceeds has been also investigated in this study.

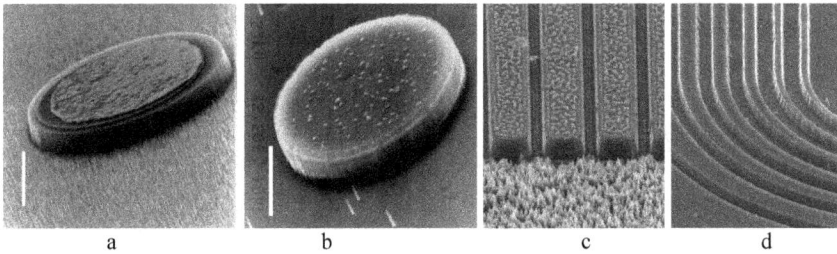

Figure 6. SEM images of SiC surfaces after etching with an Al ((a) & (b)) and Ni ((c) & (d)) mask. The Al electrode was uncovered in (a) and covered with a glass in (b). In (c) and (d) images are shown from the same sample but with a dense pattern in (c) (source fingers area of a JFET). Note the lack of micromasking in the narrow spaces between the source fingers in (c) probably due to ion loss by deflection on the sidewalls and to the shadow effect in deep trenches, which reduces the deposition of micromasking particles close to the side walls. After [42].

In conclusion, the etch rate decreases for mask openings smaller than 100 μm and 20 μm in the cases of circular and linear patterns, respectively.

4. Morphology of etched surfaces/sidewalls

4.1 Micromasking

One difficult aspect of SiC ion etching has been the formation of residues (which lead to a rougher surface – grass/micromasking effect) after long etching for various experimental conditions. The widely adopted explanation is that it is related to the need for relatively heavy ion bombardment during SiC etching due to the hardness of SiC. A predominant chemical mode of etching at room temperature is not possible in this case. Therefore, a sputtering of chamber parts (such as the electrode hosting the SiC sample) as well as of the employed metal–based hard masks is unavoidable. Some of the sputtered material is re-deposited on the surface under etching, resulting in micromasking of this surface and in significant roughness as the etching proceeds (Fig. 6).

Residue formation can be a serious problem for subsequent processes, such as metal contact (ohmic or Schottky) formation. Therefore, the re-deposition of non-volatile species (etch by-products and material sputtered from the hard mask, electrode, chamber) should be avoided. Several methods have been developed to prevent residue formation.

A general approach is to use relatively low ion-bombardment energies (or equivalently low DC self-bias values) to minimize sputtering of non-volatile materials [37], although this result in lower etch rates and secondary effects like smaller etching anisotropy [59].

Use of the appropriate electrode/platen is also necessary. Significant micromasking is observed when the electrode is from Al as it has been described in detail in [1, 21]. Similarly, aluminium-containing platens such as sapphire, aluminium nitride, anodized aluminium, and nickel-plated aluminium all create unacceptable levels of micromasking [37]. The residues are mainly Al-containing compounds and the resulting roughness effect is often called Al micromasking. This is in part because the commercial RIE systems incorporating Al electrodes are designed to accommodate multiple large Si wafers. Thus, the area of the Al electrode is usually much larger than that of the SiC samples being etched. Smooth, residue-free surfaces were obtained by covering the powered electrode [2] with a graphite [21, 37] or glass sheet [42] or even by employing a molybdenum electrode [30, 31]. This method is dependent on the reactor geometry, and side effects such as polymer formation on the non-covered electrode or flaking of the reactor walls, and a mass-loading effect in the case of the graphite or quartz sheet have been observed [3]. On the other hand, the use of quartz coverage of the cathode reduced micromasking, and when a Si wafer is placed on top of the quartz cover, the micromasking disappeared [42, 43]. The incorporation of the Si wafer results in the formation of more volatile SiF_x products thanks to the supply of Si atoms and thus, the excess F ions are not used for the formation of non-volatile species [43]. Indeed, the atomic fluorine density in the plasma, measured with OES, decreases when the Si wafer is placed on the top of the quartz [42]. However, possible loading effects can be present in this case by increasing SiC under-etching substrate size.

It has been reported that the addition of a small quantity of H_2 in the gas mixture prevents the Al micromasking effect due to the formation of Al-H volatile products,[2] but it lowers the etch rate [2, 21]. On the contrary, the addition of O_2 can increase the micromasking effect as the oxidation of Al deposits decreases their volatility [3].

Many studies have demonstrated the relation between the hard-mask-originating species and micromasking. Masks from Al and Ni have exhibited good selectivity values (see analysis below) and are employed as hard masks for the SiC etching. However, in many cases Al resulted in micromasking while the Ni-mask produces less micromasking and this is one more reason (in addition to high selectivity) for its wide use in SiC plasma etch as masking material [37, 42]. According to [60], Ni-based by-products in SF_6-based

---

[2] The gas phase reaction of H_2 and Al clusters is likely to form alane (AlH_3) volatile compound.

chemistries are characterized by higher vapour-pressure values in comparison to the Al ones and this may explain the smoother surface obtained with the Ni masks. Note, however, that the Ni-mask deposition method (evaporation, sputtering, electroplating) can result in different behaviour in terms of micromasking [42].

The areal density of the mask also plays a role, and dense Al masks or even Ni masks (Fig. 6) can result in significant micromasking [42, 60] despite the use of covered RF electrodes and/or high chamber pressure [42]. The use of less dense Ni masks [43] or the use of graphitic masks according to [60] can resolve this issue. However, the graphitic masks can be used for shallow etching as the selectivity is quite low (< 2). It has been also proposed to use SF_6 and O_2 plasma in an alternating way, not performing the etching in a single continuous step in order to remove the eventually formed F-related polymers in their initial state [42].

Plasma instabilities (momentary collapse, etc.) can also result in micromasking [48, 61] and the etch process has to be monitored to avoid these instabilities and the resulting micromasking.

A SiC surface contaminated with impurities results in micromasking too [42]. Surface treatments as those proposed in the following Section 4.2 have to be adopted prior to plasma etching to avoid micromasking due to surface contamination.

Optimum chamber pressure for avoiding micromasking, differs between RIE and ICP systems. Studies in RIE systems propose an increase in chamber pressure to reduce ionic bombardment (smaller ion energy) and the consequent micromasking effect [27, 32, 33] due to smaller masking material removal and thus re-deposition. On the contrary, in ICP systems, high-pressure (20-30 mTorr) conditions resulted in an important micromasking effect, for ICP SiC-etching with CF_4/O_2 mixtures [24] and for deep (> 5 μm) etching. Reducing the pressure or the thickness to be etched eliminated the micromasking effect. The effect of the pressure has been explained by the longer mean free path of the etching by-products at low pressure, having thus increased probability of desorption from the surface under etching. The same tendency has been apparently observed in [4] while in [58] a very low pressure (1.8 mTorr) has been employed to prevent reaction products from redepositing on an etched substrate.

The use of Ar as additive to SF_6 also has been proposed as a polymer cleaner since Ar increases the ion bombardment effect [36].

In conclusion, covering/replacing the Al-based electrode in combination with the use of the non-dense Ni mask, as well as use of low DC self-bias values, are the optimum methods for eliminating micromasking. Low chamber pressure values have to be employed in ICP systems while high pressure values are preferable in RIE systems.

4.2   Micromasking after deep (> 10 μm) etching

A problem of micromasking in the form of pillar formation has been observed in the bottom of etched areas after deep etching (mainly for via-hole formation and MEMS fabrication) even for process conditions not resulting in micromasking when the etched depth is small (< 10 μm). According to Voss *et al.* [37], this micromasking seems to be attributed to two separate causes. One is the deposition of non-volatile species including etch products and material sputtered from the chamber on the via-hole, and the other is the slow etching of defects in the SiC itself. The latter is supported by the findings of Okamoto [55] according to whom the origin of micropipes is highly reactive and is passivated with non-volatile products, which are generated by a chemical combination of the etched Ni from the metal mask and the formed SiF_x species during etching. Thus, the origin of micropipes acts as a micromask. Nevertheless, the defects acting as nuclei for pillar formation are not limited to crystal ones but also to surface contamination or surface defects like those resulting from the lapping of the wafers [37].

An indication of contamination residues is the presence of dimples on the surface under etching (Fig. 7) [4]. According to the authors, the surface of a dimple may react more readily with Al or Ni than the untextured SiC surface, thereby trapping Al or Ni to produce micromasking. If no dimples are present, Al may remain mobile on the surface until it is desorbed. Experimentally, residual hydrocarbons and water were found to greatly increase the density of dimples in the absence of in-situ plasma cleaning prior to etching [4].

Indeed, various studies [4, 37, 45, 62] reduced pillar formation by introducing an Ar pre-treatment and/or O_2 plasma prior to etching. However, the Ar sputter etches the Ni mask substantially and less so the SiC. An alternative solution to use instead, is an O_2 cleaning process [4].

In some studies [4, 35, 37] the use of Ar during the etch process has been also proposed for reducing micromasking. In [4], a mixture of 85%Ar/15%SF_6 has been proposed as optimum for deep (> 40 μm) etching without micromasking despite the reduction of the etch rate and the Ni selectivity. The authors of [30] proposed a gas mixture of $SF_6/O_2/Ar$ (5:1:5) to obtain via bottoms free of pillars without reducing the etch rate and Ni mask selectivity.

In addition to the Ar pre-treatment, the authors of [62] have added CF_4 to SF_6/He gas chemistry for eliminating completely the pillar formation.

N. Okamoto [55] tried to keep the gas pressure as low as possible, in order to avoid Ni mask by-products from depositing on the bottom of the etched area. However, a compromise has to be made in the gas pressure value because at low pressures

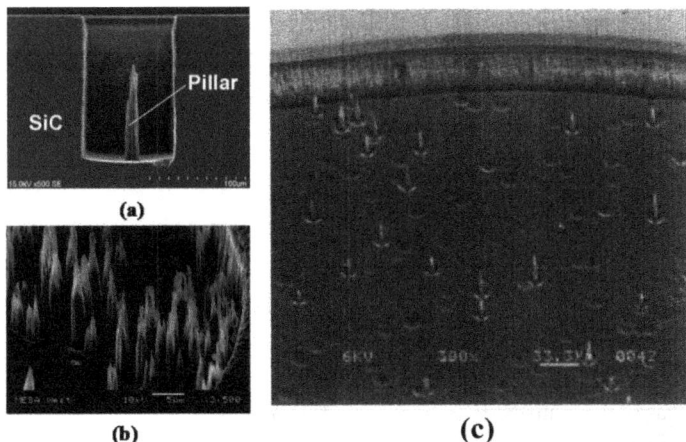

Figure 7.   SEM images of pillars (grass-like spikes) in SiC vias. The electroplated Ni has not stripped off in (c). Dimples are obvious in (c) and (b).
Images in (a) and (b) are respectively reproduced from [L. F. Voss, et al., J. Vac. Sci. Technol. B 26 (2008) 487-494] and [N. Okamoto, J. Vac. Sci. Technol. A 27 (2009) 295-300] with the permission of the American Vacuum Society. Image in (c) is reproduced from [4] by permission of Taylor & Francis.

($<$ 35 mTorr) the etch rate and the mask selectivity are reduced and micro trenching is very important.[3]

In [48], the sputter cleaning of the surface was not sufficient to prevent the formation of pillars at ICP power levels lower than 1500 W. According to the authors, at low coil RF power, the density of certain reactive ion species is somewhat reduced and pillars are formed. In addition, the authors related pillar formation with prior chamber history (opening to air, use of Cl-based chemistries, etc.) as well as high substrate temperature favouring Ni mask sputtering and re-deposition. The latter has been demonstrated by employing different SiC wafer carriers and bonding methods resulting in different heat dissipation.

In conclusion, there is unanimity that conditioning of the etching chamber and the use of substrates with low density of crystal defects, as well as a suitable preparation with Ar

---

[3] It is difficult to compare with other studies as the usual pressure range in ICP systems is up to 30 mTorr while in this study the range was up to 200 mTorr.

and O_2 plasma of the SiC surfaces are the necessary conditions to obtain a smooth bottom after very deep (> 50 µm) etching. It has also been proposed to increase the ICP coil power or reduce chamber pressure to obtain via holes free of pillars.

4.3 Ion-etching induced polishing effect of SiC surfaces

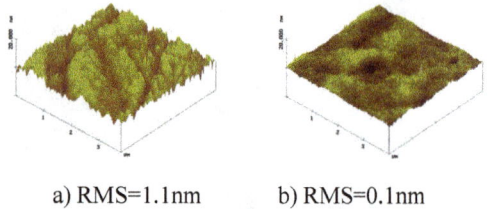

a) RMS=1.1nm   b) RMS=0.1nm

Figure 8. AFM micrographs of SiC surface: (a) prior to RIE etching and (b) after an optimum etching of 14 µm. Reprinted from [32] with permission from Elsevier.

In many cases an optimized RIE or ICP etching of SiC is characterized by a very smooth surface, even smoother than that prior to etching [13, 20, 32, 36, 63] (Fig. 8).

This smoothing of initially rough surfaces is commonly observed in ion-driven etch processes, and originates in the angular dependence of the ion mill rates [1]. This leads to faster removal rates for high aspect ratio features and creates a smoother morphology. For this reason, it has been proposed [23] to employ RIE etching as a method for smoothing the SiC surface after a polishing process.

Note that a prior RCA surface-to-be-etched cleaning is necessary as well as a BHF cleaning just prior to loading in the RIE chamber to obtain a very smooth surface after deep etching.

The combination of fluorine-containing gas with H_2 [13] or Ar [36, 64] has also been proposed to remove non-volatile or low-volatility fluorocarbon or carbon-rich etch products from the etched surface in parallel to the ion-milling action. According to [19], it is easier to remove these low-volatility reaction products, such as CF_x (CF_2, CF_3), in O_2 rich ($SF_6 + O_2$) gas mixtures resulting in lower surface roughness.

4.4 Microtrenching effect

A common problem in SiC dry etching is the microtrenching effect (Fig. 9), which refers to a V-shaped groove formed on the bottom adjacent to the sidewall due to an enhancement in local etch rates [49].

Microtrenching limits the SiC device's performance (breakdown voltage) and must be avoided. Indeed, the electric field lines have a higher density in the trenched areas, which

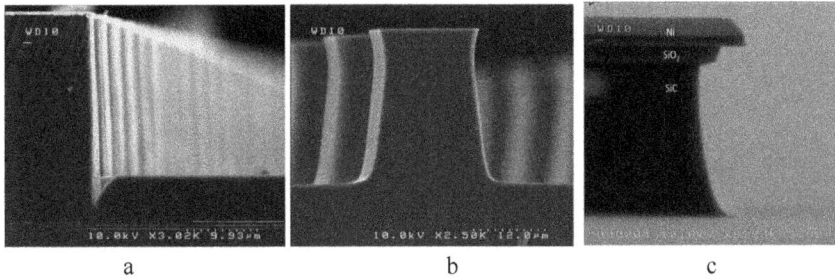

| a | b | c |

Figure 9.   *Cross-sections of SiC etched in an ICP system. P= 8 mTorr for all cases.*
(a) P_{source} = 600 W, $P_{electrode}$ = 100 W, (b) P_{source} = 1800 W, $P_{electrode}$ = 100 W, (c) P_{source} =
1800 W, $P_{electrode}$ = 100 W. Reprinted from [47].

become high-electric stress zones. For instance, it has been shown that the trenching effect may result in a local short channel effect in SiC buried-gate JFETs [65]. Moreover, it is deleterious in MEMS applications as it can significantly weaken a pressure sensor diaphragm by concentrating stress.

This phenomenon is common in other semiconductors and it has been attributed to: (i) the mask faceting by the reactants and incident plasma ions [53], (ii) the angular dispersion of the latter due to the collisions within the plasma sheath [66, 67] and (iii) the ion deflection by the charged sidewalls and the local electric field distribution formed in substrate trenches inducing enhanced ion bombardment at the bottom of the sidewall [68, 69, 70]. Furthermore, a model excluding the ion flux, and based only on the diffusion of reactive species at the surface, has been able to explain the sidewall disturbances, in particular microtrenching and bowing [71].

Different methods have been proposed and/or tested to eliminate microtrenching formation on the foot of the sidewalls, e.g., controlling the self-induced DC self-bias and the energy of incident ions [3, 68, 72], decreasing the substrate temperature to -80°C [67], growing an inhibitor film on the trench sidewalls (especially to avoid bowing) [67] or using high pressure to increase neutral scattering and obtain a less directional ion flux [72, 73] (due to the reduction of the mean free path of the ions leading to an isotropic velocity profile). The choice of the appropriate technique depends on the physical properties (isolating or conducting) of the sample substrate and of the mask, as well as on the etching setup process.

Similarly, various methods have been employed in the case of SiC etching to avoid the microtrenching phenomenon. In most cases [24, 42, 43, 45, 47, 49, 50], the trenching disappears for etching conditions resulting in a non-vertical (re-entrant/bowing) sidewall profile with sidewall angle lower than 85° (Fig. 9).

A quite drastic solution is to form a sloped etch wall by using a SiO_2 mask with wet-etched slope to transfer the same slope in the SiC underneath during dry etching [65].

Use of Cl-based or addition of Br-chemistries [74] combined with either cryogenic [18] or high temperature (900 °C) etching [75] has been reported to be efficient enough to suppress the microtrenching.

ICP source power is the more important parameter for microtrench elimination for etching with pure SF_6 chemistries, and trenching is decreased when the source power is increased [47, 79]. This is usually attributed to the chemical mode of etching, resulting in a more isotropic etching character. Note, however, that the trenching is present at high pressures and very high source power [47] rendering questionable the above explanation.

There are contradictory results about the role of platen power (and so the DC self-bias). It has been shown in [24] that the microtrenching effect in CF_4/O_2 ICP etching of SiC is strongly dependent on the substrate DC self-bias (appearing for DC self-bias values higher than 30 V). This can be attributed to the enhanced ion bombardment at the base of the sidewall due to the deflection of ions impinging on the sidewall at low angles. Similar results have been obtained in an ECR-based study [76]. This fits well with the conclusion reached from etching of other materials such as SiO_2 that any parameter increasing ion energy and/or flux results in a larger microtrench. However, contrary to these results, other studies [47, 50, 59, 79] have shown that increasing the ICP platen power, or equivalently self-bias during SiC etching, is effective in reducing micro-trenches. Similarly, microtrenching disappeared at high substrate RF power in a RIE study [77]. According to some of these studies the increased electric field results in less incident reflections off the sidewalls, and thus less bombardment at the corners of the trenches.

The effect of the ICP chamber pressure on microtrenching is also the subject of debate. In [4, 47], microtrenching is present at high pressures only, while in [37, 55, 45] microtrenching disappeared at high pressures. The explanation given for the latter is that by increasing pressure, the flux of ions impinging on the surface is reduced and so is the ion build-up to the corner of the etched feature, which promotes faster etching [37]. In the results reported in [24, 50], trenches free of microtrenching have been obtained at both low and high pressures.

Nevertheless, increase of the chamber pressure results in suppression of the microtrenching in the case of RIE systems [42].

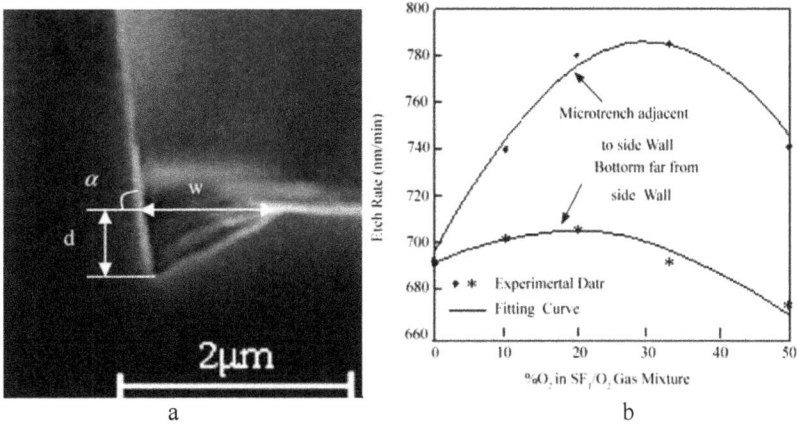

Figure 10. (a) SEM profile showing sidewall angle, microtrenching depth and width. Reprinted from [79] with permission from Elsevier. (b) SiC etch rate versus gas mixture composition. Etching is carried out at an ICP coil power of 500 W and bias voltage of −300 V. Reprinted from [49] with permission of Journal of Semiconductors Editorial Office.

In most related studies, the presence of O_2 in fluorine-based plasmas enhances the microtrench effect [12, 25, 49, 45] and it has been reported that replacing O_2 with Ar in SF_6 RIE [78] has drastically reduced microtrenching. As shown in Fig 10b, the increase of O_2 content increases the microtrench-etch-rate much more than that of the bottom trench. One proposed explanation is that the use of O_2 results in the formation of the non-volatile isolating polymer film SiF_xO_y on the sidewall increasing the trenching [12, 49] due to its greater tendency to charge than that of SiC [79]. According to the latter study, employing a factorial experimental approach, the increase of the O_2 content is the most important process parameter for the formation of the microtrench and the increase in microtrench width and depth as well as sidewall angle (Fig. 10a). On the contrary, Okamoto [55] states that the "repetition of deposition/etching of SiOF generated from oxygen and SiF_x species suppressed the ion bombardment at the bottom edge of the via-hole and the formation of microtrenches". Moreover, in a recent study [80], the Bosch process has been employed to address the trenching issue. It was demonstrated that keeping O_2 in the gas mixture was necessary to suppress trenching (see more details below in Section 8.2).

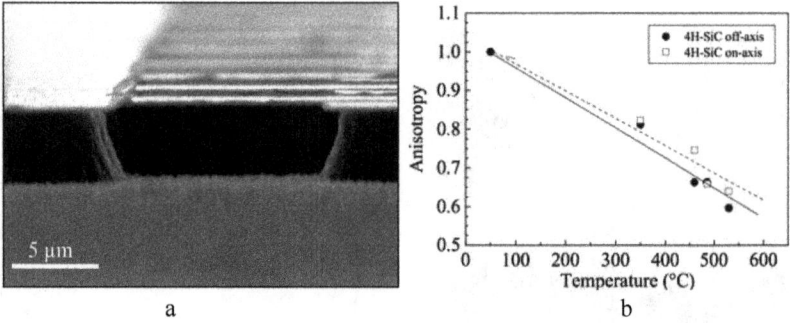

a   b

Figure 11. (a) Sloped etch sidewall geometry. SEM image of an isotropic etching profile obtained from singly clamped beams. Etching conditions: 300 mTorr, 10 sccm SF$_6$, 100 W, 450°C, 4H-SiC on axis, Ni mask. (b) Anisotropy versus temperature for on- and 8.5° off-axis Si-face (40 sccm SF$_6$, 100 W, and 150 mTorr, Ni mask). Reprinted from [56].

In summary, there are no unanimously accepted optimum conditions for the suppression of microtrenching especially in the case of the ICP etching, which is quite probably an indication that the effect is sensible to specific process environment (mask, etching setup process, etc.). Nevertheless, some trends can be established according to the majority of the studies. For instance, use of O$_2$ should be avoided when microtrenching is an issue. Increased ICP source power results in a lower sidewall angle and consequently should decrease or eliminate microtrenching. The effects of the substrate power and gas pressure are subjects of contradictory results. High DC self-bias or increasing chamber pressure in RIE systems seems to reduce/eliminate microtrenching.

4.5 Isotropic etching

The large majority of SiC plasma etching investigations are focused on anisotropic etching, while few are dedicated on isotropic etching [26, 55, 56, 57, 76, 81]. According to these studies, an increase of the substrate temperature is the necessary condition for turning from anisotropic to isotropic etching.

The effect of the temperature increase on the anisotropy has been clearly demonstrated in [56, 57] performed in an RIE [56] and alternatively in an ECR-plasma reactor [57] in which the temperature was varied up to 600 °C. The increase of temperature results in elliptically sloped sidewalls (Fig. 11a) and the anisotropy decreases linearly with the temperature increase (Fig. 11b).

The author of [55] attributed the anisotropic or isotropic character of the etching on the formation, or not, of a passivating layer on the sidewall. This passivating layer formation depends on the desorption of the etching by-products, the latter being directly related to the surface temperature.

In [56, 57], the isotropic character of the etching was attributed to the low DC self-bias (< 50 V in absolute value) and the self-annealing effects of the bottom surface. At high temperature, the higher ion-flux-induced-defects annihilation rate decreases the radiation-induced reactivity of the trench bottom (ion-energetic mode of etching), leaving a less damaged surface behind. This causes almost constant surface conditions for the arriving reactive species on the bottom and the sidewalls of the trench, reducing the anisotropy caused by the radiation-induced reactivity at low temperatures. Therefore, the etching of the substrate is purely of chemical nature not or only weakly enhanced by ion-radiation defects.

In summary, an increasing isotropic etching is obtained by raising the temperature of the SiC sample surface under etching.

4.6   Sidewall shape

Concave (bowed) sidewalls have been obtained for high-pressure/low-DC self-bias RIE etching, while for low-pressure/high-DC self-bias RIE vertical or slightly sloped sidewalls have been obtained [42]. On the contrary more vertical sidewalls are obtained by increasing the pressure in ICP chambers [47, 45] while a re-entrant/concave/bowed profile has been observed at low pressures. This difference between and ICP and RIE is quite probably related to the difference in the pressure operating range between the two systems on one hand and the diffference in sheath region extension on the other. Moreover, high ICP source power increases the sidewall angle at its bottom [47, 79] and a re-entrant profile (local bowing) is observed just below the mask level indicating as physical origin a charging of the mask material.

As previously mentioned (Section 2.3), the slope of the etched sidewalls could be controlled by employing Cl-based gas chemistry [19] and a photoresist as etch mask. The bevel is due to the beveled edges of the photoresist (see following Section 4.7). More precisely the authors propose to employ different ratio (BCl_3 + N_2) gas mixtures to control the etching angle from 40 to 80°. They attributed the formation of beveled walls in this case to an oxidation of the mask edge and to a N_2-promoted passivation. In addition, they proposed employing (Cl_2 + O_2) gas mixtures for controlling the etching angle from 7° to 17°.

Deep SiC etching, mainly for via-hole formation, resulted in various sidewall morphologies (Fig. 12). Tapering of the upper parts of the trenches (Fig. 12a) was observed in most cases and it was attributed [58] mainly to the elimination/erosion of the Ni mask edges during long periods of RIE. The mask edge elimination/erosion can be explained either by the increase of the local electric field on the edges of metal masks in the presence of strong physical sputtering [82], or simply by bearing in mind that for a long etch process the corners are eroded and attain an angle commensurate with that of the maximum sputtering yield.

In a quite unique case, concave (bowing) sidewalls have been observed in deep via-hole formation ICP etching [5] (Fig. 12b). It is unknown if such morphology is due to specific process conditions, or an observation artefact of a sample cut non perpendicularly to the wafer surface resulting in a bowed etching profile even though no actual bowing resides in the sample [83].

Moreover, the often-observed tapering down to the via bottom can be attributed either to a mask erosion or to an etched surface temperature increase during the etching. Indeed, the variation of surface temperature affects not only the etch rate (see analysis above) but also the sidewall shape. For instance, the difference between the sidewall shapes reported in Fig. 12c and Fig. 12d is due to surface temperature since the two samples were etched in the same run and the only difference was their mounting on the Ni carrier (simply mounted and In-bonded) [55]. In the case of simply mounted the high temperature resulted in a more isotropic etching as well as in a Ni-mask warpage and reaction with the SiC. In the case of the bonded sample, the lower temperature allows an anisotropic etching.

In addition, etched Ni from the mask is not only evaporated toward the ICP chamber but it is also introduced into via-holes where it combines with volatile SiF_x species. The products of this combination are non-volatile and are deposited on the sidewalls with the consequence of via-holes having vertical sidewalls without any undercut. This film accumulated up to the bottom edge of the sidewall, has blocked the ion bombardment enhancement at the vertical sidewall, so no microtrenches appeared either [55].

An interesting sidewall shape after high aspect ratio ICP etching is shown [59] in 6 μm wide trenches were etched at different duration. Microtrenching is formed after a long enough etching. The microtrenches dictate the evolution of the trench geometry. The microtrench is increasing with etching time and combined with a microloading effect at high depth etching, results in a triangular shape [59]. According to the same study, the trench sidewalls were more vertical for high DC self-bias.

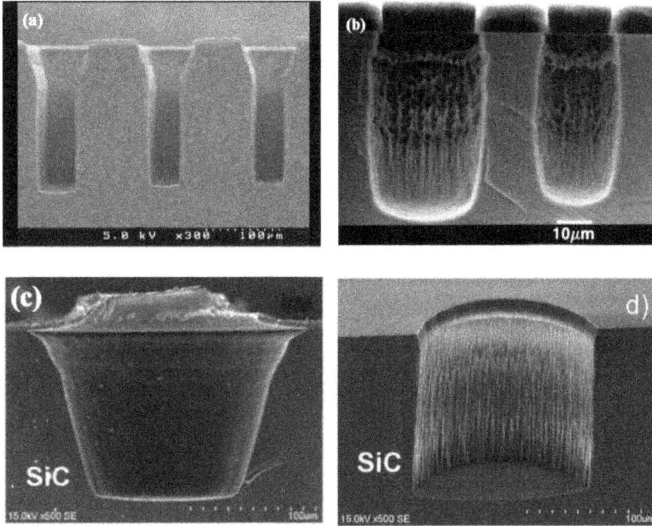

Figure 12.   SEM micrograph of via holes. (a): Mask opening widths of 55, 40 and 40 μm from the left. Reproduced from [58] with the permission of the American Vacuum Society. (b) Close-up of neighbouring via-holes of different diameter, emphasizing the role of aspect-ratio-dependent etch rates. Reprinted by permission from Springer Nature Customer Service Centre GmbH: J. Electron. Mat. [5]. Copyright 2001. Simply mounted (c) and In-bonded (d) SiC samples resulting in different sidewall shape due to the effect of temperature. Images in (c) and (d) are reproduced from [55] with the permission of the American Vacuum Society.

Finally, multi-angle etching [84] or an adapted single-crystal reactive etching and metallisation (SCREAM) like process [76] employing special designed ECR sources [85] can be used to form negatively sloped side-walls or free standing undercut structures.

In summary, the sidewall shape depends strongly on the mode of etching as well as hard mask edge following lift-off. Vertical sidewalls are obtained at high DC self-bias values while sloped sidewalls are obtained at high substrate temperatures.

4.7   Sloped walls from sloped etch mask

Contrary to the above studies in which the sloped walls are formed by the etching conditions, the usual way to form sloped etched sidewalls is by controlling the masking material geometry as in [3, 86]. In these studies, mesas are etched with a slope between

30 and 80° by using a mask of SiO_2 in an SF_6/O_2 plasma. Sloped $(30° \pm 5°)$ patterns in the SiO_2 were defined by photolithography and BHF etching (100 nm/min). By using these sloped etch mask patterns it was possible to obtain sloped SiC sidewalls with an angle given by the following formula (Fig. 13):

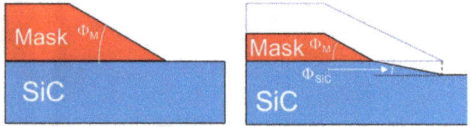

Figure 13.   Sloped etch sidewall geometry.

$$\tan(\Phi_M) = \frac{ER_{SiC}}{ER_{mask}} \tan(\Phi_{SiC}) = S \tan(\Phi_{SiC})$$

where S is the inverse of selectivity between SiC and the mask. The angle Φ_{SiC} is determined by varying the selectivity of the content of O_2 in the gas mixture.

In [87], a special bevel sidewall shape has been proposed for better edge termination of high voltage *pn* junctions. The improved bevel mesa structure, fabricated in this study, has a rounded corner at the mesa bottom, while a nearly vertical sidewall is formed at the junction edge.

4.8   Vertical scratches

Another problem reported in early stages of SiC etching was the appearance of etch sidewalls with a big density of "vertical scratches" (Fig. 14a). Very often it is related to the morphology of the negative photoresist used in the lift-off process of forming the metal mask for etching (Fig. 14b) and/or to the use of Al as masking material [88] since the deposited Al forms large grains.

Indeed, even a small mask roughness can induce an "amplified" roughness of the sidewalls due to the high selectivity of metal masks with SiC [89]. A possible solution is to avoid the lift-off process, but employ ion etching for patterning the metal hard mask resulting in vertical and smooth mask sidewalls [89].

Use of a Bosch process

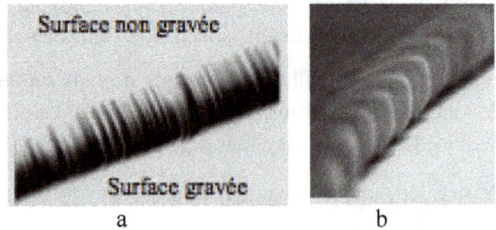

Figure 14.   (a) Etched SiC side walls, (b) Photoresist AZ 5214E after development. Reprinted from [43].

has been reported [80] to address this sidewall roughness problem since during the passivation step the sidewalls of the mask and SiC layer are protected by a polymer.

5. Mask material (adherence, micromasking, selectivity)

The main issues related to masking material in SiC etching are mask selectivity (especially for deep etching) and mask material-induced micromasking. The masks edge geometry is an issue for very deep etching or etching of small diameter (< 200 nm) features.

Generally, photoresist is etched more rapidly than SiC and thus is not a viable RIE mask for SiC etching. Therefore, other materials (mostly metals) have been studied as RIE/ICP masking materials as a high etch selectivity with masking material is required, especially for deep etching applications.

In many cases, use of an Al mask resulted in micromasking (see analysis in Section 4.1 above) and sidewall roughness (see Section 4.8 above).

AlN has also been investigated as a non-metal mask [40, 56] to etch SiC. According to [40], AlN can aid in preventing micromasking defects on the etch surface and degradation in the plasma etch tools. A selectivity of 16:1 (SiC/AlN) has been obtained. Furthermore, nitrogen released from the AlN during the etching can influence the etching profile by increasing the anisotropy and lowering the etch rate as shown in [56].

According to [44] Cu has extremely high (infinite) selectivity against SiC since under certain conditions, "deposition" instead of Cu etching is observed. Indeed, etch products such as copper fluoride are formed on the surface with fluorine in the plasma and improve selectivity. However, these by-products can result in irregular sidewalls and mask adherence problems. This is likely why Cu is not used currently as a SiC etching mask. Another point to not use Cu, as masking material, is the relative high diffusion coefficient compared to other impurities in SiC [90] and its role as deep center.

Reported results on the use of ITO as an etching mask are contradictory. Low [20] and high [3, 24] selectivity against SiC has been reported. Low (< 100 V) DC self-bias has been employed in order to have a selectivity of 50 according to [24]. Moreover, ITO is deposited by sputtering and it is difficult to optimize the mask edge quality following the lift–off.

Ni is mostly used since it has higher selectivity to SiC (~21 [20], >25 [39], 35 [58], ~100 [19]) and less micromasking problems than Al. The higher selectivity has been attributed to the smaller volatility of etching by-products [19]. A problem with the use of evaporated Ni is adherence, which is related to the induced stress between the Ni and

SiC, and this is aggravated in the case of thick and large area Ni layers. A maximum thickness of 100 and 150 nm is recommended for Ni deposited by e-beam and sputtering deposition respectively [42]. Nevertheless, a higher (250 nm) maximum thickness is reported in other studies [4]. A cleaning of the SiC surface (descum) prior to Ni deposition as well as a heating at 120 °C during deposition are mandatory to obtain the above mentioned Ni thickness without adherence problems. Electroplated Ni of several μm is necessary to perform deep etching [45, 58].

Cr also is used instead of Ni as it does not present the adherence problems of Ni and its selectivity is around half that of Ni [19, 91].

In principle, SiO_2 masks eliminate eventual metal contamination of devices and are preferable for device fabrication in industry although vertical sidewall formation can be more challenging due to SiO_2 corner rounding [92]. Etch selectivity is a problem when using SiO_2 since the use of fluorinated gas chemistry results in high SiO_2 etch rates. The etch selectivity (SiC/SiO_2) improved by increasing the ICP power from 800 W to 2 kW and/or the chamber pressure [50] reaching values above 10. Rounded SiO_2 corners and low selectivity were observed at low pressures (< 7 mTorr) while selectivity steadily increased with pressure. This is because the SiC etching rate goes up dramatically, while the SiO_2 etching rate drops with increasing pressure [50]. According to [93], the selectivity of SiC against SiO_2 can be increased to 5–10 or even higher by using slightly oxygen-rich conditions, or by increasing the self-bias. Note that under certain conditions use of SiO_2 masks results in high micromasking [42]. A possible reason for the latter is surface contamination from the CVD-deposited SiO_2 mask. Some authors [10, 47] employ a metal (Ni, Cr or Al) in combination with SiO_2 (see for instance Fig. 9c) to exploit the advantages offered by each material and obtain a better surface morphology.

A series of studies compared various hard masks for the SiC etch for example Ni, Al, Cr, ITO, SnO_2 in [11], SiO_2, Ni, Al, ITO in [20], Ni, Al and Cu in [44]. Table 2 (based on [93], [40], [91], [92]) shows the reported selectivities of various mask materials to SiC.

Table 2. Mask selectivity with 4H-SiC.

Mask	SiO_2	Al	AlN	Cr	Ni	Resist	ITO
F-based	0.8-3	5-30	16	< 40	> 40	< 0.6	10-20
Cl-based	4-15	2-10				< 0.8	3-10

An important point is that etch selectivity against masking material depends on the process conditions and any comparison should be done under identical process settings. In principle, the etch selectivity with the masking material increases when conditions favouring a chemical etching process occur since ion bombardment is not very selective [43]. Thus, the selectivity between the mask and SiC can be increased by increasing the chamber pressure and employing a more "chemical" etching mechanism in RIE-etching [43]. A similar trend has been observed in ICP-etching [40, 55]. The selectivity decreases when DC platen power rises and Ni selectivity has decreased at high DC self-bias values [4]. Reference [50] supports this conclusion.

The effect of oxygen was also investigated with respect to mask selectivity [20, 39, 46]. O_2 was added in an attempt to reduce the mask erosion rate through formation of surface oxides with low volatility [20]. However, there were unacceptably low selectivities of SiC against ITO when including O_2 in the gas mixture. According to [20] Al selectivity was infinite for 50% and higher O_2 content while Ni selectivity decreased with O_2 content. On the contrary, Ni selectivity increased by 45% when adding 5% of O_2 [46]. A similar tendency was observed in [58].

In conclusion, there is consensus that Ni has high etching selectivity against SiC and fewer problems of micromasking. The main concern with Ni is its adherence on SiC and eventual metal contamination in industry-oriented device fabrication. There is also consensus that Al promotes micromasking and exhibits smaller selectivity than Ni. Plasma parameters like pressure, DC self-bias oxygen content may act differently on the selectivity values of ITO and other metals. Selectivity increases by increasing chamber pressure and/or decreasing DC self-bias and a same tendency is likely observed when increasing O_2 content, although there are contradictory results for the latter. SiO_2 is the more industry-compatible mask but the process optimization for obtaining a selectivity higher than 3 is a challenge for any process developer.

6.   Surface conditioning prior-to / following etching

As explained above in the analysis of micromasking arising after deep SiC etching (Section 4.2), a suitable preparation with Ar and O_2 plasma is necessary to reduce or even eliminate pillar formation on the bottom of etched surfaces. The purpose of this cleaning is to remove residual hydrocarbons and water acting as nucleation sites for non-volatile species deposition.

On the other hand, post-plasma etching SiC surface cleaning is necessary as there are various deposits due to the etching process.

Indeed, XPS observations of the etched surface [13, 15] revealed that fluorine- and oxygen-based species remain on the surface. For this reason, an Ar light sputtering or sacrificial oxidation [15] or high-temperature H_2 annealing at 800 °C for 30 min [13] were performed as post-processing treatments on the etched surface. The specific post-etching wet treatments to be performed when an Al hard mask is used have been investigated in [89].

By conducting all the post-processing treatments, a reduction of fluoride atom concentration on the etched surface was observed, compared with that in the case without post-processing. Since fluorocarbon films are inherently volatile at high temperatures, high-temperature H_2 annealing is more effective than the plasma treatment, and indeed it was able to control the fluoride atomic concentration to less than 3 at.%.

7.   Carrier of the SiC sample under etching

As the SiC wafer diameter was limited to 4-inch wafers until recently (2013-2014) a wafer carrier had to be employed even in foundries for performing the plasma etching. This is not anymore an issue for semiconductor industries as they use 6-inch wafers. Nevertheless, a SiC-wafer-carrier is still used in academic laboratories as well as in via-hole formation in SiC substrates.

There are many factors to take into account when choosing the carrier that holds the sample during the etching process. For instance, in "pure" chemical or ion-enhanced-inhibitor etching modes, the wafer carrier significantly influences the chemical reactions and the resulting passivation layer. Thus, the choice of carrier wafer is often dictated by the chemistry of the process. On the other hand, a wafer carrier consisting of the same material as that to be etched is in many cases selected, as it does not induce additional reactions. However, the latter is not possible when loading effects are to be avoided and wafer carriers with low etch rate (or equivalently low reactive species consumption) are chosen in this case. A typical example in the case of Si plasma etching is a Si wafer covered with SiO_2 or Ni. Another important factor is the thermal conductivity of the carrier wafer in order to have efficient heat dissipation. Finally, the wafer carrier can affect micromasking in processes involving heavy ion bombardment.

ICP is a high ion density reactor involving high levels of RF power. Thus, SiC samples under etching, even when working with pieces, need to be mounted on a high thermal conductivity carrier wafer [37, 55] to improve heat dissipation. Otherwise, the sample is essentially thermally floating which in turn affects the etch rate and morphology. The influence of the wafer carrier in SiC ICP etching has been discussed in [48] in which Si, sapphire and quartz wafer carriers were used under identical etching conditions.

Graphite has been investigated as a carrier wafer in various studies. The choice was based on loading effects (less consumption of reactive species by using the graphite in comparison to a SiC carrier [25]) or lower micromasking [37].

Sapphire [59] and Ni-coated sapphire [48] wafers as well as various configurations of Si wafers such as bare ones [27, 33, 42], thermal-oxide-layered [50] or Ni-coated [48, 54] have also been reported.

Various adhesives have been employed such as thermoplastic polymers [27, 37, 54] or thermal release tape [59], or "heat-conductive paste" [5], or silicon-oil [50] or silicone-grease [58].

In summary, there is no unanimously accepted wafer carrier for SiC etching. Some groups use carriers with lower etch rate in comparison to SiC such as Ni- or SiO_2- coated wafers while other groups do not pay attention to this issue as they use bare Si wafers. Nevertheless, all groups agree on the necessity to bond the SiC on the carrier wafer for heat dissipation purposes.

8.   DRIE (Deep RIE) process in SiC: via-hole formation - MEMS

8.1   Continuous etching process

Many details of the SiC single DRIE process conditions and resulting effects on etch rate and morphology of etched areas were reported above (Sections 3.9, 4.2, 4.6). A summary is given below.

High density plasma systems (ICP, ECR, ...) are used for SiC DRIE. In most studies [37, 39, 55, 58, 59, 89], an SF_6/O_2 gas chemistry is used for obtaining maximum etch rate. As previously mentioned, the reason for choosing SF_6 is that this fluoride allows for high etch rates. Moreover, the addition of up to 20% of O_2 further increases the etch rate. It has also been reported that Ni mask selectivity also increases by adding O_2 [58]. Note that many manufacturers of plasma etching systems have optimized and recommend this chemistry for etching SiC. Nevertheless, the addition of oxygen can favour microtrenching and an optimization of the process conditions is necessary to address this issue.

High ICP coil power (near the upper limit of system power) is employed to obtain high radical production and high etch rates [45, 54, 55, 59]. Note, however, that a limitation in the upper ICP coil power value is set by plasma stability and self-heating of the sample under etching. Indeed, in many experimental configurations the samples are attached on carrier wafers not only for the dry etching step but for subsequent process steps too.

Attachment materials cannot withstand temperature exceeding about 80°C after long-time etching [48].

The ICP electrode/platen power seems [39, 59] to be the more important parameter for increasing the etch rate but there is a limitation due to the drop of the selectivity with the Ni mask [4] and the micromasking often observed at high platen power resulting in heavy ion bombardment [37, 59].

Microtrenching and pillar formation on the bottom of the etched hole seems to be reduced at high ICP power [37].

The etch rate drops at high pressures [37]. High pressures promote residue formation [37] while lower pressures increase microtrenching [4, 37] and reduce pillar formation on the bottom of the via-hole [55].

8.2 Bosch process

Silicon carbide is less reactive than other semiconductors (such as Si) in atomic fluorine and much less in atomic chlorine. The Bosch process has not been investigated by many research teams, even from those targeting SiC-based MEMS development, due to the highly anisotropic character of the plasma etching. Indeed it is very difficult, if not impossible, to find etching conditions allowing isotropic plasma etch at room temperature. Therefore, it is difficult to obtain tuneable etch profiles as in the case of silicon where the sidewall can be adjusted from outward to inward sloping by simply increasing the duration of the etch step relative to that of the passivation step. The conclusion is that it may prove more difficult to obtain the same aspect ratios (etch depth divided by minimum feature size) as those realized in silicon.

The [94] is the first study addressing the issue of concave sidewall (bowing - Fig. 15a) formation after the deep etch that creates the long source-finger stripes in SiC-VJFETs by employing a Bosch process. The process utilized alternating steps of polymer deposition from the C_4F_8 source gas, and anisotropic ICP etching by CF_4/O_2. The polymer deposited on the mesa sidewalls provides protection against side etching and therefore makes the Bosch process suitable for deep trench applications. As apparent from Fig. 15d, the concavity of sidewalls is substantially reduced in comparison to the "non-Bosch" etching process.

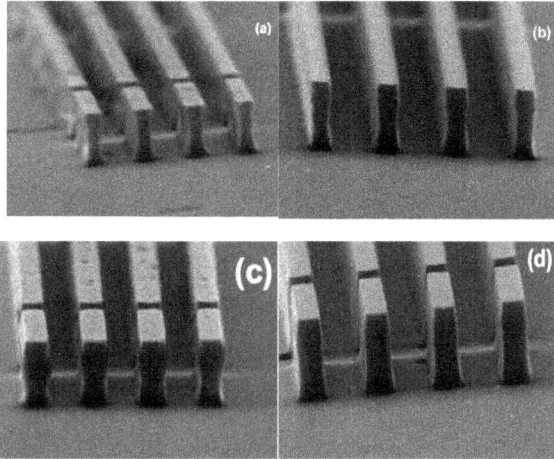

Figure 15. 4 μm deep trench etching in SiC: (a) Regular CF_4/O_2 (40/20 sccm) ICP process 50 V/700 W/7 mTorr, (b) Regular CF_4/O_2 ICP process 100 V/700 W, (c) Bosch process with etching 9 cycles with 100 V DC bias, followed by 50 V DC bias, (d) Bosch process with an etching process including 9 cycles with 100 V DC bias, followed by regular process at 100 V DC bias. Reprinted from [94].

The second study [80] used a Bosch process approach for addressing the issue of microtrenching in the case of ICP SiC etching employing SF_6/O_2 gas chemistry. C_4F_8 was again used for the passivation steps of the cycles and the effects of C_4F_8 flow rate as well as of etch/passivation step duration were investigated. The microtrenching disappears as xthe C_4F_8 flow rate increases. Above the optimum value of 40 sccm, microtrenching appears again probably due to the thick passivation layer formed on the sidewalls. By using a fixed optimum C_4F_8 flow rate (40 sccm), the microtrenching is entirely eliminated when the shortest etching time and passivation time (i.e., t_e = 5 s and t_p = 3 s) were used.

9. Nanopillar/nanowire formation

Few papers have reported the fabrication of nanometre scale SiC structures by the top-down approach [95, 96, 97, 98, 99].

The authors of [95] have demonstrated nano-electromechanical system (NEMS) devices based on top-down formation, and horizontal SiC nanowires (NWs) with a diameter of 55 nm by etching (ECR or ICP) heteroepitaxially grown 3C-SiC layers on silicon. Note that the axis of NWs was parallel to the substrate surface. A second study [96] has

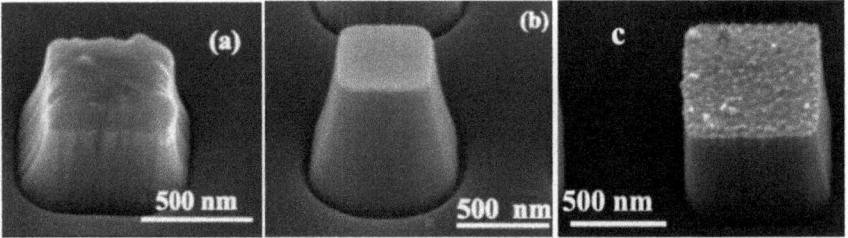

Figure 16.   SEM images of SiC nanopillars formed under H_2 free (a, b) and H_2 adding conditions (c). Etching with Al (a) and Ni (b, c) mask. Adding H_2 results in vertical sidewalls. Fig. 16a and 16b reprinted from [98]. Fig. 16c reprinted from [99].

reported nanopillar structures of SiC obtained by dry ICP-RIE with a self-assembled SiO_2 mask. The maximum achieved pillar height cannot exceed 600 nm due to the low selectivity of SiO_2 with respect to SiC. In addition, as is typical with self-assembled methods, it was not possible to control the pitch distance, diameter of mask, and mask materials.

Other studies [97, 98, 99] have followed a typical top-down approach to form SiC nanopillars with their axis vertical to the surface of the SiC substrate (Fig. 16). $3C$-SiC nanopillars have been formed in [98, 99] while many polytype nanopillars have been formed in [97]. E-beam lithography has been employed for masking the SiC with Ni (30-70 nm in [98, 99] and 110nm in [97] mask diameter). Common tendencies have been found in these studies. A major problem was the low aspect ratio values, which did not exceed 10 (Fig. 17a) for H_2 free chemistries [97, 98], while this value increases to 20 (Fig. 17b) by employing a H_2 containing gas mixture (Ar 20 sccm, SF_6 5 sccm, H_2 10 sccm) [99].

The angle of the sidewall slope [98] or equivalently the aspect ratio [97] increases when the induced DC self-bias increases or when the gas pressure decreases. A noticeable result is that the maximum sidewall slope is around 82 - 84° for H_2 free chemistries, and according to [97] this value fits with the sidewall angle of the masking material after the lift-off (Fig. 18). However, when adding H_2 the slope reached a value close to 90°, which indicates that passivation of sidewalls is crucial for having high slope values or equivalently high anisotropy [99]. A columnar structure with top diameter of 20 nm and a height of 400 nm has been obtained in this case (Fig. 18b). Note that a special chamber incorporating an ECR source (max power 800 W) has been used in [98, 99].

On the other hand, gas additions such as oxygen, as well as other gases like CF_4, CHF_3 and C_2H_4, do not increase the sidewall-slope substantially [98].

It was also observed that the sidewall-slope of the Ni etch mask increases as the etching proceeds (Fig. 18). Various reasons have been proposed for that, such as an increase of the local electric field from the used metal mask and the subsequent strong physical sputtering [97, 82].

In addition, use of Al masks led to strongly rippled edges of the mesa sidewalls [97, 98] quite probably due to the larger grain size of the deposited Al layer, while nanopillars formed by use of Ni masks were free of such sidewall roughness (Fig. 17a and 16b).

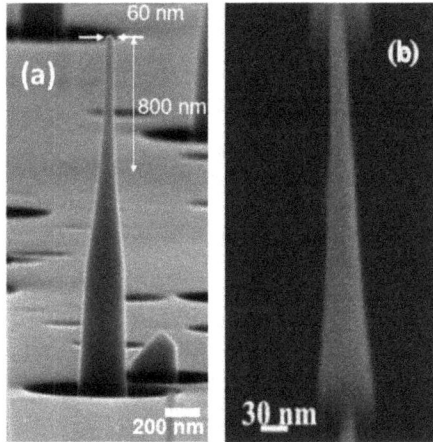

Figure 17. SEM images of SiC nanopillars formed under (a) H_2 free and (b) H_2 adding conditions. Maximum anisotropy under H_2 free conditions (SF_6/O_2 80:20, 300 V DC bias, 6 mTorr, 1500 W RF power). Maximum anisotropy for H_2 based conditions: 20 sccm Ar, 5 sccm SF_6, 10 sccm H_2, bias 200 V, platen power 20 W, 1 mTorr. Fig. 17a reprinted from [97] © IOP Publishing. All rights reserved. Fig. 17b reprinted from [99].

10. Electrical properties after etching

Various studies were devoted to the electrical properties of RIE etched surfaces. Towards this purpose, a Schottky contact was formed on the SiC etched surface in most cases and its electrical characteristics (barrier height, reverse breakdown voltage, leakage current, ideality factor) were employed for the evaluation of the etched surface.

High-energy ion bombardment is considered as the main cause for electrical degradation of etched surfaces and most studies have investigated the effect of DC self-bias. At high DC self-bias, a reduction in the breakdown voltage and ideality factor was observed, which was attributed to increased lattice damage [100]. On the other hand, at low DC self-bias, increased leakage currents have been observed [100] due to a C-rich surface or polymer on the surface not efficiently removed by the chemistry or ion bombardment at reduced DC self-bias. An optimum DC self-bias of -50 V, corresponding to ion energies of ~75 eV, resulted in less electrical degradation [100]. If higher values of DC self-bias

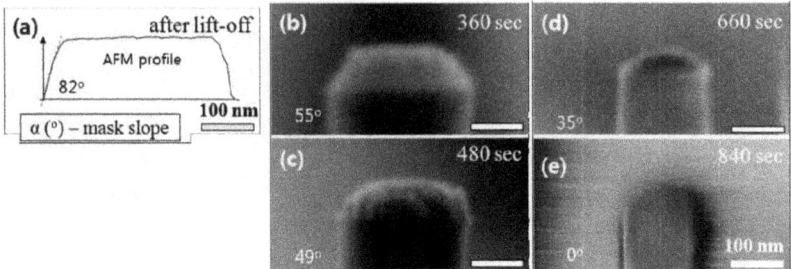

Figure 18. (a) AFM profile after lift-off, and (b-e) magnified SEM images of Ni mask after etching for (b) 360 sec, (c) 480 sec, (d) 660 sec and (e) 840 sec. Same scale from (a) to (e). Reprinted from [97] © *IOP Publishing. All rights reserved.*

are necessary for getting higher etch rates, then the majority of the etching can be performed at higher DC self-bias, and the latter can be reduced toward the end of the process.

The degradation of the electrical characteristics of etched surface has been attributed to a non-stoichiometric surface resulting from excess carbon because of the difference in volatility of the Si-related and C-related etch products [12, 101]. An interesting result was the connection of the conductive character of the etched surface with the type of the C-F bonds (semi-ionic or covalent) created on it after the etching [12]: at higher etch rate, the etched SiC surface becomes less conductive since covalent bonds are dominating in this case.

A group from KTH, Sweden, performed a systematic study related to the electrical properties of etched SiC surfaces [102, 103, 104]. They employed both MOS capacitors and BJTs fabricated on etched surfaces to evaluate best performance according to method of etching (RIE vs. ICP) and post-etching sacrificial oxidation process (dry O_2 vs nitrous (N_2O) ambient). The best performance was achieved by performing ICP etching and dry sacrificial oxidation between 900 and 1250 °C after the etching.

Plasma etching seems also to contribute to the SiC bipolar devices' forward-bias degradation induced by stacking faults as has been shown in a study directly comparing the effect of RIE etching conditions to the electrical characteristics of mesa-isolated 4*H*-SiC *pn* diodes [33, 105]. Two different sets of RIE conditions were employed. The first one is a rather "physical" process with pressure, 200 W power, and a 50/50 SF_6/Ar gas mixture. These conditions resulted in a very energetic 550 V DC self-bias and 100 nm/min of etch rate. The second set is a "chemical process" in pure SF_6, again at

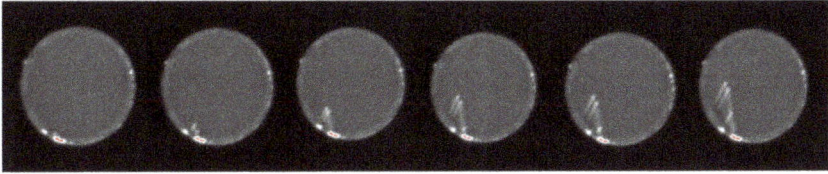

Figure 19.   Defect propagation from mesa wall of 100 μm diodes at 2.8 V forward voltage corresponding to a forward current of 1 mA. Time evolution of photoemission from left to right. Reprinted from [33].

200 W, and under a much higher pressure (> 150 mTorr) in order to keep the DC self-bias at 100 V. The etch rate was 160 nm/min. In the beginning of the electrical stress, luminous spots on the diode's edges were only observed (Fig. 19). However, these spots are the starting points for the glide of the SFs. Only 10% of the diodes processed with the "physical" RIE were able to reach 220 V instead of 280 V, which is the expected value for a bulk avalanche breakdown (Fig. 20). On the contrary, for the diodes processed with a "chemical" RIE, the avalanche breakdown appears now at 280 V for most of the diodes (75%). In addition, the reverse leakage current is reduced by around 1 order of magnitude and more homogeneous *I–V* curves were obtained (Fig. 20). Therefore, the process at high chamber pressure induces much less dislocation formation on the mesa wall leading to the glide of Stacking Faults (SFs) which are responsible for a forward-bias degradation effect.

In summary, low DC self-bias have to be used if the etched surface is part of the device's active area to avoid degradation of electrical characteristics of devices such as premature breakdown, lower ideality factor and forward-bias degradation. Otherwise, a sacrificial oxidation and subsequent oxide removal is necessary. Hopefully, there are experimental conditions to combine low DC self-bias with a high degree of anisotropy in the case of SiC plasma etching.

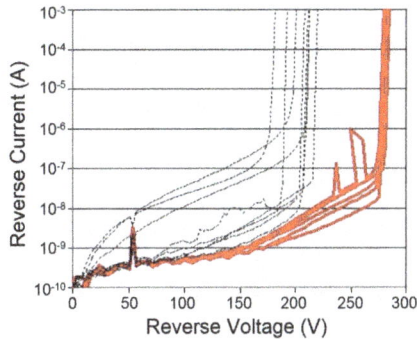

Figure 20.   I–V curves in reverse regime until the avalanche for the diodes fabricated by employing a "physical" RIE (black dashed lines) and a "chemical" RIE (red full lines). Reprinted from [105] © Wiley Materials.

11. Main conclusions

1) The widely accepted belief that compounds tend to be etched atomically seems to apply in the case of SiC, resulting in the formation of a C-rich thin layer (quite probably C-F bonds) on the etched surface due to lower removal rate of C atoms.

2) Cl-based chemistry is not as useful in SiC etching as F-chemistry due to its lower etch rates (etching by-products SiF_x and CF_x are much more volatile than $SiCl_x$ and CCl_x).

3) The SiC etch rate increase for fluorinated gases with smaller bond energy. For reference, the average bound energies of CHF_3, CF_4, SF_6, and NF_3 are decreasing in that order.

4) SF_6-based plasmas are mainly used due to their high etch rates and to the fact that the same gas is mostly used in Si-related technology. The abundance of F atoms in the SF_6 molecule contributes also in the high SiC etch rate.

5) CF_4 is rarely used as it results in lower etch rates than SF_6, and has the issue of polymer formation.

6) NF_3 is not used, as it is highly toxic and expensive.

7) A higher etch rate is typically obtained when a relatively low oxygen percentage is added to the fluorinated gas, but the etch rate value increase is moderate. The optimum oxygen percentage depends on the employed F-containing molecule. The richer in F, the less percentage in oxygen is needed. For very high O_2 percentages, the resulting dilution effects reduce the etch rate.

8) The role of Ar and its effect on the etch rate value is a subject of debate. It seems that the role of Ar is more complicated than the initially considered increase of the sputtering effect with a simultaneous increase of the etch rate. Some studies observed a substantial increase of the etch rate with the addition of Ar while others did not.

9) Fluorine production controls the etch rate in most cases especially in RIE systems. For high-enough fluorine production the etch rate-limiting factor is a surface-based reaction most probably the C-F reaction that produces CF_2 volatiles.

10) Most studies on RIE etching of SiC agree that the increase of the gas pressure induces an increase of the etch rate up to a certain value, and then a decrease beyond that. This behaviour is directly related to the atomic fluorine production in the RIE chamber while surface reactions, increased collisions, and low ion energy have all been proposed as possible causes for the drop of etch rate above a certain pressure.

Only at low pressures, where the sputtering effect is dominant, is a high DC self-bias value necessary for increasing the etch rate value.

11) It seems impossible to extract a single unanimous behaviour of the etch rate versus pressure in the case of ICP systems, as contradictory and specific-to-process-conditions results have been reported.

12) A steady increase of the etch rate with the RF substrate power (or equivalently DC self-bias) has been observed in most studies of SiC RIE etching

13) An increase of the etch rate with the RF coil power has been observed in many studies of SiC ICP etching with the reported factor of increase varying in the 0.6-1.7 range for constant pressure.

14) Decreasing the plasma length by moving the substrate electrode results in an etch rate decrease in the case of RIE and ICP systems, and an etch rate increase in the case of ECR/helicon systems. The reason for this difference is not clear.

15) A substrate holder temperature increase results in a more isotropic etching with an increase of both vertical and lateral etch rates.

16) Covering the Al-based electrode in combination with the use of a Ni mask, which covers relatively small areas of the SiC sample to be etched, as well as use of low DC self-bias values, are the optimum conditions for eliminating micromasking. Low chamber pressure values have to be employed in ICP systems while high-pressure values are preferable in RIE systems.

17) Use of substrates with low density of crystal defects and suitable preparation with Ar and O_2 plasma of the SiC surfaces to be deeply (> 50 μm) etched are necessary conditions to obtain smooth bottoms.

18) There are no unanimously accepted optimum conditions for the suppression of microtrenching especially in the case of the ICP etching. Use of O_2 should be avoided when microtrenching is an issue. Increased ICP source power results in a lower sidewall angle and consequently should decrease or eliminate microtrenching. Increasing chamber pressure in RIE systems eliminates microtrenching.

19) There is consensus that Ni has high etching selectivity and few problems of micromasking. Its main drawback is its adherence on SiC. There is also consensus that Al promotes micromasking and exhibits smaller selectivity than Ni. There are contradictory results on the selectivity values of ITO and other metals. SiO_2 is the more industry-compatible mask but the process optimization for obtaining a

selectivity higher than 3 is a challenge for any process developer. Cl-chemistry can be a solution in the latter case.

20) Low DC self-bias has to be used if the etched surface is part of the device's active area to avoid degradation of electrical characteristics of devices such as premature breakdown, lower ideality factor and forward-bias degradation.

21) Obtaining low diameter (< 80 nm) vertical (top-down) SiC pillars/nanowires by employing plasma etching is still an issue of the related technology.

22) Most of the above conclusions have been drawn from experiments performed on pieces of SiC wafers. Therefore, wafer-level etch effects have not been addressed. These include, but are not limited to, loading effects (the amount of exposed material to etch), and wafer-level variation due to processes like diffusion of the plasma gases.

Acknowledgements

The authors would like to thank Konstantin Vasilevskiy, Nicolas Camara, Antonis Stavrinidis, George Konstantinidis, Thomas Stauden, Lars Hiller and Florentina Niebelschütz for performing plasma etching experiments and corresponding analysis of the results. They also acknowledge Mihai Lazar, Richard Gaisberger, Thierry Chevolleau and Evangelos Gogolides, for valuable comments on the content of the present chapter.

K. Zekentes acknowledges the support of European Commission through the Marie-Curie project SICWIRE.

References

[1]   S.J. Pearton, Dry Etching of SiC, in: R. Cheung (Ed.), Silicon Carbide Micro Electromechanical Systems for Harsh Environments, Imperial College Press, London, 2006, pp. 102-127. https://doi.org/10.1142/9781860949098_0004

[2]   P.H Yih, A.J. Steckl, V. Saxena. A review of SiC Reactive Ion Etching in Fluorinated Plasmas, Phys. Stat. Sol. (b) 202 (1997) 605-642. https://doi.org/10.1002/1521-3951(199707)202:1<605::AID-PSSB605>3.0.CO;2-Y

[3]   J.R. Flemish, Dry Etching of SiC, in: S.J. Pearton (Ed.), Processing of Wide Gap Semiconductors, William Andrew Pub, New York, (2000), pp. 151-177. https://doi.org/10.1016/B978-081551439-8.50006-7

[4]   G. M. Beheim, Deep Reactive Ion Etching for Bulk Micromachining of Silicon Carbide, in: M. Gad-el-Hak (Ed.), The MEMS Handbook, CRC Press, Boca Raton, (2002), pp. 21-1 - 21-12. https://doi.org/10.1201/9781420050905.ch21

[5]   K.P. Leerungnawarat. P. Lee, S.J. Pearton, F. Ren, S.N.G. Chu, Comparison of F_2 plasma chemistries for deep etching of SiC, J. Electron. Mater. 30 (2001) 202-206. https://doi.org/10.1007/s11664-001-0016-0

[6]   A. Miotello, L. Calliari, R. Kelly, N. Laidani, M. Bonelli, L. Guzman, Composition changes in Ar^+ and e^--bombarded SiC: An attempt to distinguisch ballistic and chemical guided effects, Nucl. Instr. Meth. Phys. Res. B 80-81 (1993) 931-937. https://doi.org/10.1016/0168-583X(93)90712-F

[7]   J. Pezoldt, B. Stottko, G, Kupris, G. Ecke, Sputtering effects in hexagonal silicon carbide, Mater. Sci. Eng. B 29 (1995) 94-98. https://doi.org/10.1016/0921-5107(94)04005-O

[8]   G. Ecke, R. Kosiba, J. Pezoldt H. Rößler, The influence of ion beam sputtering on the composition of the near-surface region of silicon carbide, Fresenius J. Anal Chem. 365 (1999) 195-198. https://doi.org/10.1007/s002160051471

[9]   J. Hong, R.J. Shul, L. Zhang, L. F. Lester, H. Choi, Y. B. Cho, Y. B. Hahn, D. C. Hays, K. B. Jung, ,1 S. J. Pearton, C.-M. Zetterling, M. Östling, Plasma Chemistries for High Density Plasma Etching of SiC, J. Electron. Mater. 28 (1999) 196-201, and in J.J Wang, E.S Lambers, S.J Pearton, M Ostling, C.-M Zetterling, J.M Grow, F Ren, R.J Shul, ICP etching of SiC, Solid-State Electron. 42 (1998) 2283-2288. https://doi.org/10.1016/S0038-1101(98)00226-3

[10]   F.A. Kahn, I. Adesida, High rate etching of SiC using inductively coupled plasma reactive ion etching in SF_6-based gas mixtures, Appl. Phys. Lett. 75 (1999) 2268-2270. https://doi.org/10.1063/1.124986

[11]   P. Chabert, Deep etching of silicon carbide for micromachining applications: Etch rates and etch mechanisms, J. Vac. Sci. Technol. B 19 (2001) 1339-1345. https://doi.org/10.1116/1.1387459

[12]   L. Jiang, R. Cheung, R. Brown, A. Mount, Inductively coupled plasma etching of SiC in SF_6/O_2 and etch-induced surface chemical bonding modifications, J. Appl. Phys. 93 (2003) 1376-1383. https://doi.org/10.1063/1.1534908

[13]   H. Mikami, T. Hatayama, H. Yano, Y. Uraoka, T. Fuyuki, Role of Hydrogen in Dry Etching of Silicon Carbide Using Inductively and Capacitively Coupled Plasma, Jpn. J. Appl. Phys. 44 (2005) 3817–3821. https://doi.org/10.1143/JJAP.44.3817

[14] G. Ecke, H. Rößler, V. Cimalla, J. Pezoldt, Interpretation of Auger depth profiles
of thin SiC Layers on Si, Mikrochim. Acta 125 (1997) 219-222.
https://doi.org/10.1007/BF01246186

[15] M. Imaizumi, Y. Tarui, H. Sugimoto, J. Tanimura, T. Takami, T. Ozeki, Reactive
Ion Etching in CF_4 / O_2 Gas Mixtures for Fabricating SiC Devices, Mater. Sci.
Forum, 338-342 (2000) 1057-1060.
https://doi.org/10.4028/www.scientific.net/MSF.338-342.1057

[16] E. Niemann, A Boos and D. Leidich, Chloride-based dry etching process in 6H-
SiC, Inst. Phys. Conf. Ser. 137 (1994) 695-698.

[17] F.A. Khan, B. Roof, L. Zhou, I. Adesida, Etching of silicon carbide for device
fabrication and through via-hole formation, J. Electron. Mater. 30 (2001) 212-219.
https://doi.org/10.1007/s11664-001-0018-y

[18] L. Jiang, N O V Plank, M A Blauw, R Cheung, E van der Drift, Dry etching of
SiC in inductively coupled Cl_2/Ar plasma, J. Phys. D: Appl. Phys. 37 (2004)
1809–1814. https://doi.org/10.1088/0022-3727/37/13/012

[19] H.-K. Sung, T. Qiang, Z. Yao, Y. Li, Q. Wu, H-K. Lee, B-D. Park, W-S. Lim,
K-H. Park, C. Wang, Vertical and bevel-structured SiC etching techniques
incorporating different gas mixture plasmas for various microelectronic
applications, Sci. Rep. 7 (2017) 3915. https://doi.org/10.1038/s41598-017-04389-y

[20] P. Leerungnawarat, D. C. Hays, H. Cho, S. J. Pearton, R. M. Strong,
C.-M. Zetterling, M. Ostling, Via-hole etching for SiC, J. Vac. Sci. Technol. B 17
(1999). 2050-2054. https://doi.org/10.1116/1.590870

[21] P.H. Yih, A.J. Steckl, Effects of Hydrogen Additive on obtaining Residue-Free
Reactive Ion Etching of β-SiC in Fluorinated Plasmas. J. Electrochem. Soc. 140
(1993) 1813-1824. https://doi.org/10.1149/1.2221648

[22] J. Sugiura, W.J. Lu, K.C. Cadien, A.J. Steckl, Reactive ion etching of SiC thin
films using fluorinated gases, J. Vac. Sci. Technol. B 4 (1986) 349-354.
https://doi.org/10.1116/1.583329

[23] R. Wolf, R. Helbig, Reactive Ion Etching of 6H-SiC in SF_6/O_2 and CF_4/O_2 with N_2
Additive for Devise Fabrication, J. Electrochem. Soc. 143 (1996) 1037-1042.
https://doi.org/10.1149/1.1836578

[24] L. Cao, B. Li, J. H. Zhao, Etching of SiC using inductively coupled plasma,
J. Electrochem. Soc. 145 (1998) 3609–3612. https://doi.org/10.1149/1.1838850

[25] J. R. Bonds, SiC Etch development in a LAM TCP 9400SE II System, MSc thesis, Mississippi State Uninversity, USA, 2002.

[26] B.P. Luther, J. Ruzyllo, D.L. Miller. Nearly isotropic etching of 6H-SiC in NF_3 and O_2 using a remote plasma, Appl. Phys.Lett. 63 (1993) 171-173. https://doi.org/10.1063/1.110389

[27] J. B. Casady, E. D. Luckowski, M. Bozack, D. Sheridan, R. W. Johnson, J. R. Williams, Etching of 6H-SiC and 4H-SiC using NF_3 in a Reactive Ion Etching system, J. Electrochem. Soc. 143 (1996) 1750-1753 and in: J. B. Casady, E. D. Luckowski, M. Bozack, D. Sheridan, R. W. Johnson, J. R. Williams, Reactive Ion Etching of 6H-SiC using NF_3, Inst. Phys. Conf. Ser. 142 (1996) 624-627. https://doi.org/10.1149/1.1836711

[28] G. McDaniel, Comparison of dry etch chemistries for SiC. J. Vac. Sci. Technol. A 15 (1997) 885-889. https://doi.org/10.1116/1.580726

[29] J. Bonds, G. E. Carter, J. B. Casady, J. D. Scofield, Effect of electrode spacing on reactive ion etching of 4H-SiC, Mater. Res. Soc. Symp. Proc. 622 (2000) T8.8.1-T8.8.6. https://doi.org/10.1557/PROC-622-T8.8.1

[30] J.D. Scofield, P. Bletzinger, B.N. Ganguly. Oxygen-free dry etching of α-SiC using dilute SF_6: Ar in an asymmetric parallel plate 13.56 MHz discharge, Appl. Phys. Lett. 73 (1998) 76-78. https://doi.org/10.1063/1.121728

[31] J.D. Scofield, B.N. Ganguly, P. Bletzinger, Investigation of dilute SF6 discharges for application to SiC reactive ion etching, J. Vac. Technol. A 18 (2000) 2175-2184. https://doi.org/10.1116/1.1286361

[32] N. Camara, K. Zekentes, Study of the reactive ion etching of 6H-SiC and 4H-SiC in SF_6/Ar plasmas by optical emission spectroscopy and laser interferometry, Sol. St. Electron. 46 (2002) 1959-1963 and in: N. Camara, G. Constantinidis, K. Zekentes, Use of Laser Interferometry and Optical Emission Spectroscopy for Monitoring the Reactive Ion Etching of 6H - and 4H-SiC, Mater. Sci. Forum Vols. 433-436 (2003) 693-696. https://doi.org/10.1016/S0038-1101(02)00129-6

[33] N. Camara, Ph.D. Thesis, 2006, INPG (Grenoble, France - Crete Univ., Heraklion, Greece)

[34] S. Rauf, P.L.G. VentzekIon, C. Abraham, G.A. Hebner, J.R. Woodworth, Charged species dynamics in an inductively coupled Ar/SF_6 plasma discharge, J. Appl. Phys. 92 (2002) 6998-7007. https://doi.org/10.1063/1.1519950

[35] M. S. Brown, J. D. Scofield, B. N. Ganguly, Emission, thermocouple, and electrical measurements in $SF_6/Ar/O_2$ SiC etching discharges, J. Appl. Phys. 94 (2003) 823-830. https://doi.org/10.1063/1.1580197

[36] M. S. So, S. G. Lim, T. N. Jackson, Fast, smooth and anisotropic etching of SiC using SF_6/Ar. J. Vac. Sci. Technol. B 17 (1999) 2055-2057. https://doi.org/10.1116/1.590871

[37] L. F. Voss, K. Ip, S. J. Pearton, R. J. Shul, M. E. Overberg, A. G. Baca, C. Sanchez, J. Stevens, M. Martinez, M. G. Armendariz, G. A. Wouters, SiC via fabrication for wide-band-gap high electron mobility transistor/microwave monolithic integrated circuit devices, J. Vac. Sci. Technol. B 26 (2008) 487-494. https://doi.org/10.1116/1.2837849

[38] J. Xia, Study of plasma etching of silicon carbide, PhD dissertation, (2010), Nanyang University, China.

[39] H. Cho, K. P. Lee, P. Leerungnawarat, S. N. G. Chu, F. Ren, S. J. Pearton, C.-M. Zetterling, High density plasma via hole etching in SiC, J. Vac. Sci. Technol. A 19 (2001) 1878-1881. https://doi.org/10.1116/1.1359539

[40] D. G. Senesky, A. P. Pisano, Aluminium nitride as a masking material for plasma etching of silicon carbide structures, Proc. IEEE 23rd Int Conf on MEMS, (2010) pp 352-355. https://doi.org/10.1109/MEMSYS.2010.5442492

[41] F. A. Khan, I. Adesida, High rate etching of SiC using inductively coupled plasma reactive ion etching in SF_6 - based gas mixtures, Appl. Phys. Lett. 75 (1999) 2268-2270. https://doi.org/10.1063/1.124986

[42] K. Vassilevski, N. Camara, A. Stavrinidis, Report on FORTH's SiC RIE technology, unpublished.

[43] H. Vang, PhD Dissertation, (2006), INSA Lyon, France.

[44] D.W. Kim, H.Y. Lee, B.J. Park, H.S. Kim, Y.J. Sung, S.H. Chae, Y.W. Ko, G.Y. Yeom, High rate etching of 6H–SiC in SF6-based magnetically-enhanced inductively coupled plasmas, Thin Solid Films 447–448 (2004) 100–104. https://doi.org/10.1016/j.tsf.2003.09.030

[45] G. M. Beheim, L. J. Evans, Control of Trenching and Surface Roughness in Deep Reactive Ion Etched 4H and 6H SiC, Mater. Res. Soc. Symp. Proc. 911 (2006) 0911-B10-15. https://doi.org/10.1557/PROC-0911-B10-15

[46] K. Robb, J. Hopkins, G. Nicholls, L. Lea, Plasma sources for high-rate etching of SiC, Solid State Technol. 48(5) (2005) 61-67.

[47] J. Biscarrat, PhD dissertation, (2015), Univ. Tours, France.

[48] Ju-Ai Ruan, Sam Roadman, Cathy Lee, Cary Sellers, Mike Regan, SiC Substrate Via Etch Process Optimization, Proc. CS MANTECH 2009 Conference and in, Ju-Ai Ruan, Sam Roadman, Wade Skelton, Low RF power SiC Substrate Via Etch, Proc. CS MANTECH 2010 Conference.

[49] Ding Ruixue, Yang Yintang, Han Ru, Microtrenching effect of SiC ICP etching in SF_6/O_2 plasma, J. Semiconductors 30 (2009), 016001-1 - 016001-3. https://doi.org/10.1088/1674-4926/30/1/016001

[50] H. Oda, P. Wood, H. Ogiya, S. Miyoshi, O. Tsuji, Optimizing the SiC Plasma Etching Process For Manufacturing Power Devices, Digest CS MANTECH, (May 2015), Scottsdale, Arizona, USA, p.126.

[51] M. Lazar, Technologie pour l'intégration de composants semiconducteurs à large bande interdite, HDR dissertation, Université Claude Bernard Lyon I , (2018)

[52] J. H. Choi, L. Latu-Romain, T. Baron, T. Chevolleau, E. Bano, Hexagonal faceted SiC nanopillars fabricated by inductively coupled SF_6/O_2 plasma method, Mater. Sci. Forum 717-720 (2012) 893-896. https://doi.org/10.4028/www.scientific.net/MSF.717-720.893

[53] J. J. Wang, E.S. Lambers, S.J. Pearton, M. Ostling, C.M. Zetterling, J. M. Grow, F. Ren, R. J. Shul, J. Vac. Sci Technol. A 16 (1998) 2204-2209. https://doi.org/10.1116/1.581328

[54] N. Okamoto, Differential etching behavior between semi-insulating and n-doped 4H-SiC in high-density SF6/O2 inductively coupled plasma, J. Vac. Sci. Technol. A 27 (2009) 456-460. https://doi.org/10.1116/1.3100215

[55] N. Okamoto, Elimination of pillar associated with micropipe of SiC in high-rate inductively coupled plasma etching, J. Vac. Sci. Technol. A 27 (2009) 295-300. https://doi.org/10.1116/1.3077297

[56] Th. Stauden, F. Niebelschütz, K. Tonisch, V. Cimalla, G. Ecke, Ch. Haupt, J. Pezoldt, Isotropic etching of SiC, Mater. Sci. Forum 600-603 (2009) 651-654. https://doi.org/10.4028/www.scientific.net/MSF.600-603.651

[57] F. Niebelschütz, Th. Stauden, K. Tonisch, J. Pezoldt, Temperature facilitated ECR-etching for isotropic SiC structuring, Mater. Sci. Forum 645-648 (2010) 849-852. https://doi.org/10.4028/www.scientific.net/MSF.645-648.849

[58] S. Tanaka, K. Rajanna, T. Abe, M. Esashi, Deep reactive ion etching of silicon carbide, J. Vac. Sci. Technol. B 19 (2001) 2173-2176. https://doi.org/10.1116/1.1418401

[59] K.M. Dowling, E.H. Ransom, D.G. Senesky, Profile Evolution of High Aspect Ratio Silicon Carbide Trenches by Inductive Coupled Plasma Etching, J. Microelectromech. Syst. 26 (2017) 135-142. https://doi.org/10.1109/JMEMS.2016.2621131

[60] M. Lazar, F. Enoch, F. Laariedh, D. Planson, P. Brosselard, Influence of the masking material and geometry on the 4H-SiC RIE etched surface state, Mater. Sci. Forum. 679-680 (2011) 477-480. https://doi.org/10.4028/www.scientific.net/MSF.679-680.477

[61] M. Lazar, INSA Lyon, France (private communication).

[62] S. H. Kuah, P. C. Wood, Inductively coupled plasma etching of poly-SiC in SF_6 chemistries, J. Vac. Sci. Technol. A 23 (2005) 947-952. https://doi.org/10.1116/1.1913682

[63] G.R. Yazdi, K. Vassilevski, J.M. Córdoba, D. Gogova, I.P. Nikitina, M. Syväjärvi, M. Odén, N.G. Wright, R. Yakimova, Free standing AlN single crystal growth on pre-patterned 4H-SiC substrates, Mater. Sci. Forum 645-648 (2010) 1187-1190. https://doi.org/10.4028/www.scientific.net/MSF.645-648.1187

[64] G.R. Yazdi, K. Vassilevski, J.M. Córdoba, D. Gogova, I.P. Nikitina, M. Syväjärvi, M. Odén, N.G. Wright, R. Yakimova, Free standing AlN single crystal growth on pre-patterned 4H-SiC substrates, Mater. Sci. Forum 645-648 (2010) 187-1190. https://doi.org/10.4028/www.scientific.net/MSF.645-648.1187

[65] S. M. Koo, S.-K. Lee, C.M. Zetterling, M. Ostling, U. Forsberg, E. Janzén, Influence of the trenching effect on the characteristics of buried-gate SiC junction field effect transistors, Mater. Sci. Forum 389-393 (2002) 1235-1238. https://doi.org/10.4028/www.scientific.net/MSF.389-393.1235

[66] R.A. Gottscho, C.W. Jurgensen, D.J. Vitkavage, Microscopic uniformity in plasma etching, J. Vac. Sci. Technol. B 10 (1992) 2133-2147. https://doi.org/10.1116/1.586180

[67] I. W. Rangelow, P. Hudek, F. Shi, Bulk micromachining of Si by Lithography and Reactive ton Etching (LIRIE),Vacuum 46 (1995) 1361-1369. https://doi.org/10.1016/0042-207X(95)00027-5

[68] A.C. Westerheim, A.H. Labun, J.H. Dubash, J.C. Arnold, H.H. Sawin, V.Yu-Wang, Substrate bias effect in high-spect ratio SiO_2 contact etching using an inductively coupled plasma reactor, J. Vac. Sci. Technol. A 13 (1995) 853-858. https://doi.org/10.1116/1.579841

[69] S.G. Ingram, The influence of substrate topography on ion bombardment in plasma etching, J. Appl. Phys. 68 (1990) 500-504. https://doi.org/10.1063/1.346819

[70] G. Memos, E. Lidorikis, G. Kokkoris, The interplay between surface charging and microscale roughness during plasma etching of polymeric substrates, J Appl Phys. 123 (2018) 073303-1 – 0733303-9. https://doi.org/10.1063/1.5018313

[71] F. Gerodolle, J. Pelletier, Two-dimensional implications of a purely reactive model for plasma etching, IEEE Trans. Electon Dev. 38 (1991) 2025-2032. https://doi.org/10.1109/16.83725

[72] M. A. Vyvoda H. Lee, M. V. Malyshev F. P. Klemens, M. Cerullo, V. M. Donnelly, D. B. Graves, A. Kornblit, J. T. C. Lee, Effects of plasma conditions on the shapes of features etched in Cl_2 and HBr plasmas. I. Bulk crystalline silicon etching, J. Vac. Sci. Technol. A 16 (1998) 3247-3258. https://doi.org/10.1116/1.581530

[73] A. Burtsev, Y.X. Li, H.W. Zeijl, C.I.M. Beenakker, An anisotropic U-shape SF_6-based plasma silicon trench etching investigation, Microelectron. Eng. 40 (1998) 85-97. https://doi.org/10.1016/S0167-9317(98)00149-X

[74] Y. Nakano, R. Nakamura, H. Sakairi, S. Mitani, T. Nakamura, 690 V, 1.00 mΩ cm^2 4H-SiC double-trench MOSFETs, Mater. Sci. Forum 717–720 (2012) 1069–1072. https://doi.org/10.4028/www.scientific.net/MSF.717-720.1069

[75] H. Koketsu, T. Hatayama, H. Yano, T. Fuyuki, Clearance of 4H-SiC sub-trench in hot chlorine treatment, Mater. Sci. Forum 717–720 (2012) 881–884. https://doi.org/10.4028/www.scientific.net/MSF.717-720.881

[76] J.R. Flemish, K. Xie, Profile and Morphology Control during Etching of SiC Using Electron Cyclotron Resonant Plasmas, J. Electrochem. Soc. 143 (1996), 2620-2623. https://doi.org/10.1149/1.1837058

[77]   F. Simescu, D. Coiffard, M. Lazar, P. Brosselard, D. Planson, Study in trench
       formation during SF_6/O_2 reactive ion etching of 4H-SiC. J. Optoelectron. Adv.
       Mater. 2 (2010) 766-769.

[78]   K. W. Chu, C. T. Yen, P. Chung, C. Y. Lee, Tony Huang, C. F. Huang, An
       Improvement of Trench Profile of 4H-SiC Trench MOS Barrier Schottky (TMBS)
       Rectifier, Mater. Sci. Forum 740-742, (2013) 687-690.
       https://doi.org/10.4028/www.scientific.net/MSF.740-742.687

[79]   Han Ru, Yang Yin-Tang, Fan Xiao-Ya, Microtrenching geometry of 6H–SiC
       plasma etching, Vacuum 84 (2010) 400–404.
       https://doi.org/10.1016/j.vacuum.2009.09.001

[80]   C. Han, Y. Zhang, Q. Song, Y. Zhang, X. Tang, F. Yang, Y. Niu, An Improved
       ICP Etching for Mesa-Terminated 4H-SiC p-i-n Diodes, IEEE Trans. Electron
       Dev. 62 (2015) 1223-1229. https://doi.org/10.1109/TED.2015.2403615

[81]   B.P. Luther, J. Ruzyllo, D.L. Miller, Nearly isotropic etching of 6H-SiC in NF_3
       and O_2 using a remote plasma, Appl. Phys.Lett. 63 (1993) 171-173.
       https://doi.org/10.1063/1.110389

[82]   D.A. Zeze, R.D. Forrest, J.D. Carey, D.C. Cox, I.D. Robertson, B.L. Weiss, S.R.P.
       Silva, Reactive ion etching of quartz and Pyrex for microelectronic applications, J.
       Appl. Phys. 92 (2002) 3624-3629. https://doi.org/10.1063/1.1503167

[83]   S.-N. Son, S.J. Hong, Quantitative Evaluation Method for Etch Sidewall Profile of
       Through-Silicon Vias (TSVs), ETRI Journal 36 (2014) 617-624.
       https://doi.org/10.4218/etrij.14.0113.0828

[84]   X.M.H. Hang, X.L. Feng, M.K. Prakash, S. Kumar, C.A. Zorman, M. Mehregany,
       M.L. Fabrication of suspended nanomechanical structures from bulk 6H-SiC
       substrates, Mater. Sci. Forum 457-460 (2004) 1531-1534.
       https://doi.org/10.4028/www.scientific.net/MSF.457-460.1531

[85]   J. Asmussen, Electron cyclotron resonance microwave discharges for etching and
       thin-film deposition, J. Vac. Sci. Technol. A 7 (1989) 883-893.
       https://doi.org/10.1116/1.575815

[86]   F. Lanois, P. Lassagne, D. Planson, M. L. Locatelli, Angle etch control for silicon
       carbide power devices, Appl. Phys. Lett. 69 (1996) 236-238.
       https://doi.org/10.1063/1.117935

[87]   T. Hiyoshi, T. Hori, J. Suda, T. Kimoto, Bevel Mesa Combined with Implanted Junction Termination Structure for 10 kV SiC PiN Diodes, Mater. Sci. Forum 600-603 (2009) 995-998. https://doi.org/10.4028/www.scientific.net/MSF.600-603.995

[88]   P. Godignon, SiC Materials and Technologies for Sensors Development, Mater. Sci. Forum 483-485 (2005) 1009-1014. https://doi.org/10.4028/www.scientific.net/MSF.483-485.1009

[89]   H. Stieglauer, J. Noesser, G. Bödege, K. Drüeke, H. Blanck, D. Behammer, Evaluation of through wafer via holes in SiC substrates for GaN HEMT technology, Proc. CS MANTECH 2012 Conference, 2012.

[90]   A. Suino, Y. Yamazaki, H. Nitta, K. Miura, H. Seto, R. Kanno, Y. Ijiima, H. Sato, S. Takeda, E. Toya, T. Ohtsuki, J. Phys. Chem. Solids 69 (2008) 311-314. https://doi.org/10.1016/j.jpcs.2007.07.007

[91]   J. Boussey, C. Gourgon, M. Cottat, E. Bano, K. Zekentes (unpublished).

[92]   V. Veliadis, presentation in ECSCRM'16 Tutorial Day.

[93]   T. Kimoto, J.A. Cooper, Fundamentals of silicon carbide technology: growth, characterization, devices and applications, John Wiley & Sons, Singapore, 2014, p. 211. https://doi.org/10.1002/9781118313534

[94]   Y. Li, Design, Fabrication and Process Developments of 4H-Silicon Carbide TIVJFET, Ph. D. Dissertation, Rutgers University, 2008 but the process is better described in M. Su, Power devices and integrated circuits based on 4H-SiC lateral JFETs, Ph. D. Dissertation, Rutgers University, 2010.

[95]   X.L. Feng, M.H. Matheny, C.A. Zorman, M. Mehregany, M.L. Roukes, Low voltage nanoelectromechanical switches based on silicon carbide nanowires, Nano Lett. 10 (2010) 2891-2896. https://doi.org/10.1021/nl1009734

[96]   A. Kathalingam, M.R. Kim, Y.S. Chae, S. Sudhakar, T. Mahalingam, J.K. Rhee, Self assembled micro masking effect in the fabrication of SiC nanopillars by ICP-RIE etching, Appl. Surf. Sci. 257 (2011) 3850-3855. https://doi.org/10.1016/j.apsusc.2010.11.053

[97]   J.H. Choi, L. Latu-Romain, E. Bano, F. Dhalluin, T. Chevolleau, T. Baron, Fabrication of SiC nanopillars by inductively coupled SF_6/O_2 plasma etching, J. Phys. D: Appl. Phys. 45 (2012) 235204-1 - 235204-9 and in: Jihoon Choi, PhD Disseration (2013) Grenoble INP, France. https://doi.org/10.1088/0022-3727/45/23/235204

[98] L. Hiller, T. Stauden, R. M. Kemper, J. K. N. Lindner, D. J. As and J. Pezoldt,
ECR-Etching of Submicron and Nanometer Sized 3C-SiC(100) Mesa Structures,
Mater. Sci. Forum 717-720 (2012) 901-904.
https://doi.org/10.4028/www.scientific.net/MSF.717-720.901

[99] L. Hiller, T. Stauden, R.M. Kemper, J.K.N. Lindner, D.J. As, J. Pezoldt, Hydrogen
Effects in ECR-Etching of 3C-SiC(100) Mesa Structures, Mater. Sci. Forum 778-
780 (2014) 730-733. https://doi.org/10.4028/www.scientific.net/MSF.778-780.730

[100] B. Li, L. Cao, J. Zhao, Evaluation of damage induced by inductively coupled
plasma etching of 6H-SiC using Au Schottky barrier diodes, Appl. Phys. Lett. 73
(1998) 653-655. https://doi.org/10.1063/1.121937

[101] B.S. Kim, J.K. Jeong, M.Y. Um, H.J. Na, I.B. Song, H.J. Kim, Electrical
Properties of 4H-SiC Thin Films Reactively Ion-Etched in SF_6/O_2 Plasma, Mater.
Sci. Forum 389-393 (2002) 953-956.
https://doi.org/10.4028/www.scientific.net/MSF.389-393.953

[102] S.M. Koo, S. K. Lee, C. M. Zetterling, M. Ostling, Electrical characteristics of
metal-oxide-semiconductor capacitors on plasma eth-damaged silicon carbide,
Solid State Electron. 46 (2002) 1375-1380.
https://doi.org/10.1016/S0038-1101(02)00068-0

[103] E. Danielsson, S.K. Lee, C. M. Zetterling, M. Ostling, Inductively coupled plasma
etch damage in 4H-SiC investigated by Schottky diode characterization, J.
Electron. Mater. 30 (2001) 247-252. https://doi.org/10.1007/s11664-001-0024-0

[104] L. Lanni, B. G. Malm, M. Östling, C.M Zetterling, SiC etching and sacrificial
oxidation effects on the performance of 4H-SiC BJTs, Mater. Sci. Forum 778-780
(2014) 1005-1008. https://doi.org/10.4028/www.scientific.net/MSF.778-780.1005

[105] N. Camara, A. Thuaire, E. Bano, K. Zekentes, Forward-bias degradation in 4H-
SiC p^+nn^+ diodes: Influence of the mesa etching, Phys. Stat. Sol. (a) 202 (2005)
660-664. https://doi.org/10.1002/pssa.200460469

Advancing Silicon Carbide Electronics Technology II   Materials Research Forum LLC
Materials Research Foundations **69** (2020) 233-275   https://doi.org/10.21741/9781644900673-5

CHAPTER 5

Fabrication of Silicon Carbide Nanostructures and Related Devices

M. Bosi[1], K. Rogdakis[2], K. Zekentes[3]*

[1] IMEM-CNR, Area delle Scienze 37A, 43124 Parma, Italy

[2]Department of Electrical & Computer Engineering, Hellenic Mediterranean University, Heraklion, 71410, Greece

[3]MRG-IESL/FORTH, Vassilika Vouton, PO Box 1385, Heraklion, Greece and Grenoble INP, IMEP-LAHC, 38000 Grenoble, France

* zekentesk@iesl.forth.gr

Abstract

SiC nanostructures combine the physical properties of bulk SiC with that induced by the reduction of their spatial dimensionality and thus can be considered as a new material offering concrete advantages for various applications. The main effort on SiC nanocrystals (0D) is dedicated towards light emission and reinforcing agent applications. A large variety of methods were employed for SiC nanocrystal fabrication. SiC nanowires (1D) have been investigated for various applications (reinforcing agent, various types of sensors, transistors, field emitters). Both bottom and top down approaches have been used with concrete advantages and disadvantages for each approach. The fabrication methods of SiC nanocrystals and nanowires are described in a comprehensive way. The main effort has been dedicated for both structures on cubic polytype material as it is the more stable at low growth (< 1900 °C) temperature. Applications and device technology are also included in the present review.

Keywords

Nanocrystal, Nanowire, Nanotube, Nanopillar, NanoFET, Catalyst, Bottom-Up Process, Core/Shell Heterostructure, Luminescence, Photoluminescence, Nanoelectromechanical Systems (NEMS), NWFET

Contents

1. **Introduction**..**235**

2. **SiC nanoparticles** **237**

    2.1   Si to SiC conversion based fabrication of SiC nanocrystals 237

    2.2   Chemical vapor based fabrication of SiC nanocrystals......................237

    2.3   Electrochemical and chemical etching based methods for
            SiC nanocrystals formation 238

    2.4   Chemical Synthesis of SiC nanocrystals 239

    2.5   Formation of SiC nanocrystals by laser ablation...............................239

    2.6   Other methods for formation of SiC nanocrystals..............................239

    2.7   Other (non-cubic) polytype SiC nanocrystals formation 240

    2.8   Formation of SiC hollow nanospheres, nanocages and core-
            shell nanospheres 240

    2.9   Luminescence of SiC nanocrystals..240

    2.10   Applications of SiC 0D nanostructures 242

3. **Bottom-up growth of SiC nanowires and nanotubes**...............................**242**

    3.1   General description of bottom-up NW growth....................................242

    3.1.1 Vapor-Liquid-Solid process 244

    3.1.2 Vapor-Solid process..244

    3.1.3 Solid-Liquid-Solid process 245

    3.2   SiC nanowire growth without a template 245

    3.2.1 Catalysts..246

    3.2.2 Precursors and processes for 3*C*-SiC nanowire growth 246

    3.3   Template assisted SiC nanowire growth 248

    3.3.1 CNTs conversion into SiC NW and NT..248

    3.3.2 Si NW conversion into Si/SiC core/shell NW, SiC NT and
            SiC NW..249

    3.4   Conclusions on SiC NW bottom-up formation 251

4. **Top-down formation of SiC NWs**...**253**

5. **Processing technology of SiC NW based devices** **255**

6. **Functionalization of SiC nanostructures** **256**

7. **Applications of SiC nanowires**...**256**

Conclusions..**259**

Materials Research Forum LLC
https://doi.org/10.21741/9781644900673-5

Acknowledgements **260**

References **260**

1.   Introduction

Nano-materials represent a broad class of materials in which at least one of their dimensions is confined at the nanoscale (<100 nm). They may be characterized as 0D (quantum dots, nanocrystals, nanospheres), 1D [nanotubes (NTs), nanowires (NWs)] or 2D (flakes, ribbons) depending on the number of dimensions extending beyond the nanometer range. For instance, NWs are wire shaped materials with diameter typically in the range of 10-100 nm, while being spatially unconstrained along the axial direction. Over the past few decades, nanomaterials have received much interest because of their novel characteristics including magnetic, optoelectronic and thermal properties resulting from quantum spatial confinement at low diameters as well as the high surface to volume atoms ratio. At such scales, the surface states of the material embark on having an important role on the electronic properties such as charge transport and trapping, energy band-structure, density of states etc.

SiC is known [1] for its advantageous physical properties such as the wide energy bandgap, the high thermal conductivity, the high breakdown electric field, the high Young's modulus and hardness, the high melting temperature and the excellent chemical and physical stability. Moreover, various in vitro and in vivo studies [2, 3, 4] have shown that this material is suitable for use in biomedical devices as it exhibits an excellent biocompatibility, better than that of Si.

Therefore, devices based on SiC nanostructures could present concrete advantages compared to bulk structures, as a result of the superior properties of SiC which are expected to be significantly enhanced at the nanoscale.

Bulk SiC crystallizes under many different forms called polytypes. The cubic (β or $3C$-SiC) polytype is the most stable at low growth temperatures [5] despite the fact that it is not the most stable from the thermodynamic equilibrium point of view. Therefore, $3C$-SiC nanostructures have been reported in most bibliographic studies and in the following text the non-mention of the polytype will correspond to the cubic polytype. Otherwise the polytype will be explicitly mentioned.

In this chapter, the state of the art of SiC nanostructures will be presented in terms of fabrication, processing and device integration technology. The targeted readership is firstly, new-comers in this very active research field and secondly, already active researchers in the field wishing to have an overview of the corresponding literature to better plan their future research.

SiC 0D structures have been the subject of intensive research due to their potential applications in optoelectronics, and are especially used as nanoscale UV light emitters. SiC is an indirect band gap semiconductor with a consequently weak optical emission. However, the emission intensity is significantly enhanced in the case of SiC nanoparticles. Indeed, when the dimensions are reduced to the quantum limits radiative recombination rates are highly enhanced while non-radiative recombination rates are substantially suppressed [6]. The SiC nanoparticles can be distinguished in solid nanocrystals, hollow nanospheres, hollow nanocages and core–shell nanospheres. SiC nanoparticles with a dimension smaller than 10 nm are called SiC quantum dots, because of the presence of quantum effects like the carriers quantum confinement. Multiple methods have been employed for the fabrication of SiC 0D structures. Detailed reviews on SiC nanocrystal fabrication and characterization can be found in [7, 8].

SiC NWs have been widely investigated as they are suitable for many applications [9]. Therefore, a substantial part of the present chapter is dedicated to their fabrication methods that can be categorized in two main branches: top-down and bottom-up, depending on the structure of the initial material being either bulk (top-down) or nano-scaled/molecular (bottom-up).

The main distinction among the bottom-up techniques is whether they incorporate or not a nano-scaled template. The original approach for preparing SiC NWs was based on using carbon NTs as a template for converting them into SiC NWs under the presence of Si vapor. Later on, Si NWs were also used as templates towards their conversion into SiC NWs by high temperature reaction with carbon sources. Apart from the template-assisted methods, extensive studies were performed using standard NW bottom-up growth techniques (owing to the knowledge acquired from Si NW growth) such as Vapor Solid (VS) or Vapor Liquid Solid (VLS) growth. These non-templated techniques rely on the reaction of Si and carbon sources on the molecular level with (VLS) or without (VS) catalyst presence. Throughout this chapter, for simplicity reasons, we will refer to the template-assisted bottom-up growth techniques as conversion experiments (Section 2.3), and the non-templated as bottom-up growth (Section 2.2).

There are substantially fewer studies on top-down fabrication of SiC NWs. E-beam lithography and subsequent dry etching is the commonly employed technology approach in this case. A pyramidal shape has been obtained when a low diameter (<80 nm) and/or long enough (>500 nm) NWs were targeted. Possible reasons for the resulting pyramidal shape are ion divergence, inclination of hard-mask-sidewalls and hard-mask-lateral-etch-rate.

2.   SiC nanoparticles

2.1   Si to SiC conversion based fabrication of SiC nanocrystals

Cubic SiC nanoparticles with size varying from 5 to 200 nm, depending on the growth conditions, have been formed by conversion of Si substrates to SiC in a molecular beam epitaxy system [10]. A flux of solid carbon has been employed for the conversion of the heated Si substrate to SiC.

On the contrary, C_{60} or methanol has been employed in [11] and [12], respectively, for the conversion process.

A linear alignment of self-assembled, cubic SiC dots grown by molecular beam epitaxy on Si substrates is demonstrated [13] by exposing Si (111) to a solid carbon flux under suitable conditions (Fig. 1).

However, this conversion-process (or carbonization, or carburization) approach for fabricating SiC nanocrystals (NCs) has not been intensively studied as simpler methods, described in the next paragraphs, have been developed.

2.2   Chemical vapor based fabrication of SiC nanocrystals

Cubic SiC powder (10–20 nm in size) have been synthesized by radial injection of silane, methane, and hydrogen into the tail flame of a radiofrequency plasma [14].

Similarly, SiC powder smaller than 10 nm have been obtained by thermal decomposition of silicon organic precursors at reduced pressure [15].

The growth of 3C-SiC NCs on various substrates was realized from a mixture of C_2H_2 and SiH_4 via laser pyrolysis by various groups [16, 17]. Flame temperatures as high as 1100 °C have been used. Ultra-low diameter (around 3 nm) cubic SiC NCs are produced, but in mixture with larger diameter NCs of SiC and also Si.

A gas phase method, based on atmospheric pressure plasma and Tetramethylsilane (TMS) as precursor, resulted in the synthesis of highly crystalline ultra-small (down to 1.5 nm), free-standing SiC NCs [18].

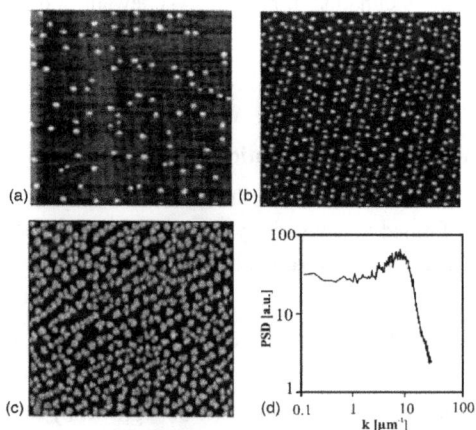

*Figure 1.   AFM images (5×5 μm²) of SiC layers grown on (111)Si with 0.06 °
offcut angle at 925 °C, (a) after 2 s, (z-scale 2.8 nm), (b) after 11 s (z-scale 8.6 nm),
and (c) after 180 s (z-scale 31 nm), and (d) power spectral density of image (b).
Reprinted with permission from John Wiley and Sons [13].*

2.3   Electrochemical and chemical etching based methods for SiC nanocrystals formation

The electrochemical etching of SiC polycrystalline wafers is characterized by monodispersity and regular shape with small sizes of the resulting SiC NCs [19] (Fig. 1). In this case, a polycrystalline 3C-SiC wafer is electrochemically etched using HF and ethanol under UV light illumination. Sonication in water removed the amorphous coverage and yielded a solution of suspended SiC NCs with around 3.5 nm diameter. The SiC NCs prepared in this way exhibited excellent luminescent properties (see section below). The disadvantages of the above method are the relatively high price of the involved polycrystalline SiC wafers and the low yield of NCs limited by the surface areas of the wafers.

A simpler method is the direct chemical etching of cubic SiC powder with micrometer grain size [20]. A mixture of HNO_3 and HF is used for the etching. An ultrasonic treatment is necessary at the end of the process to remove the amorphous SiC shell covering the cubic SiC NCs with diameters smaller than 6.5 nm. Thanks to its simplicity, various groups have adopted this method for the production of SiC NCs.

2.4   Chemical Synthesis of SiC nanocrystals

The chemical synthesis of SiC NCs presents an inherent difficulty due to the high bond strength between Si and C atoms. Conversely, very high temperature is required for performing such a task.

SiC NCs around 10 nm of diameter have been obtained by dissociation of triethylsilane at 1000 °C [21]. Similar diameter SiC NCs have been synthesized by nanocasting and carbothermal reduction [22].

Henderson and Veinot [23] proposed a straightforward method for SiC NCs preparation. Phenylsiloxane polymer has been prepared by heating $C_6H_5SiCl_3$ and $SiCl_4$, followed by the polymer hydrolysis and the co-condensation to form a C-rich matrix with embedded SiC NCs. The final 10 nm SiC NCs have been obtained by oxidation and etching of the matrix.

2.5   Formation of SiC nanocrystals by laser ablation

Laser ablation in two different configurations has been employed for the production of SiC NCs.

In the first configuration [24], Si wafers immersed in a solution of ethanol and toluene (7:1 volume ratio) have been laser ablated to obtain a mixture of 3C-SiC and Si NCs of few nanometers diameter. The latter have been removed by immersion in HF and H_2O_2 solution. Ring-like $3C$-SiC nanostructures have also been observed in samples prepared with the same procedure, however in small proportion with respect to quasi spherical SiC nanostructures.

Laser ablation of polycrystalline SiC in water resulted in the formation of 10 nm in size SiC NCs [25].

2.6   Other methods for formation of SiC nanocrystals

A variety of other methods have been proposed for the fabrication of SiC NCs.

Room temperature, high-energy ball milling of elemental Si and C mixtures resulted in the synthesis of nanosized (8 nm) crystalline SiC powders [26].

For all the above-mentioned methods, the obtained SiC solid NCs were usually pseudo-spherical. Dasog *et al.* have obtained SiC NCs with well spherical shapes by the method of solid-state metathesis reaction among SiO_2, Mg and C powders [27]. The diameter of the spheres ranged from 50 up to 300 nm.

Materials Research Forum LLC
https://doi.org/10.21741/9781644900673-5

2.7 Other (non-cubic) polytype SiC nanocrystals formation

Since the cubic polytype is the most stable at low growth temperatures, the vast majority of the above-mentioned studies dealt with the formation 3C-SiC NCs. Obviously for obtaining other polytype NCs, the most suitable approach is to etch crystals of the same polytype [28, 29]. An electrochemical etching followed by an ultrasonic treatment has been employed in this case. The optical properties (photoluminescence, PL) have been mainly investigated.

2.8 Formation of SiC hollow nanospheres, nanocages and core-shell nanospheres

SiC hollow nanospheres have been investigated due to their potential applications (drug delivery cells, lightweight fillers and catalysts). Various methods have been employed towards their fabrication.

3C-SiC hollow cubic nanocages were formed by a simplified Yajima process [30]. The obtained polycrystalline nanocages had edge lengths of 60–400 nm.

Hollow nanospheres have been obtained by sodium [31] or Na-K [32] reduction of $SiCl_4$ and C_6Cl_6 at 600–700 °C and 130 °C, respectively.

Hollow spherical nanocrystals have been obtained by a template-mediated VS reaction in [33] and in [34] by employing C NCs and SiO vapor in the first case, as well as solid C and Si in the latter one.

Core-shell nanospheres are produced either by surface modification of SiC nanospheres [35,36] or by forming simultaneously cores and shells [37].

2.9 Luminescence of SiC nanocrystals

Quantum confinement effects have been observed in colloidal 3C-SiC NCs of 1 to 6 nm diameter [19]. Fig. 2 shows the main results of this study. The confinement quantum-effect is expressed through the blue-shift of the energy band-gap (E_G) PL peak when the excitation source energy is enhanced. With decreasing excitation wavelengths from 490 to 320 nm, the PL spectra were shifted from 540 to 450 nm. This shift has been explained as a consequence of NCs size distribution. The E_G widens and the luminescence blue-shifts as SiC NCs become smaller. The emission intensities were very strong, enabling emission spots to be visible even by naked eyes (Fig. 2b).

As in the Si case, two PL mechanisms were proposed for the SiC NCs: conduction-to-valence band radiative transitions and radiative transitions via surface states. The two mechanisms can be distinguished by variation of the NC size. The mechanism of radiative transitions via surface states will exhibit a relatively fixed emission wavelength.

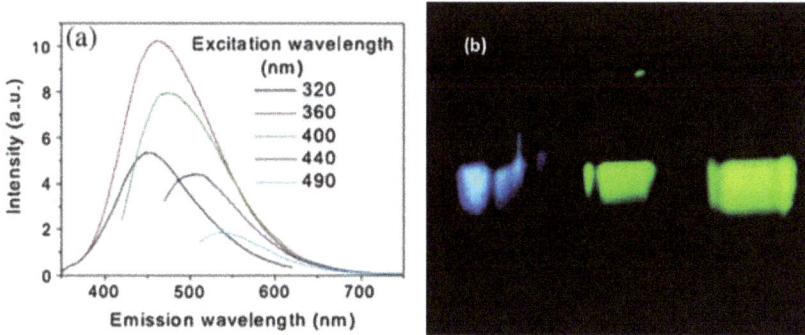

Figure 2.   *(a) PL spectra of SiC NCs taken under five different excitation wavelengths, and (b) emission photos of SiC NCs under excitation with three different wavelengths (from left to right): 320, 400 and 450 nm. Reprinted with permission from [19]. Copyright 2005 by the American Physical Society.*

Many results on PL of 3C-SiC NCs have been reported. For instance, close-packed 3C-SiC NC film exhibited blue to near-ultraviolet (UV) emission due to the quantum confinement effect [38].

The observation of quantum confinement effect in the case of hexagonal SiC NCs is subject of contradictory reports. According to [7] and [29] there is no definite evidence of quantum confinement effect due to the very low value of the Bohr radius. Indeed, the exciton Bohr radii are evaluated to be 2.0, 0.7, and 1.2 nm for 3C-, 6H-, and 4H-SiC, respectively [7]. Thus, very small diameter hexagonal SiC NCs are needed in order to observe the quantum confinement effect in the PL of hexagonal SiC NCs. However, according to Botsoa *et al.* [39] the quantum confinement has been observed in their measurements by above-E_G PL and the observation of energy sub-bands in absorption spectrum. Inline with these observations, the surface states non-radiative and radiative PL has been screened in the case of colloidal solution by employing a suitable polar wetting medium (e.g. ethanol) and by selecting the smaller diameter (< 2 nm) NCs through centrifugation.

The potential of SiC NCs as light emitters is the main application targeted for this class of materials. Indeed, the quantum yield of 3C-SiC quantum dots suspended in water is 17% [40], which is comparable to that of some direct E_G semiconductors. Light Emitting Diodes (LEDs) have been fabricated by using porous SiC [41] and polymer-encapsulated

3C-SiC quantum dots [42]. However, the quantum efficiency was very low, revealing the premature character of the related technology.

The above-mentioned tunable emission and high fluorescence in combination with the stability of SiC in water solutions as well as its high biocompatibility and low cytotoxicity renders SiC NCs ideal candidates for bioimaging probes. Various studies have investigated SiC towards this application [40, 43, 44, 45]. The results are very promising but the corresponding technology is far away from the production of commercial probes.

2.10   Applications of SiC 0D nanostructures

Apart from the aforementioned applications related to the PL properties of SiC 0D nanostructures, there is a series of other applications where they can be applied.

SiC is a well-known tough material and SiC powders are used since many decades in applications requiring hard materials. SiC NCs have been used as reinforcing agents of Mg [46], Al_2O_3 [47], nanocrystalline Ni matrix [48] and epoxies [49].

SiC NCs have been also investigated as material for biosensors [50, 51] and for photocatalyst as well as electrocatalyst [52]. The chemically inert, nontoxic, and biocompatible character of SiC is the main factor for the former application while the large E_G, the excellent thermal and chemical stability as well as the large specific surface area of NCs are the advantages for the latter application.

SiC exhibits high dielectric polarization in the microwaves regime, resulting in excellent microwave absorption performance. In addition, the combination with the high thermal and chemical stability and low density it makes it ideal for being used as microwave absorber. However, only room temperature related investigation of SiC NCs as microwave absorbers has been performed until now with promising results [53, 54].

3.   Bottom-up growth of SiC nanowires and nanotubes

3.1   General description of bottom-up NW growth

Bottom-up growth of NWs refers to a synthesis process in which the nano-sized structures are assembled starting from the constituent atoms, which are delivered either by solid or gaseous precursors.

During a bottom-up process, a NW generally grows along the crystal direction that minimizes the surface free energy of the semiconductor. For a zinc-blend and diamond crystal, as is the case of cubic SiC, this is usually the [111] direction [55], although other directions such as [001], [110] and [112] were reported, depending mainly on the NW

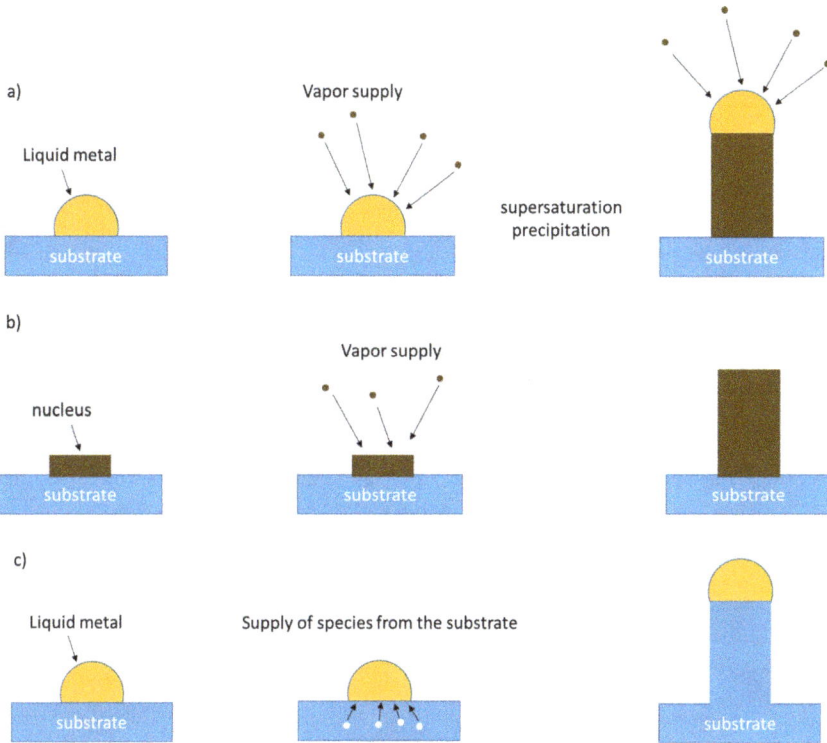

Figure 3.   *Schematic illustration for nanowire growth by the Vapor-Liquid-Solid (a), Vapor-Solid (b), Solid-Liquid-Solid (c) mechanisms.*

diameter [55]. The most common processes for bottom-up NW growth are the VLS, VS and solid-liquid-solid (SLS), depending on the phases present during the growth (Fig. 3). In the following section, a brief description of these basic mechanisms will be given.

While the top-down approach can in principle lead to a higher degree of freedom in NW engineering (strict control of density, position, integration with other fabrication processes, etc.), the bottom-up approach allows a facile and rapid synthesis of NW at high density and bulk quantity (usually in a felt form). It should be also considered that bottom-up grown of NW do not suffer from crystal structure damage caused by violent etching processes or e-beam lithography used in top-down methods. Moreover, the size of the nanostructures obtained by the bottom up approach is not limited by the resolution of the lithographic or etching techniques employed. Besides, the high substrate coverage of this cheap and easy process could potentially open the path for a scalable mass production with high yield.

3.1.1 Vapor-Liquid-Solid process

VLS is the most widespread technique used for the synthesis of semiconducting NW of various materials such as Si, Ge, III-V, SiC and III-N. The NW growth is assisted by a metal deposited on the substrate, which forms a eutectic with the substrate itself and thus remains in the liquid phase even at temperatures, below the bulk metal melting temperature. As the substrate temperature increases, the alloy formed by the metal/substrate system melts and forms small liquid droplets, due to the surface tension of the liquid: this process is usually called "dewetting". The precursors for the NW growth are delivered in the gas phase and are absorbed inside the liquid droplet until the saturation point is reached. Once the supersaturation is achieved, a solid-phase compound precipitates and nucleates at the liquid/substrate interface, thus initiating the growth of the NW. The NW elongates starting from the liquid/substrate interface, as more precursors are absorbed and precipitate in the liquid. The metal droplet always remains at the top of the growing NW. At the end of the growth, once the temperature is decreased, the liquid droplet solidifies and remains at the top of the NW. In this process the NW diameter is mainly controlled by the dimension of the liquid droplet, which is determined by the thickness of the initial catalyst metallic layer and by the dewetting parameters. The detailed description of the VLS growth mechanism is beyond the scope of this book, but a comprehensive review could be found in [56].

3.1.2 Vapor-Solid process

For this growth mechanism the presence of a metallic catalyst on the substrate is not needed, since the gaseous precursors react at or near the substrate to directly synthesize the NW. The catalyst for the NW growth is usually given by nanometric nuclei that are created in situ during the process, originating to the various reactions in the vapor phase. The shape of the nuclei is determined by the competition between the surface energy minimization and the kinetic conditions. For this reason, the control of the NW nucleation, density and spatial position may be difficult. In VS process the diameter and the morphology of the NW could be controlled by tuning the temperature and the precursors partial pressure.

3.1.3 Solid-Liquid-Solid process

SLS growth is another process mediated by the presence of a metallic catalyst, but in this case it always remains on the surface of the substrate. One of the precursor atoms needed for the NW growth, usually Si, is provided by the substrate itself. A formation of a metal-Si eutectic is observed, similarly with VLS. The main difference between SLS and VLS is that Si atoms continuously diffuse through the substrate–liquid (S–L) interface from the substrate side, providing the Si atoms needed for the growth. In this case the typical metal droplet at the end of the NW observed in VLS process is not present. Different orientation and polytypes are obtained with SLS mechanism [9] and usually the resulting NWs have a core/shell structure since the growth is mediated by the intermediate production of an oxide layer.

3.2   SiC nanowire growth without a template

A wide variety of $3C$-SiC NWs morphologies have been obtained by means of non-templated bottom up growth methods [9]. This includes nanoneedles with felted, curly, and straight shape [57], flowers [58], SiC hierarchial nanostructures [59], ultralong NWs [60] and $3C$-SiC / SiO$_2$ core/shell NW [61]. Most of the reported work is based on VLS or SLS techniques, employing vapor phase technique and different precursors such as silane, propane, methytrichloro silane, carbon monoxide and carbon tetrachloride.

The non-templated bottom-up growth of $3C$-SiC NW can be divided in two categories:

a) growth of pure $3C$-SiC NW the surface of which may be covered by a very thin layer of native oxide

b) growth of core/shell $3C$-SiC/SiO$_2$ NWs, where the crystalline $3C$-SiC core is covered by a SiO$_2$ shell some tens of nanometres thick. It is important to point out that the oxide is formed during the growth process, and it is not due to an oxidation process occurring after the growth or ex-situ. In some cases the SiO$_2$ shell could also contain C impurities and, in order to remove it, a simple HF etching could not be effective.

Pure $3C$-SiC or core/shell NWs can be obtained by choosing the appropriate precursors and catalyst. The nature of the nanostructure is mainly determined by the kinetic of the reactions and by the supersaturation conditions in the gas phase. In the following section a brief description of the most common catalyst used for the $3C$-SiC NWs growth will be given, followed by details regarding the most common precursors and growth processes adopted.

Advancing Silicon Carbide Electronics Technology II   Materials Research Forum LLC
Materials Research Foundations **69** (2020) 233-275   https://doi.org/10.21741/9781644900673-5

3.2.1 Catalysts

The catalyst is playing a major role in the NW growth process. Considering that $3C$-SiC is usually obtained at temperatures between 1100 °C and 1380 °C, the most common metals used are Ni and Fe, because their eutectic temperature with Si (966 °C and 1207 °C, respectively) is close to the growth temperature of the NWs [62]. A layer between 2-5 nm of Ni or Fe is usually deposited on the Si substrates by means of evaporation or sputtering. Achieving a good, uniform, controlled and reproducible thickness of the catalyst layer is the first condition towards a control of the VLS or SLS process. The catalyst could also be deposited on the substrate from a solution, such as nickel nitrate ($Ni(NO_3)_2$) or iron nitrate ($Fe(NO_3)_3$) dissolved in ethanol [61, 63, 64, 65]. In this case, a surfactant (oleylamine) can be added to the solution in order to achieve a better wettability on the Si substrate [64]. Since the NW growth with VLS or SLS processes occurs only in the presence of a catalyst, it is possible to pattern the substrate and deposit the metal only in selected regions of the surface. In this way, a selective growth of the NWs only in certain area of the substrate is possible. Moreover, selective growth of SiC NWs was also demonstrated on Si microelectromechanical structure (MEMS). In this case, the metal catalyst was deposited onto the Si MEMS structure after their release [66].

As it has been mentioned, $3C$-SiC is biocompatible and hemocompatible and is therefore a promising material for in vivo bio-devices. In order to promote and foresee biocompatible applications of $3C$-SiC NWs, it is mandatory to develop a system in which every material employed is non-toxic. Nickel used as catalyst for NWs growth could pose some problems due to its cytotoxicity for certain in-vivo application in human body. For this reason it is important to consider alternative materials as catalyst, such as iron, since it is reported to be less cytotoxic with respect to Ni [67, 68, 69].

3.2.2 Precursors and processes for $3C$-SiC nanowire growth

Gaseous precursors are the most convenient ones for vapor phase growth, since their flow and partial pressures could be controlled with great accuracy by standard mass flow meters. Usually, the carrier gas employed for $3C$-SiC growth is H_2 but also inert gas such as N_2 are used, especially for the growth of core/shell structures. On the other hand, growth of NWs by means of powders could bring some degree of irreproducibility mainly due to powder granularity difference between different batches of reagents or milling techniques. The most common precursors for the $3C$-SiC thin film growth are silane and propane. They were also used in a standard Vapor Phase Epitaxy (VPE) reactor to synthesize pure $3C$-SiC NWs with VLS method, using Ni and Fe as catalysts [65, 70]. In this process, the procedure and the reactions are quite straightforward: the substrate is first heated to 1100-1200 °C (depending on the eutectic temperature of the substrate/metal system) and then, after some minutes in which the dewetting occurs, SiH_4 and C_3H_8 are injected with a fixed ratio, depending on the reactor geometry. Growth time lasts typically some minutes, since the growth of the NWs is usually an "explosive" process and the NW grow up rapidly to the final length. The final dimension of the NWs is usually independent from the growth time.

The use of two gaseous precursors for the SiC synthesis (SiH_4 and C_3H_8) allows the optimization of the C/Si ratio. In order to simplify the process and to reduce the parameters, single precursors with fixed C:Si ratio such as methyltrichlorosilane (MTS, CH_3Cl_3Si) [71], Dichloromethylvinylsilane ($CH_2CHSi(CH_3)Cl_2$) or diethylmethylsilane ($CH_3SiH(C_2H_5)_2$) [72] were proposed. Polycrystalline SiC NWs were obtained by using diethylmethylsilane, while dichloromethylvinlysilane permitted to get narrow $3C$-SiC NWs, with smooth surface and good crystallinity. Carbon monoxide was employed as precursor to synthesize crystalline $3C$-SiC/SiO_2 by VLS [61]. At 1100 °C an ensemble of roughly aligned rods was obtained, with diameters up to 2 μm and lengths of about 40 μm. At 1050 °C a dense network of long, interwoven fibers was obtained, with length of several hundreds and uniform diameters below 80 nm. In all the cases, the SiC NWs were coated by a shell of about 20 nm of SiO_2. Alternatively, CO could also be obtained by carbothermal reduction of WO_3 by C in a reductive environment [73]. The result of this process is crystalline SiC NWs covered with an amorphous SiO_2 shell. In all these cases Ni was used as catalyst.

Si atoms could also provided by the substrate but the use of Si powders with expandable graphite or a mixture of milled Si and SiO_2 powders was also demonstrated. In the first case [74], a catalyst-free process is described: a C source with porous structures is first intercalated by oxidants compound such as sulfuric acid, potassium permanganate. The resulting "expandable graphite" can increase its volume hundreds of times when heated, thus forming a porous structure with excellent absorption capacity for reactant gas. Expandable graphite and Si powders were then used to prepare SiC NWs by a VS method, without the need of a catalyst. This method is potentially scalable to large scale, for industrial synthesis of pure SiC NWs and was capable of producing crystalline $3C$-SiC NWs.

Si and SiO_2 powders were mixed and used as starting precursor for the VLS growth of SiC/SiO_2 core/shell NWs, along with C_3H_6 and using Fe as a catalyst [75]. The observed NWs had a 30-50 nm core of crystalline SiC, covered by a 10 nm SiO_2 shell.

An alternative method for SiC NWs synthesis is by using carbon tetrachloride (CCl_4) with Ni as catalyst [76]. CCl_4 decomposes by forming chlorine, either Cl_2 or Cl. Silicon is

Advancing Silicon Carbide Electronics Technology II   Materials Research Forum LLC
Materials Research Foundations **69** (2020) 233-275   https://doi.org/10.21741/9781644900673-5

then etched from substrate and form different species of Si chlorides, from SiCl to SiCl$_4$. These species then react with the C formed by the decomposition of CCl$_4$, giving pure SiC NWs with diameter up to 50 nm and length of 10 μm, by means of VLS method. Another alternative method for the synthesis of 3C-SiC NWs is by means of pyrolysis of liquid polysilacarbosilane (l-PS) at 1300 °C by chemical vapor deposition [77]. l-PS is the decomposition product of polydimethylsilane, a crystalline polymer. l-PS contains Si-Si and Si-C bonds and is decomposed at high temperature into small species such as cyclic silanes, silane fragments, H$_2$, and CH$_4$. The silanes then act as gaseous sources for the synthesis of 3C-SiC NWs through the VLS. With this method, large amount of light-green cotton-like fibres were produced, that were identified as pure 3C-SiC NW with lengths of several centimetres and diameters of 100-200 nm.

3.3   Template assisted SiC nanowire growth

The first growth attempt of SiC NWs was historically achieved by Dai *et al.* in 1995 [78] through the silicidation of carbon nanotubes (CNTs), which were used as a template for the local confinement of the chemical reaction. Later on, Si NWs were also used as templates for their conversion into SiC NWs, NTs as well as Si/SiC core/shell nanostructures [79, 80, 81]. Comprehensive topical review of this procedure was reported in [82] and in [9]. In this section we will provide a short summary of the method, the most important issues and some new results on the subject.

3.3.1 CNTs conversion into SiC NW and NT

This process is based on the wide knowledge obtained from CNTs growth experiments already from early 90s, and their further silicidation technique, which is a well-studied mechanism in microelectronics technology [83]. Starting from single-wall or multiwall CNTs, the silicidation process consists of a Si-rich gas reaction with C atoms. The Si source is generally provided by either a Si-rich powder such as Si or SiO powder [84], a gas such as silane [85] or even by a Si substrate itself (Fig. 4). SiC NWs with very narrow diameters (2–50 nm [86]) and long enough (10μm [87]) can be easily achieved. However, the control of the produced nanostructure type and its crystalline quality are difficult to be achieved and properly tuned. The first problem originates from the difficulty in filling CNTs, and very often both SiC-NTs and SiC-NWs are simultaneously formed in the experiment (Fig. 5) [88]. Experimental studies on the synthesis of SiC NTs are however rare. It should be underlined that these SiC-NTs are not single walled. The main principle that is used in the literature to grow SiC-NTs is the silicidation of CNTs using the shape memory synthesis [89, 90, 91]. The principle of this synthesis is based on the gas–solid reaction between SiO and C that leads to a SiC material having the same shape as the initial template thus the carbon material.

Figure 4.   SEM images for (a) CNTs grown on metal catalyst loaded anodized Si substrate at 700 °C for 30 min under C_2H_2/H_2, (b) SiC nanorods grown at 1250 °C for 60 min from the reaction of CNTs with $SiH_4/C_3H_8/H_2$, and (c) SiC microcrystals grown at 1250 °C for 60 min from the reaction of CNTs with a gas mixture of TMS/H_2. Reprinted from [85] with permission. Copyright Elsevier (2004).

3.3.2 Si NW conversion into Si/SiC core/shell NW, SiC NT and SiC NW

The surface of bottom-up or top-down grown Si NWs could be in principle converted into $3C$-SiC by a standard carbonization method [92]. By this technique, few nm of the Si surface is converted into $3C$-SiC, the thickness being essentially limited by Si and C atoms inter-diffusion through the formed SiC layer. However, the structural quality of the obtained SiC is quite poor, and its thickness is difficult to be controlled.

Later on, experiments to totally carburize Si-NWs have been performed [79, 80, 81, 82]. The SiC-NWs were crystalline but contained a high density of planar defects preventing the wide use of this approach for SiC NWs production. Nevertheless, this method allows

Materials Research Forum LLC
https://doi.org/10.21741/9781644900673-5

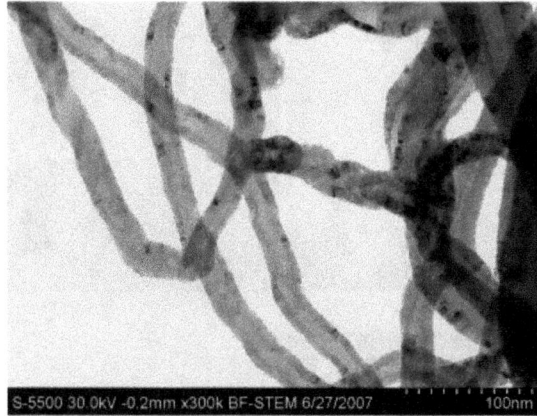

Figure 5. High resolution SEM picture of SiC nanotubes obtained after 1 h of sintering at 1200 °C. Thickness of the SiC nanotubes walls is about 6 nm while the outer diameter is about 22 nm. Diagonal lines in the SiC phase are stacking faults. Reproduced with permission from [88]. © IOP Publishing. All rights reserved.

the controlled growth of Si/3C-SiC core/shell heterostructures, which are expected to combine the high electrical properties of Si with that of a protective and possibly biocompatible coating composed by SiC [79].

Si–SiC core–shell nano-composites have been synthesized via a two-step process [93]: in a first step, Si nano-towers were grown by VLS mechanism followed by a second step, where 3C-SiC NCs were deposited using a thermal plasma-based deposition technique. This rough SiC shell exhibited a poor crystalline quality and has been grown in order to generate compressive stress inside the Si core for a study of plasticity. Si-NWs have also been coated with SiC in order to fabricate a robust electrode material for aqueous micro-supercapacitors [94].

The first study to prepare single crystalline Si–SiC core–shell NWs was reported in [80] starting from bulk Si and using a top-down approach. The doping of the Si core can be easily controlled by the preliminary Si etching. A simple chemical KOH etching on the Si–SiC core–shell NWs has proved that the SiC shell covers perfectly the nanostructures [80] and ensures a good protection for future use as a bio-nano-sensor [95].

A common problem of Si/3C-SiC epitaxy of thin films is the presence of interfacial voids due to the tendency of Si to out-diffuse from the substrate at the high temperature [96, 97]. This drawback was actually exploited to form 3C-SiC NTs starting from Si NWs

[98]. The main advantage of SiC-NTs compared to SiC-NWs is the higher surface to volume ratio, which could be exploited in nano-sensors or actuators functioning in harsh environments. Latu-Romain *et al.* have developed a way to grow SiC-NTs with controlled dimensions and good crystalline quality by using top-down grown Si-NWs carburization [81]. The growth method can be divided into three main steps. First, Si-NWs are obtained by plasma etching on Si(100) substrate (Fig. 6), then Si-NWs are heated and carburized at atmospheric pressure to form a SiC shell to prevent any damage of the nanostructures at high temperature.

Finally, carburization is conducted at a lower pressure to increase Si out-diffusion leading to a hollow structure. In this way, Si-NWs are converted into {2 0 0} faceted SiC-NTs with a high crystalline quality and with dense sidewalls. Moreover, the size of the final SiC-NTs in terms of external diameter is nearly the same as that of the initial Si-NWs. It is interesting to notice that, whatever the experimental conditions explored in that work, Si out-diffusion still controls the carburization reaction and the Si diffusion can be monitored via the pressure. This original process has several advantages, such as the very high crystalline quality of the SiC-NTs and the possibility to tune the final SiC-NTs dimension by controlling the size of the Si-NWs template.

3.4   Conclusions on SiC NW bottom-up formation

Different approaches are implemented for the realization of SiC based nanostructures. The bottom-up methods described in this chapter have the advantage of being very simple, with high yield and potentially with lower crystallographic defects and lattice damage with respect to the top down techniques. However, several issues are still to be addressed in order to increase the quality of bottom-up 3*C*-SiC nanostructures, amongst them to lower the density of stacking faults and defects. Indeed, the quality of bottom-up 3*C*-SiC NW in terms of electronic properties has been poor, mainly because of the high density of stacking faults generation during the growth, due to their very low formation energy, as well as the high intrinsic carrier concentration resulting from unintentional doping (mainly coming from nitrogen).

For the optimization of electronic devices based on SiC NW, the study and the control of intrinsic carrier concentration has to be addressed. It is recognized that 3*C*-SiC NWs are *n*-type because of the high residual (unintentional) doping probably originating in nitrogen contamination. Efforts to minimize the defects and carriers density are being pursued by achieving a better control on the growth parameters (control the atmosphere) and a better understanding of the defects formation.

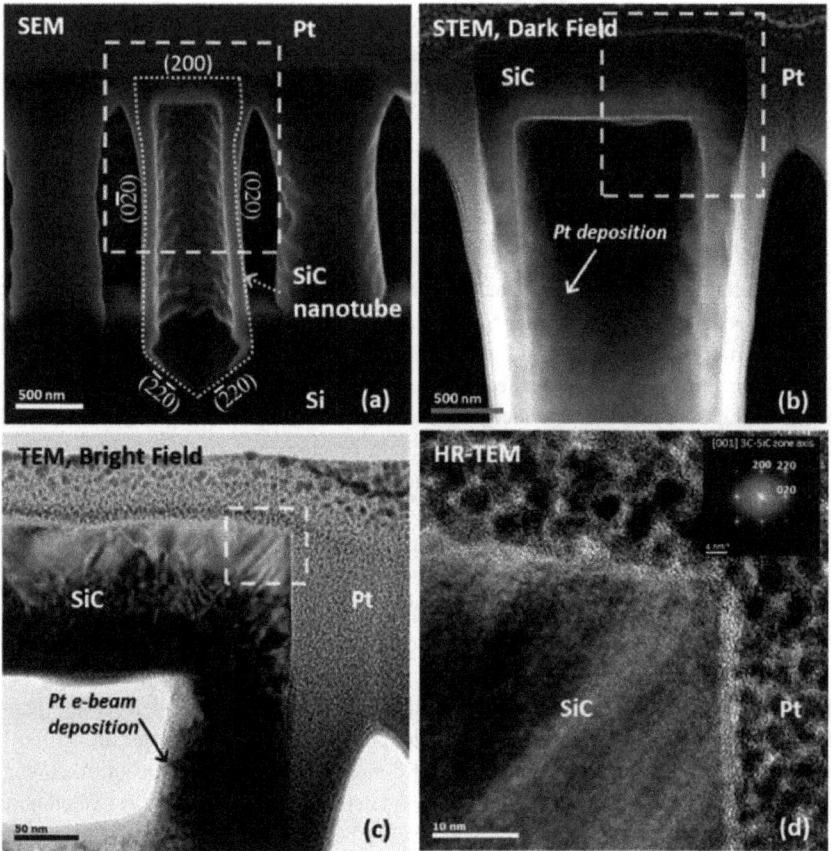

Figure 6.   SiC-NT, after carburization (a) SEM image of the transversal thin lamella where morphology of the SiC-NT could be observed. Indexation of the crystallographic planes of the SiC-NT has been added thanks to the TEM study. (b) The STEM image in dark field at 200 keV of the SiC-NT, (c) The TEM image in bright field at 200 keV of the SiC-NT, (d) The HR (High Resolution)-TEM image of the top corner of the SiC-NT. Inset: fast Fourier transform of the HR-TEM image. Reproduced with permission from [81]. IOP Publishing. All rights reserved.

On the contrary, top-down approach demonstrated improved device behaviour mainly because the much lower carrier concentration of the bulk starting SiC material. Indeed, using SiC nano pillars (NP) it has been possible to greatly increase electrical performance, benefited from the low doping level of the SiC, the high crystalline quality of SiC and the optimization of the contact annealing. An apparent mobility of ~150 cm^2/(V·s) has been obtained for SiC-NPFETs, which is the best mobility measured on SiC nanoFETs [102].

Conversion of Si nanostructured templates to SiC is considered a promising route to achieve a better control of the nanostructure. The combination of top down methods to fabricate the Si nanostructure and a conversion to SiC following principles used in thin films epitaxy or in bottom up method could allow a better, controlled growth of Si/3C-SiC core/shell heterostructures that can combine the high electrical properties of Si with that of a protective and biocompatible SiC coating.

4.   Top-down formation of SiC NWs

The reported work on top-down SiC NWs growth is very limited, in contrast with the bottom up growth approach [99, 100, 101, 102, 103, 104, 105, 106]. By incorporating e-beam lithography and subsequent plasma etching methods, horizontal or vertical nanostructures with respect to the substrate have been formed. Cubic and hexagonal SiC NWs have been fabricated depending on the initial substrate employed.

Horizontal NWs with the respect to the substrate have been formed targeting the fabrication of nanoelectromechanical systems (NEMS) [99] and JFETs [105, 106]. The p-type channel JFETs have been fabricated by ion implantation in selected areas of a 4H-SiC n-type epilayer and subsequent plasma etching of surrounding areas and part of the implanted areas [105]. Alternatively, n-type channel has been formed by growing an n-type epilayer on top of a p-type substrate and subsequent plasma etching for defining channel and pad areas [106].

The vertical top-down SiC NWs suffer from a low aspect ratio due to the important sidewall slope (Fig. 7). The proposed reasons for this feature are the scattering of impinging plasma ions and/or the higher etch of the hard-mask edge. A columnar structure with top diameter of 20 nm and a height of 400 nm (aspect ratio of 20) has been obtained in the best case [101].

An additionnal effect of a very long-time etch is the faceted morphology of the etched wires/pillars (Fig. 7) [101]. The cross section of the SiC pillars shows a rhombus, pentagonal or hexagonal morphology depending on polytypes and crystallographic orientations. The favored morphologies of SiC NPs originate from a complex interplay between their polytypes and crystal orientations, which reflects the so-called Wulff's rule [101].

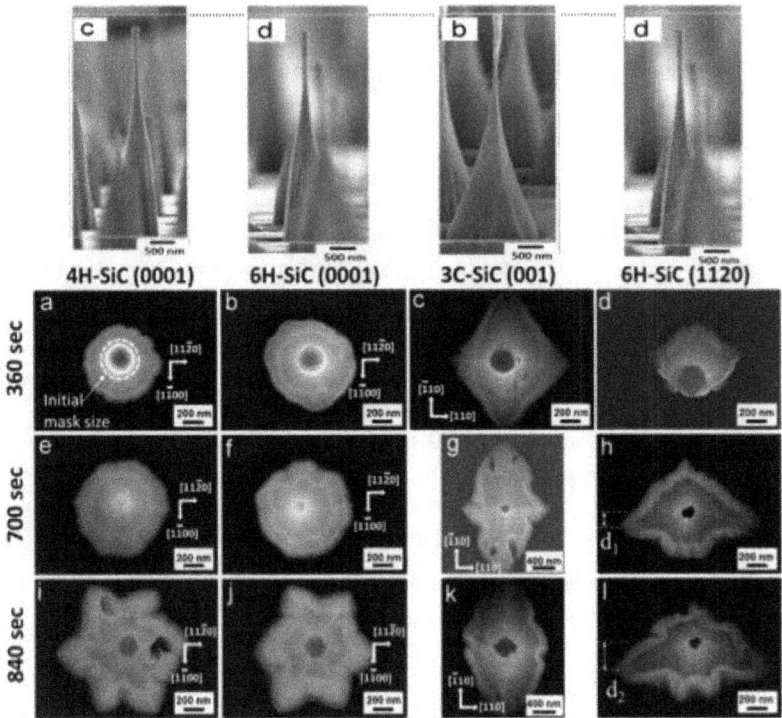

Figure 7. (a) Schematic of SiC NW top down formation and (b) corresponding NW morphology. Reproduced with permission from [101]. IOP Publishing. All rights reserved.

Recently [107], a new method resolving the above issue and simultaneously relaxing the need for high-resolution lithography was proposed. Initially, NanoImprint Lithography (NIL) has been employed for defining the nanopillar network. NIL is in many aspects capable of producing results comparable to those of e-beam lithography, but at a considerably lower cost and with a much higher throughput. Then, hard mask etch instead of lift-off has been employed for obtaining vertical hard mask sidewall. Finally the diameter of the NPs has been reduced by sacrificial oxidation. Massively parallel, dense, vertical 4H-SiC NW arrays with controlled diameter from 300 nm down to 70 nm have been obtained with this method (Fig. 8).

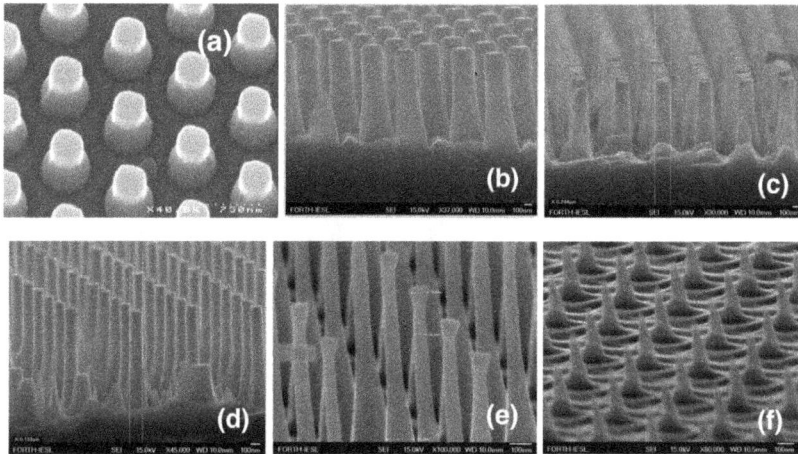

Figure 8.   4H-SiC pillars before ((a) and (b)) and after 30 (c), 60 (d), 90 (e) and 120 (f) min of oxidation and oxide removal. Reprinted with permission [107].

5.   Processing technology of SiC NW based devices

A review of the plasma etching technology involved in top-down formation of SiC NWs is presented in another chapter of the present volume [108] and will not be discussed in this section, since the interested reader will have all necessary information in the chapter on SiC plasma etching.

In a first study of ohmic contacts on SiC NWs [109] nickel was identified to present the lower contact resistance to $3C$–SiC NWs by comparing Ni/Au and Ti/Au metallizations. The specific contact resistances for Ni/Au ohmic contacts ($5.9\times10^{-6} \pm 8.8\times10^{-6}$ $\Omega \cdot cm^2$) were roughly 40 times lower than those for Ti/Au ohmic contacts formed on SiC NWs. For achieving lower contact resistances, all samples were processed by rapid thermal annealing (RTA) at 700 °C for 30 s with ambient N_2 [109]. These results have been corroborated by another study [110], in which it has been additionally noticed a self-heating annealing of the contacts due to the Joule effect. This was observed only for devices with Ti/Au contacts by revealing the role of conductive titanium oxide that is easily formed at the interface of Ti and $3C$-SiC. The origin of the low contact resistance in Ni-based ohmic contacts to SiC NWs after high-temperature annealing is assumed to originate from the formation of polycrystalline nickel silicide phase (Ni_2Si, $NiSi_2$, $NiSi$, etc.) at the interface between the Ni and $3C$-SiC NWs [111], as seen in the bulk SiC materials.

An additional conclusion of the study of Rogdakis [110] was that annealing above 600 °C resulted in many cases in non-operational devices or in damaged contacts containing voids. This effect has been systematically investigated by J. Choi [102]. After an annealing step at 700 °C, Ni silicide phases emerging from the metal electrode begin to intrude into the SiC NW channel partially converting the SiC into Ni silicide phases by solid state reaction (Fig. 9). Consequently, a SiC/Ni silicide heterostructure or a fully Ni silicidized SiC NW depending on the channel length is formed.

6.   Functionalization of SiC nanostructures

In recent years there is an increased effort towards the surface functionalization of SiC nanostructures, in order to tune properties such as e.g. conductivity, E_G and hydrophobicity [8]. In addition, for various bio-related applications a surface functionalization is necessary as an interface with the biological medium.

A series of studies have been dedicated to modify the surface charge of SiC NCs and in this way improve their dispersion in aqueous and non-aqueous media as well as protecting them from potential chemical reactions in physiological environment [112, 113, 114, 115].

In other studies, the surface of SiC NCs has been functionalized to modify their hydrophobic [116] character or their conductivity [117]. For example, the increase of SiC NWs/NCs conductivity by surface functionalization is necessary for being used as support material for fuel cells [8].

The surface of SiC NWFETs has been functionalized in order to have a label-free electrical DNA detection in [118].

7.   Applications of SiC nanowires

A detailed review of SiC NW applications has been performed in [9] and later on in [7]. A summary updated with recent data is presented below.

The high strength values (53.4 GPa of bending strength and 660 GPa of Young's modulus for 20–30 nm NWs [119]) and the super plasticity [120, 121], in combination with the thermal stability and chemical inertness of SiC NWs indicate their suitability as reinforcing material in composite structures. A variety of SiC NW-based composites have been demonstrated including polymer-matrix composites [122, 123], metal-matrix composites [124], ceramic-matrix composites [125, 126] and C/C composites [127].

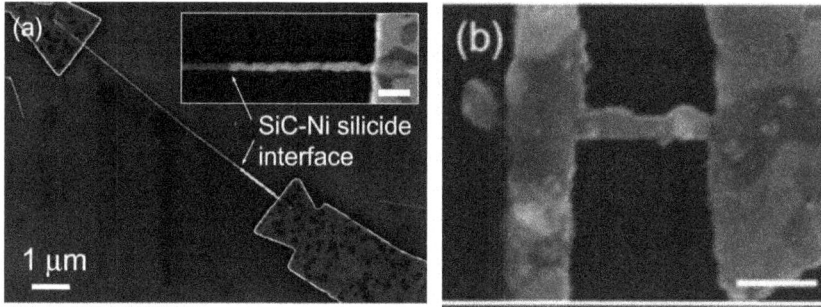

Figure 9.   *SEM images of a silicided SiC nano FET (a) partial Ni silicide intrusion (b) complete salicidation of SiC NW. Reprinted with permission from [102].*

The excellent mechanical properties of SiC NWs in combination with the wide E_G are attracting features for the development of NEMS [99, 128].

SiC NWs exhibit superhydrophobicity under proper surface treatment [129, 130] and therefore can be employed for the development of self-cleaning glass.

SiC NWs can be employed for gas sensors [131] and stress sensors [132, 133] as their conductance varies by application of stress or by adsorption of molecules on their surface. The transverse piezoresistance coefficient of the 6H-SiC NWs has been estimated around 10^{-9} Pa^{-1} [132] a value high enough for stress sensor applications. In [131] SiC NW-based capacitive structures have been employed for measuring the humidity. In addition SiC NWs are excellent candidates for the development of biosensors [134].

Theoretical studies on the 3C-SiC NW FETs, in various transport regimes [110, 135, 136, 137], have shown that the SiC NW FETs have similar performance to the Si-based ones while they offer, in addition, the advantage of high temperature operation. Only one experimental study [109] on SiC NW FETs has used the top-gate geometry while in the others a back-gated 3C-SiC NWFETs has been employed [135, 138, 139, 140, 141, 142, 143]. In all cases a MIS (Metal Insulator Semiconductor) geometry has been used. The electrical characterization revealed devices with either ohmic or rectifying contacts leading to two different operation modes. The transistors with ohmic-like contacts exhibit very weak gating effect and the device switching off is not achievable even for high negative gate voltages due to the high electron concentration of the NWs [109, 135, 138, 141]. In contrast, the devices with Schottky contact barrier at source / drain regions demonstrate a well determined switching-off and a better performance thanks to the

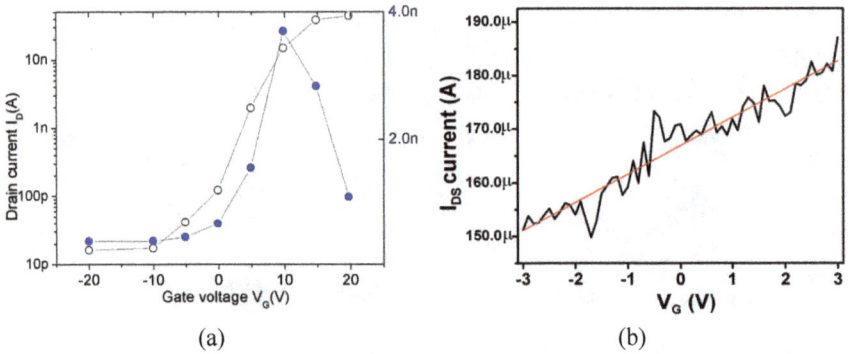

Figure 10. Transfer characteristics of SiC NWFETs with: (a) back-gated, bottom-up formed 3C-SiC NW with rectifying contacts (b) back-gated, top-down formed 3C-SiC NW with ohmic contacts. Reprinted, with permission, from [140] © (2011) IEEE, [143] © TTP.

modulation of the drain current by the gate voltage, through the control of Schottky barriers transparency at the source and drain regions [139, 140, 142]. Nevertheless, ohmic contact devices are expected to demonstrate enhanced performance when the quality of both the NW material and of the interface between the gate oxide and the NW will be substantially improved.

The 3C-SiC NW channels have been mostly formed by a bottom up approach except one case [143] where a top down formation has been used. The transistors with top-down formed NWs exhibited three orders of magnitude higher current and transconductance values with respect to SiC NWFETs with bottom-up grown NWs. However, in both cases it was not possible to switch-off the transistors revealing the negative impact of the high unintentional doping and the interface traps density on the device performance. The above conclusion has been corroborated by the results of studies using top-down formed 4H-SiC NWs as channel in NWFETs of JFET type [105, 106]. Indeed, the best results in terms of transistor operation ($I_{ON}/I_{OFF} > 10^5$, $g_m = 8.8$ mS) have been obtained in this case.

The low electron affinity, the excellent chemical and physical stability and the high aspect ratio make SiC NWs ideal candidates for the development of field emitters. Towards this purpose, it is necessary to optimize the tip geometry [144, 145], the alignment and density of SiC NWs [146, 145], their coating [147, 148] and their doping [149]. Excellent results have been obtained in terms of turn-on field (well below 2 V/μm under various current densities and low voltage values) [145, 146, 147, 148, 149] and high field operation [150].

SiC NWs have been also investigated for their potential application as photocatalysts [151], supercapacitors [152] and microwave absorbers [153].

Conclusions

In the present review, the state-of-the-art formation technology of SiC 0D and 1D structures with an emphasis on separate NCs and NWs has been presented. These two types of SiC nanostructures have been widely investigated and present a real potential for applications, which are also examined in the present book. Porous SiC and nanostructured films are other classes of SiC nanostructures, which are however not examined in the present work as they present less interest the last years. The interested reader is addressed to other books [7] and review articles [8] for further information on these SiC nanostructures.

The formation of 0D nanostructures has been mainly investigated by chemical methods, having the advantage of obtaining narrow (<10 nm) diameter SiC NCs at a low cost, a crucial factor for all the applications considered for this material. Chemical etching of SiC powder is widely used to obtain SiC 0D nanostructures. The disparity in size distribution is the main problem of this method and the reason to investigate other methods such as gas-phase based ones. SiC NCs exhibit an efficient and stable PL in the range from ultra-violet to yellow thanks to the quantum confinement effect. Thus, SiC NCs are ideal candidates for tunable light emitters for multiple applications, including biomedical ones. SiC NCs are also suitable as catalysts, reinforcing material in composites, biosensors and microwave absorbers.

SiC NWs are the most widely investigated SiC nanostructures as they are suitable for charge transport related applications, in addition to the aforementioned one related to SiC NCs. Bottom up formation techniques have been mainly used for fabricating SiC NWs, again for low cost purposes. SiC NWs have been widely investigated for field-emitter and NW FET applications. In addition, SiC NWs are suitable for applications related to self-cleaning glass, catalysts, reinforcing material in composites, biosensors and microwave absorbers.

A general conclusion for both 0D and 1D SiC structures is that the widely used bottom up formation methods results in cubic material with planar defects. On the contrary, top-down formed SiC NWs are free from planar defects and with a clearly better crystalline quality.

SiC nanostructures are very promising for addressing the requirements of various applications. An important research effort has been devoted towards this purpose by demonstrating various devices and systems. However, the related technology is not mature enough for commercialization.

Acknowledgements

The authors would like to thank Edwige Bano for valuable discussions. K. Zekentes acknowledges the support of European Commission through the Marie-Curie project SICWIRE.

References

[1] Special Issue on Silicon Carbide Devices and Technology, IEEE Trans. Electron Dev. 55 (2008) 1795-2065. https://doi.org/10.1109/TED.2008.926685

[2] R. Yakimova, R.M. Petoral, G.R. Yazdi, C. Vahlberg, A. Lloyd Spetz, and K. Uvdal, Surface functionalization and biomedical applications based on SiC, J. Phys. D: Appl. Phys. 40 (2007) 6435–6442. https://doi.org/10.1088/0022-3727/40/20/S20

[3] S.E. Saddow, C.L. Frewin, C. Coletti, N. Schettini, E. Weeber, A. Oliveros, and M. Jarosezski, Single-crystal silicon carbide: A biocompatible and hemocompatible semiconductor for advanced bio-medical applications, Mater. Sci. Forum, 679–680 (2011) 824–830. https://doi.org/10.4028/www.scientific.net/MSF.679-680.824

[4] C. Coletti, M.J. Jaroszeski, A. Pallaoro, M. Hoff, S. Iannotta, and S.E. Saddow, Biocompatibility and wettability of crystalline SiC and Si surfaces. In 29th Annual International Conference of the IEEE Engineering in Medicine and Biology Society, Lyon, France, 2007; pp. 5849–5852.17. https://doi.org/10.1109/IEMBS.2007.4353678

[5] T. Kimoto, A. Itoh and H. Matsunami, Step-Controlled Epitaxial Growth of High-Quality SiC Layers, Phys. Status Solidi b 202 (1997) 247–62. https://doi.org/10.1002/1521-3951(199707)202:1<247::AID-PSSB247>3.0.CO;2-Q

[6] L. E. Brus, P. F. Szajowski, W.L. Wilson, T. D. Harris, S. Schupler and P.H. Citrin, Electronic spectroscopy and photophysics of Si nanocrystal: relationship to bulk c-Si and porous Si, J. Am. Chem. Soc., 117 (1995) 2915. https://doi.org/10.1021/ja00115a025

[7]   Jiyang Fan Paul K. Chu, Silicon Carbide Nanostructures Fabrication, Structure, and Properties, Springer (2014) 9.

[8]   R. Wu, K. Zhou, C. Y. Yue, J. Wei, Y. Pan, Recent progress in synthesis, properties and potential applications of SiC nanomaterials, Progress in Materials Science 72, (2015) 1-60. https://doi.org/10.1016/j.pmatsci.2015.01.003

[9]   K. Zekentes, K. Rogdakis, SiC nanowires: Material and devices, J. Phys. D. Appl. Phys. 44 (2011) 133001. https://doi.org/10.1016/0022-0248(95)00330-4

[10]   K. Zekentes, V. Papaioannou, B. Pecz and J. Stoemenos, Early stages of growth of β-SiC on Si by MBE, J. Cryst. Growth, 157 (1995) 392-399. https://doi.org/10.1016/0022-0248(95)00330-4

[11]   J. Yang, X. Wang, G. Zhai, N. Cue, and X. Wang, J. Cryst. Growth, 224, (2001) 83. https://doi.org/10.1016/S0022-0248(01)00749-7

[12]   V. Palermo, A. Parisini, D. Jones, Silicon carbide nanocrystals growth on Si(100) and Si (111) from a chemisorbed methanol layer, Surface Science, 600 (2006) 1140–1146. https://doi.org/10.1016/j.susc.2005.12.048

[13]   V. Cimalla, J. Pezoldt, Th. Stauden, Ch. Förster, and O. Ambacher, A. A. Schmidt, K. Zekentes, Linear alignment of SiC dots on silicon substrates, J. Vac. Sci. Technol. B. Letters, B22 (2004) L20-L23 and V. Cimalla, J. Pezoldt, Th. Stauden, A.A. Schmidt, K. Zekentes, and O. Ambacher, Lateral alignment of SiC dots on Si, Phys. Stat. Sol. 1(2004) 337-340. https://doi.org/10.1002/pssc.200303951

[14]   C.M. Hollabaugh, D. E. Hull, L. R. Newkirk, J. J. Petrovic, R.F.-plasma system for the production of ultrafine ultrapure silicon carbide powder, J Mater Sci, 18 (1983) 3190–3194. https://doi.org/10.1007/BF00544142

[15]   S. Klein, M. Winterer, H. Hahn, Reduced-pressure chemical vapor synthesis of nanocrystalline silicon carbide powders. Chem. Vap. Deposition 4 (1998) 143–149.
https://doi.org/10.1002/(SICI)1521-3862(199807)04:04<143::AID-CVDE143>3.0.CO;2-Z

[16]   F. Huisken, B. Kohn, R. Alexandrescu, S. Cojocaru, A. Crunteanu, G. Ledoux and C. Reynaud, Silicon carbide nanoparticles produced by CO_2 laser pyrolysis of SiH_4/C_2H_2 gas mixtures in a flow reactor, J. Nanopart. Res., 1 (1999) 293–303. https://doi.org/10.1023/A:1010081206959

[17]   F. Lomello, G. Bonnefont, Y. Leconte, N. Herlin-Boime and G. Fantozzi, Processing of nano-SiC ceramics: densification by SPS and mechanical

characterization, J. Eur. Ceram. Soc. 32 (2012)633–41.
https://doi.org/10.1016/j.jeurceramsoc.2011.10.006

[18]   S. Askari, A. U. Haq, M. Macias-Montero, I. Levchenko, F. Yu, W. Zhou,
       K. Ostrikov, P. Maguire, V. Svrcek and D. Mariotti, Nanoscale 8 (2016) 17141-
       17149. https://doi.org/10.1039/C6NR03702J

[19]   X.L. Wu, J.Y. Fan, T. Qiu, X. Yang, G.G. Siu and P. K. Chu, Experimental
       evidence for the quantum confinement effect in 3C-SiC nanocrystallites. Phys.
       Rev. Lett. 94 (2005) 026102. https://doi.org/10.1103/PhysRevLett.94.026102

[20]   J. Zhu, Z. Liu, X. L. Wu, L. L. Xu, W. C. Zhang, P. K. Chu, Luminescent small-
       diameter 3C-SiC nanocrystals fabricated via a simple chemical etching method,
       Nanotechnology 18 (2007) 365603–5. https://doi.org/10.1088/0957-
       4484/18/36/365603

[21]   V. G Pol, S. V. Pol and A. Gedanken, Novel synthesis of high surface area silicon
       carbide by RAPET (reactions under autogenic pressure at elevated temperature) of
       organosilanes, Chem. Mater. 17 (2005) 1797–1802.
       https://doi.org/10.1021/cm048032z

[22]   A-H. Lu, W. Schmidt, W. Kiefer, F. Schüth, High surface area mesoporous SiC
       synthesized via nanocasting and carbothermal reduction process, J Mater Sci 40
       (2005) 5091–5093. https://doi.org/10.1007/s10853-005-1115-8

[23]   E. J. Henderson and J. G. C. Veinot, From phenylsiloxane polymer composition to
       size-controlled silicon carbide nanocrystals. J. Am. Chem. Soc. 131 (2009)809–15.
       https://doi.org/10.1021/ja807701y

[24]   S. Yang, W. Cai, H. Zeng and X. Xu, Ultra-fine b-SiC quantum dots fabricated by
       laser ablation in reactive liquid at room temperature and their violet emission, J.
       Mater. Chem. 19 (2009) 7119–7123. https://doi.org/10.1039/b909800c

[25]   Y. Zakharko, D. Rioux, S. Patskovsky, V. Lysenko, O. Marty, J. M. Bluet and
       M. Meunier, Direct synthesis of luminescent SiC quantum dots in water by laser
       ablation, Phys. Status Solidi-R 5 (2011) 292–294.
       https://doi.org/10.1002/pssr.201105284

[26]   Z-G. Yang and L. L. Shaw, Synthesis of nanocrystalline SiC at ambient
       temperature through high energy reaction milling. Nanostruct. Mater 7 (1996)
       873–886. https://doi.org/10.1016/S0965-9773(96)00058-X

[27] M. Dasog, L. F. Smith, T. K. Purkait and J. G. C. Veinot, Low temperature synthesis of silicon carbide nanomaterials using a solid-state method. Chem. Commun. 49 (2013) 7004–7006. https://doi.org/10.1039/c3cc43625j

[28] A. M. Rossi, T. E. Murphy and V. Reipa, Ultraviolet photoluminescence from 6H silicon carbide nanoparticles. Appl. Phys. Lett. 92 (2008) 253112. https://doi.org/10.1063/1.2950084

[29] J. Fan, H. Li, J. Wang, M. Xiao, Fabrication and photoluminescence of SiC quantum dots stemming from 3C, 6H, and 4H polytypes of bulk SiC, Appl. Phys. Lett. 101 (2012) 131906. https://doi.org/10.1063/1.4755778

[30] C.H. Wang, Y. H. Chang, M. Y. Yen, C. W. Peng, C. Y. Lee, H. T. Chiu, Synthesis of silicon carbide nanostructures via a simplified Yajima process-reaction at the vapor-liquid interface, Adv. Mater. 17 (2005) 419–422. https://doi.org/10.1002/adma.200400939

[31] G. Shen, D. Chen, K. Tang, Y. Qian, S. Zhang, Silicon carbide hollow nanospheres, nanowires and coaxial nanowires, Chem. Phys. Lett. 375 (2003) 177–184. https://doi.org/10.1016/S0009-2614(03)00877-7

[32] P. Li, L. Xu and Y. Qian, Selective synthesis of 3C-SiC hollow nanospheres and nanowires, Cryst. Growth, 8 (2008) 2431–2436. https://doi.org/10.1021/cg800008f

[33] Z. Liu, L. Ci, N. Y. Jin-Phillipp and M. Rühle, Vapor-solid reaction for silicon carbide hollow spherical nanocrystals, J. Phys. Chem. C 111 (2007) 12517–12521. https://doi.org/10.1021/jp073012g

[34] Y. Zhang, E. W. Shi, Z. Z. Chen, X. B. Li and B. Xiao, Large-scale fabrication of silicon carbide hollow spheres. J. Mater. Chem. 16 (2006) 4141–5. https://doi.org/10.1039/b610168b

[35] A. Kassiba, W. Bednarski, A. Pud, N. Errien, M. Makowska-Janusik, L. Laskowski, et al., Hybrid core–shell nanocomposites based on silicon carbide nanoparticles functionalized by conducting polyaniline: electron paramagnetic resonance investigations, J. Phys. Chem. C 111 (2007) 11544–51. https://doi.org/10.1021/jp070966y

[36] A. Peled and J. P. Lellouche, Preparation of a novel functional SiC at polythiophene nanocomposite of a core–shell morphology. J. Mater. Chem. 22 (2012)2069–73. https://doi.org/10.1039/C2JM14506E

[37] L. Z. Cao, H. Jiang, H. Song, Z. M. Li and G. Q. Miao, Thermal CVD synthesis and photoluminescence of SiC/SiO$_2$ core–shell structure nanoparticles. J. Alloy Compd. 489 (2010) 562–5. https://doi.org/10.1016/j.jallcom.2009.09.109

[38] J. Y. Fan, H. X. Li, Q. J. Wang, D. J. Dai and P. K. Chu, UV-blue photoluminescence from close-packed SiC nanocrystal film, Appl. Phys. Lett. 98 (2011) 08913–3. https://doi.org/10.1063/1.3556657

[39] J. Botsoa, J. M. Bluet, V. Lysenko, L. Sfaxi, Y. Zakharko, O. Marty and G. Guillot, Luminescence mechanisms in 6H-SiC nanocrystals. Phys. Rev. B 80 (2009) 155317. https://doi.org/10.1103/PhysRevB.80.155317

[40] J. Fan, H. Li, J. Jiang, L. K.Y. So, Y. W. Lam and P. K. Chu 3C-SiC nanocrystals as fluorescent biological labels, Small 4 (2008) 1058–1062. https://doi.org/10.1002/smll.200800080

[41] H. Mimura, T. Matsumoto and Y. Kanemitsu, Blue electroluminescence from porous silicon carbide, Appl. Phys. Lett. 65 (1994) 3350–3352. https://doi.org/10.1063/1.112388

[42] B. Xiao, X. L. Wu, W. Xu and P. K. Chu, Tunable electroluminescence from polymer- passivated 3C-SiC quantum dot thin films, Appl. Phys. Lett. 101 (2012) 123110. https://doi.org/10.1063/1.4753995

[43] J. Botsoa, V. Lysenko, A. Géloën, O. Marty, J. M. Bluet and G. Guillot, Application of 3C-SiC quantum dots for living cell imaging, Appl. Phys. Lett. 92 (2008) 173902–3. https://doi.org/10.1063/1.2919731

[44] Y. Zakharko, T. Serdiuk, T. Nychyporuk, A. Géloën, M. Lemiti and V. Lysenko, Plasmon-enhanced photoluminescence of SiC quantum dots for cell imaging application, Plasmonics 7 (2012) 725–32. https://doi.org/10.1007/s11468-012-9364-2

[45] D. Beke, Z. Szekrenyes, D. Palfi, G. Rona, I. Balogh, P. A. Maak, et al., Silicon carbide quantum dots for bioimaging, J. Mater. Res. 28 (2013)205–9. https://doi.org/10.1557/jmr.2012.296

[46] H. Ferkel and B. L. Mordike, Magnesium strengthened by SiC nanoparticles, Mat. Sci. Eng. A 298 (2001) 193–199. https://doi.org/10.1016/S0921-5093(00)01283-1

[47] J. Zhao, L. C. Stearns, M. P. Harmer, H. M. Chan and G. A. Miller, Mechanical behavior of alumina-silicon carbide "nanocomposites", J. Am. Ceram. Soc. 76 (1993) 503–510. https://doi.org/10.1111/j.1151-2916.1993.tb03814.x

Advancing Silicon Carbide Electronics Technology II   Materials Research Forum LLC
Materials Research Foundations **69** (2020) 233-275   https://doi.org/10.21741/9781644900673-5

[48]   A. F. Zimmerman, D. G. Clark, K. T. Aust and U. Erb, Pulse electrodeposition of Ni-SiC nanocomposite, Mater. Lett. 52 (2002) 85–90. https://doi.org/10.1016/S0167-577X(01)00371-8

[49]   N. Chisholm, H. Mahfuz, V. K. Rangari, A. Ashfaq and S. Jeelani, Fabrication and mechanical characterization of carbon/SiC-epoxy nanocomposites, Compos. Struct. 67 (2005) 115–124. https://doi.org/10.1016/j.compstruct.2004.01.010

[50]   N. J. Yang, H. Zhang, R. Hoffmann, W. Smirnov, J. Hees, X. Jiang, et al., Nanocrystalline 3C-SiC electrode for biosensing applications, Anal. Chem. 83 (2011) 5827–30. https://doi.org/10.1021/ac201315q

[51]   E. H. Williams, J. A. Schreifels, M. V. Rao, A. V. Davydov, V. P. Oleshko, N. J. Lin, et al., Selective streptavidin bioconjugation on silicon and silicon carbide nanowires for biosensor applications, J. Mater. Res. 28 (2013) 68–77. https://doi.org/10.1557/jmr.2012.283

[52]   X. F. Liu, M. Antonietti and C. Giordano, Manipulation of phase and microstructure at nanoscale for SiC in molten salt synthesis, Chem. Mater. 25 (2013) 2021–7. https://doi.org/10.1021/cm303727g

[53]   H. L. Zhu, Y. J. Bai, R. Liu, N. Lun, Y. X. Qi, F. D. Han, et al., In-situ synthesis of one-dimensional MWCNT/SiC porous nanocomposites with excellent microwave absorption properties, J. Mater. Chem. 21 (2011)13581–7. https://doi.org/10.1039/c1jm11747e

[54]   S. Xie, G. Jin, S. Meng, Y. W. Wang, Y. Qin, X. Y. Guo, Microwave absorption properties of in situ grown CNTs/SiC composites, J. Alloy. Compd. 520 (2012) 295–300. https://doi.org/10.1016/j.jallcom.2012.01.050

[55]   S.A. Fortuna, X. Li, Metal-catalyzed semiconductor nanowires: a review on the control of growth directions, Semicond. Sci. Technol. 25 (2010) 24005. https://doi.org/10.1088/0268-1242/25/2/024005

[56]   J.M. Redwing, X. Miao, X. Li, Vapor-Liquid-Solid Growth of Semiconductor Nanowires, in: Handb. Cryst. Growth, Elsevier, (2015) 399–439. https://doi.org/10.1016/B978-0-444-63304-0.00009-3

[57]   Z.J. Li, W.P. Ren, A.L. Meng, Morphology-dependent field emission characteristics of SiC nanowires, Appl. Phys. Lett. 97 (2010) 263117. https://doi.org/10.1063/1.3533813

[58]   G.W. Ho, A.S.W. Wong, D.-J. Kang, M.E. Welland, Three-dimensional crystalline SiC nanowire flowers, Nanotechnology. 15 (2004) 996–999. https://doi.org/10.1088/0957-4484/15/8/023

[59]   Guozhen Shen, Yoshio Bando and D. Golberg, Self-Assembled Hierarchical Single-Crystalline β-SiC, Nanoarchitectures, 7 (2007) 35-38. https://doi.org/10.1021/cg060224e

[60]   J. Chen, Q. Shi, L. Gao, H. Zhu, Large-scale synthesis of ultralong single-crystalline SiC nanowires, Phys. Status Solidi. 207 (2010) 2483–2486. https://doi.org/10.1002/pssa.201026288

[61]   M. Negri, S.C. Dhanabalan, G. Attolini, P. Lagonegro, M. Campanini, M. Bosi, F. Fabbri and G. Salviati, Tuning the radial structure of core–shell silicon carbide nanowires, Cryst. Eng. Comm. 17 (2015) 1258–1263. https://doi.org/10.1039/C4CE01381F

[62]   J.M. Redwing, X. Miao, X. Li, Vapor-Liquid-Solid Growth of Semiconductor Nanowires, in: Handb. Cryst. Growth, Elsevier, (2015) 399–439. https://doi.org/10.1016/B978-0-444-63304-0.00009-3

[63]   G. Attolini, F. Rossi, M. Bosi, B.E. Watts, G. Salviati, Synthesis and characterization of 3C–SiC nanowires, J. Non. Cryst. Solids, 354 (2008) 5227–5229. https://doi.org/10.1016/j.jnoncrysol.2008.05.064

[64]   G.A. Bootsma, W.F. Knippenberg and G. Verspui, Growth of SiC whiskers in the system SiO2-C-H2nucleated by iron, J. Cryst. Growth. 11 (1971) 297–309. https://doi.org/10.1016/0022-0248(71)90100-X

[65]   X. Zhou, H.. Lai, H. Peng, F.C.. Au, L. Liao, N. Wang, I. Bello, C. Lee, S. Lee, Thin β-SiC nanorods and their field emission properties, Chem. Phys. Lett. 318 (2000) 58–62. https://doi.org/10.1016/S0009-2614(99)01398-6

[66]   B.E. Watts, G. Attolini, F. Rossi, M. Bosi, G. Salviati, F. Mancarella, M. Ferri, A. Roncaglia, A. Poggi, β-SiC NWs grown on patterned and MEMS silicon substrates, Mater. Sci. Forum. 679–680 (2011) 508–511. https://doi.org/10.4028/www.scientific.net/MSF.679-680.508

[67]   S. Zhu, N. Huang, L. Xu, Y. Zhang, H. Liu, H. Sun, Y. Leng, Biocompatibility of pure iron: In vitro assessment of degradation kinetics and cytotoxicity on endothelial cells, Mater. Sci. Eng. C. 29 (2009) 1589–1592. https://doi.org/10.1016/j.msec.2008.12.019

[68] M. Assad, N. Lemieux, C.H. Rivard, L.H. Yahia, Comparative in vitro biocompatibility of nickel-titanium, pure nickel, pure titanium, and stainless steel: Genotoxicity and atomic absorption evaluation, Biomed. Mater. Eng. 9 (1999) 1–12.

[69] P. Lagonegro, M. Bosi, G. Attolini, M. Negri, S.C. Dhanabalan, F. Rossi, F. Boschi, P.P. Lupo, T. Besagni, G. Salviati, SiC NWs Grown on Silicon Substrate Using Fe as Catalyst, Mater. Sci. Forum. 806 (2014) 39–42. https://doi.org/10.4028/www.scientific.net/MSF.806.39

[70] G. Attolini, F. Rossi, M. Negri, S.C. Dhanabalan, M. Bosi, F. Boschi, P. Lagonegro, P. Lupo, G. Salviati, Growth of SiC NWs by vapor phase technique using Fe as catalyst, Mater. Lett. 124 (2014). https://doi.org/10.1016/j.matlet.2014.03.061

[71] H.-J. Choi, H.-K. Seong, J.-C. Lee, Y.-M. Sung, Growth and modulation of silicon carbide nanowires, J. Cryst. Growth. 269 (2004) 472–478. https://doi.org/10.1016/j.jcrysgro.2004.05.094

[72] J.-S. Hyun, S.-H. Nam, B.-C. Kang, J.-H. Boo, Growth of 3C-SiC nanowires on nickel coated Si(100) substrate using dichloromethylvinylsilane and diethylmethylsilane by MOCVD method, Phys. Status Solidi. 6 (2009) 810–812. https://doi.org/10.1002/pssc.200880621

[73] B. PARK, Y. RYU, K. YONG, Growth and characterization of silicon carbide nanowires, Surf. Rev. Lett. 11 (2004) 373–378. https://doi.org/10.1142/S0218625X04006311

[74] J. Chen, Q. Shi, L. Xin, Y. Liu, R. Liu, X. Zhu, A simple catalyst-free route for large-scale synthesis of SiC nanowires, J. Alloys Compd. 509 (2011) 6844–6847. https://doi.org/10.1016/j.jallcom.2011.03.131

[75] A. Meng, Z. Li, J. Zhang, L. Gao, H. Li, Synthesis and Raman scattering of β-SiC/SiO$_2$ core–shell nanowires, J. Cryst. Growth. 308 (2007) 263–268. https://doi.org/10.1016/j.jcrysgro.2007.08.022

[76] G. Attolini, F. Rossi, F. Fabbri, M. Bosi, B.E. Watts, G. Salviati, A new growth method for the synthesis of 3C-SiC nanowires, Mater. Lett. 63 (2009). https://doi.org/10.1016/j.matlet.2009.09.012

[77] G. Li, X. Li, Z. Chen, J. Wang, H. Wang, R. Che, Large Areas of Centimeters-Long SiC Nanowires Synthesized by Pyrolysis of a Polymer Precursor by a CVD

Route, J. Phys. Chem. C. 113 (2009) 17655–17660.
https://doi.org/10.1021/jp904277f

[78] H. Dai, E. W. Wong, Y. Z. Lu, S. Fan and C. M. Lieber, Synthesis and characterization of carbide nanorods, Nature 375 (1995) 769–772. https://doi.org/10.1038/375769a0

[79] L. Tsakalakos, J. Fronheiser, L. Rowland, M. Rahmane, M. Larsen, Y. Gao, SiC Nanowires by Silicon Carburization, MRS Proc. 963 (2006) 963-Q11-3. https://doi.org/10.1557/PROC-0963-Q11-03

[80] M. Ollivier, L. Latu-Romain, M. Martin, S. David, A. Mantoux, E. Bano, V. Souliere, G. Ferro and T. Baron, Si–SiC core–shell nanowires, J. Cryst. Growth 363 (2013) 158–63. https://doi.org/10.1016/j.jcrysgro.2012.10.039

[81] L. Latu-Romain, M. Ollivier, V. Thiney, O. Chaix-Pluchery and M . Martin, Silicon carbide nanotubes growth: an original approach, J. Phys. D: Appl. Phys. 46 (2013) 092001. https://doi.org/10.1088/0022-3727/46/9/092001

[82] L Latu-Romain and M Ollivier, Silicon carbide based one-dimensional nanostructure growth: towards electronics and biology perspectives, J. Phys. D: Appl. Phys. 47 (2014) 203001. https://doi.org/10.1088/0022-3727/47/20/203001

[83] A. H. Reader, A. H. van Ommen, P. J. W. Weijs, R. A. M. Wolters and D. J. Oostra, Transition metal silicides in silicon technology, Rep. Prog. Phys. 56 (1993)1397. https://doi.org/10.1088/0034-4885/56/11/002

[84] W. Tan, P. Hunley and I. S. Omer, Properties of silicon carbide nanotubes formed via reaction of SiO powder with SWCNTs and MWCNTs, *Proc. IEEE Southeastcon 2009, Technical Proc. (Atlanta, GA, 5–8 March)* pp 230–5. https://doi.org/10.1109/SECON.2009.5174082

[85] Y. H. Mo, M. D. Shajahan, Y. S. Lee, Y. B. Hahn and K. S. Nahm, Structural transformation of carbon nanotubes to silicon carbide nanorods or microcrystals by the reaction with different silicon sources in rf induced CVD reactor, Synth. Met. 140 (2004) 309–15. https://doi.org/10.1016/S0379-6779(03)00381-3

[86] E. Munoz, A. B. Dalton, S. Collins, A. A. Zakhidov, R. H. Baughman, W. L. Zhou, J. He, C. J. O'Connor, B. McCarthy and W. J. Blau, Synthesis of SiC nanorods from sheets of single-walled carbon nanotubes, Chem. Phys. Lett. 359 (2002) 397–402. https://doi.org/10.1016/S0009-2614(02)00745-5

[87] H. Liu, G. A. Cheng, C. Liang and R. Zheng, Fabrication of silicon carbide nanowires/carbon nanotubes heterojunction arrays by high-flux Si ion

implantation, Nanotechnology 19 (2008) 245606. https://doi.org/10.1088/0957-4484/19/24/245606

[88]   K. L. Wallis, J. K. Patyk, and T. W. Zerda, Reaction kinetics of nanostructured silicon carbide, J. Phys.: Condens. Matter 20 (2008) 325216. https://doi.org/10.1088/0953-8984/20/32/325216

[89]   C-H. Pham, N. Keller, G. Ehret and M. J. Ledoux, The first preparation of silicon carbide nanotubes by shape memory synthesis and their catalytic potential, J. Catal. 200 (2001) 400–10. https://doi.org/10.1006/jcat.2001.3216

[90]   J. M. Nhut, R. Vieira, L. Pesant, J. P. Tessonnier, N. Keller, G. Ehret, C. H. Pham and M. J. Ledoux, Synthesis and catalytic uses of carbon and silicon carbide nanostructures, Catal. Today 76 (2002) 11–32. https://doi.org/10.1016/S0920-5861(02)00206-7

[91]   X. H. Sun, C. P. Li, W. K. Wong, N. B. Wong, C. S. Lee, S. T. Lee and B. K. Teo, Formation of silicon carbide nanotubes and nanowires via reaction of silicon (from disproportionation of silicon monoxide) with carbon nanotubes, J. Am. Chem. Soc. 124 (2002) 14464–71. https://doi.org/10.1021/ja0273997

[92]   S. Nishino, Production of large-area single-crystal wafers of cubic SiC for semiconductor devices, Appl. Phys. Lett. 42 (1983) 460. https://doi.org/10.1063/1.93970

[93]   A. R. Beaber, S. L. Girshick and W. W. Gerberich, Dislocation plasticity and phase transformations in Si–C core–shell nanotowers, Int. J. Fract. 171 (2011) 177–83. https://doi.org/10.1007/s10704-010-9566-6

[94]   J. P. Alper, M. Vincent, C. Carraro and R. Maboudian, Silicon carbide coated silicon nanowires as robust electrode material for aqueous micro-supercapacitor, Appl. Phys. Lett. 100 (2012)163901. https://doi.org/10.1063/1.4704187

[95]   L. Fradetal, V. Stambouli, E. Bano, B. Pelissier, J. H. Choi, M. Ollivier, L. Latu-Romain, T. Boudou and I. Pignot-Paintrand, Bio-functionalization of Silicon Carbide nanostructures for SiC nanowire-based sensors realization, J. Nanosci. Nanotechnol, 14 (2014)3391–7. https://doi.org/10.1166/jnn.2014.8223

[96]   K. Zekentes, V. Papaioannou, B. Pecz and J. Stoemenos, Early stages of growth of β-SiC on Si by MBE, J. Cryst. Growth 157(1995) 392-399. https://doi.org/10.1016/0022-0248(95)00330-4

[97]   K.C. Kim, C. Il Park, J. Il Roh, K.S. Nahm, Y.H. Seo, Formation mechanism of interfacial voids in the growth of SiC films on Si substrates, J. Vac. Sci. Technol. A Vacuum, Surfaces, Film. 19 (2001) 2636. https://doi.org/10.1116/1.1399321

[98]   L. Latu-Romain, M. Ollivier, A. Mantoux, G. Auvert, O. Chaix-Pluchery, E. Sarigiannidou, E. Bano, B. Pelissier, C. Roukoss, H. Roussel, F. Dhalluin, B. Salem, N. Jegenyes, G. Ferro, D. Chaussende, T. Baron, From Si nanowire to SiC nanotube, J. Nanopart. Res. 13 (2011) 5425–33. https://doi.org/10.1007/s11051-011-0530-9

[99]   X. L. Feng, M. H. Matheny, C. A. Zorman, M. Mehregany and M. L. Roukes, Low Voltage Nanoelectromechanical Switches Based on Silicon Carbide Nanowires, Nano Lett. 10 (2010) 2891. https://doi.org/10.1021/nl1009734

[100] A. Kathalingam, M. R. Kim, Y. S. Chae, S.Sudhakar, T. Mahalingam and J. K. Rhee, Appl. Surf. Sci. 257 (2011) 3850. https://doi.org/10.1016/j.apsusc.2010.11.053

[101] J. H. Choi, L. Latu-Romain, E. Bano, F. Dhalluin, T. Chevolleau and T. Baron, Fabrication of SiC nanopillars by inductively coupled SF_6/O_2 plasma etching, J. Phys. D: Appl. Phys. 45 (2012) 235204. https://doi.org/10.1088/0022-3727/45/23/235204

[102] J. Choi, SiC Nanowires: from growth to related devices, PhD dissertation, Grenoble INP, France, (2013)

[103] L. Hiller, T. Stauden, R. M. Kemper, J. K. N. Lindner, D. J. As and J. Pezoldt, ECR-Etching of Submicron and Nanometer Sized 3C-SiC (100) Mesa Structures, Materials Science Forum, 717-720 (2012) 901-904. https://doi.org/10.4028/www.scientific.net/MSF.717-720.901

[104] L. Hiller, T. Stauden, R. M. Kemper, J. K. N. Lindner, D. J. As and J. Pezoldt, Hydrogen Effects in ECR-Etching of 3C-SiC(100) Mesa Structures, Materials Science Forum, 778-780 (2014) 730-733. https://doi.org/10.4028/www.scientific.net/MSF.778-780.730

[105] M. S. Kang, J.H. Lee, W. Bahng, N. K. Kim and S. M. Koo, Top-Down Fabrication of 4H–SiC Nano-Channel Field Effect Transistors, J. Nanosci. Nanotechnol. 14 (2014) 7821-7823. https://doi.org/10.1166/jnn.2014.9387

[106] M. S. Kang, S. Yu and S. M. Koo, Elevated Temperature Operation of 4H-SiC Nanoribbon Field Effect Transistors, J. Nanosci. Nanotechnol. 15 (2015) 7551-7554. https://doi.org/10.1166/jnn.2015.11166

[107] M. Cottat, A. Stavrinidis, C. Gourgon, C. Petit-Etienne, M. Androulidaki, E. Bano, G. Konstantinidis, J. Boussey, K. Zekentes, 4H-SiC Nanowire arrays formation by nanoimprint lithography, plasma etching and sacrificial oxidation, Proc. WOCSDICE 2019, Cabourg, France.

[108] K. Zekentes, V. Veliadis, Plasma etching of SiC, chapter in the present book.

[109] C. O. Jang, T. H. Kim, S. Y. Lee, D. J. Kim and S. K. Lee, Low-resistance ohmic contacts to SiC nanowires and their applications to field-effect transistors, Nanotechnology, 19 (2008) 345203. https://doi.org/10.1088/0957-4484/19/34/345203

[110] K. Rogdakis, Experimental and theoretical study of 3C-Silicon Carbide Nanowire Field Effect Transistors, PhD dissertation, Univ. of Crete-Grenoble INP, (2010).

[111] J. Eriksson, F. Roccaforte, F. Giannazzo, R. Lo Nigro, V. Raineri, J. Lorenzzi, and G. Ferro, Improved Ni/3C-SiC contacts by effective contact area and conductivity increases at the nanoscale, Appl. Phys. Lett., 94(2009)112104-3. https://doi.org/10.1063/1.3099901

[112] S. Baklouti, C. Pagnoux, T. Chartier, J. F. Baumard, Processing of aqueous a-Al_2O_3, a- SiO_2 and a-SiC suspensions with polyelectrolytes, J. Eur. Ceram. Soc. 17 (1997) 1387–1392. https://doi.org/10.1016/S0955-2219(97)00010-1

[113] P. Tartaj, M. Reece, J. S. Moya, Electrokinetic behavior and stability of silicon carbide nanoparticulate dispersions, J. Am. Ceram. Soc. 81 (1998) 389–394. https://doi.org/10.1111/j.1151-2916.1998.tb02345.x

[114] B.P. Singh, J. Jena, L. Besra and S. Bhattacharjee, Dispersion of nano-silicon carbide (SiC) powder in aqueous suspensions, J. Nanopart. Res. 9 (2007) 797–806. https://doi.org/10.1007/s11051-006-9121-6

[115] J. Che, X. Wang, Y. Xiao, X.Wu, L. Zhou and W. Yuan, Effect of inorganic-organic composite coating on the dispersion of silicon carbide nanoparticles in non-aqueous medium, Nanotechnology 18 (2007) 135706. https://doi.org/10.1088/0957-4484/18/13/135706

[116] V. Médout-Marère, A. El. Ghzaoui, C. Charnay, J. M. Douillard, G. Chauveteau and S. Partyka, Surface heterogeneity of passively oxidized silicon carbide particles: hydrophobic- hydrophilic partition, J. Colloid. Interf. Sci. 223 (2000) 205–214. https://doi.org/10.1006/jcis.1999.6625

[117] M. Iijima and H. Kamiya, Surface modification of silicon carbide nanoparticles by azo radical initiators, J. Phys. Chem. C 112 (2008) 11786–11790. https://doi.org/10.1021/jp709608p

[118] L. Fradetal, E. Bano, G. Attolini, F. Rossi, and V. Stambouli, A silicon carbide nanowire field effect transistor for DNA detection, Nanotechnology 27 (2016) 235501. https://doi.org/10.1088/0957-4484/27/23/235501

[119] E. W. Wong, P. E. Sheehan and C. M. Lieber, Science 277 (1997) 1971. https://doi.org/10.1126/science.277.5334.1971

[120] X. D. Han, Y. F. Zhang, K. Zheng, X. N. Zhang, Z. Zhang, Y. J. Hao, Y. Guo, J. Yuan, Z. L. Wanget, Low-temperature in situ large strain plasticity of ceramic SiC nanowires and its atomic-scale mechanism, Nano Lett., 7 (2007) 452–7. https://doi.org/10.1021/nl0627689

[121] Y. F. Zhang, X. D. Han, K. Zheng, Z. Zhang, X. N. Zhang, J. Y. Fu, Y. Ji Y. Hao X. Guo Z. L. Wang, Direct observation of super-plasticity of beta-SiC nanowires at low temperature, Adv, Funct. Mater. 17 (2007) 3435–40. https://doi.org/10.1002/adfm.200700162

[122] W. Nhuapeng, W. Thamjaree, S. Kumfu, P. Singjai, T. Tunkasiri, Fabrication and mechanical properties of silicon carbide nanowires/epoxy resin composites, Curr. Appl. Phys. 8 (2008) 295–9. https://doi.org/10.1016/j.cap.2007.10.074

[123] S. Meng, G. G. Jin, Y. Y. Wang and X. Y. Guo, Tailoring and application of SiC nanowires in composites, Mater Sci. Eng. A 527 (2010) 5761–5. https://doi.org/10.1016/j.msea.2010.05.045

[124] M. Pozueloa, W. H. Kao and J. M. Yang, High-resolution TEM characterization of SiC nanowires as reinforcements in a nanocrystalline Mg-matrix, Mater. Charact. 77 (2013) 81–8. https://doi.org/10.1016/j.matchar.2013.01.003

[125] W. Yang, H. Araki, C. C. Tang, S. Thaveethavorn, A. Kohyama, T. Noda., Single-crystal SiC nanowires with a thin carbon coating for stronger and tougher ceramic composites, Adv. Mater. 17 (2005) 1519–23. https://doi.org/10.1002/adma.200500104

[126] W. Yang, H. Araki, S. Thaveethavorn, H. Suzuki, T. Noda, Process and mechanical properties of in situ silicon carbide nanowire-reinforced chemical vapor infiltrated silicon carbide/silicon carbide composite, J. Am. Ceram. Soc. 87 (2004) 1720–5. https://doi.org/10.1111/j.1551-2916.2004.01720.x

[127] Q. G. Fu, B. L.Jia, H. J. Li, K. Z. Li, Y. H. Chu, SiC nanowires reinforced MAS joint of SiC coated carbon/carbon composites to LAS glass ceramics, Mater. Sci. Eng. A 532 (2012) 255–9. https://doi.org/10.1016/j.msea.2011.10.088

[128] T. Barois, A. Ayari, P. Vincent, S. Perisanu, P. Poncharal and S. T. Purcell, Ultra low power consumption for self-oscillating nanoelectromechanical systems constructed by contacting two nanowires, Nano Lett. 13 (2013) 1451–6. https://doi.org/10.1021/nl304352w

[129] J. J. Niu and J. N.Wang, A novel self-cleaning coating with silicon carbide nanowires, J Phys Chem B 113 (2009) 2909–2912. https://doi.org/10.1021/jp808322e

[130] G. Kwak, M. Lee, K. Senthil and K. Yong, Wettability control and water droplet dynamics on SiC–SiO2 core–shell nanowires, Langmuir 26 (2010) 12273–7. https://doi.org/10.1021/la101234p

[131] H. Y. Wang, Y. Q. Wang, Q. F. Hu and X. J. Li, Capacitive humidity sensing properties of SiC nanowires grown on silicon nanoporous pillar array, Sens. Actuator B – Chem. 166 (2012) 451–6. https://doi.org/10.1016/j.snb.2012.02.087

[132] F. M. Gao, J. J. Zheng, M. F. Wang, G. D. Wei, W. Y. Yang, Piezoresistance behaviors of p-type 6H-SiC nanowires, Chem. Commun. 47 (2011) 11993–5. https://doi.org/10.1039/c1cc14343c

[133] R. W. Shao, K. Zheng, Y. F. Zhang, Y. J. Li, Z. Zhang and X. D. Han, Piezoresistance behaviors of ultra-strained SiC nanowires, Appl. Phys. Lett. 101, (2012) 233109–4. https://doi.org/10.1063/1.4769217

[134] L Fradetal, E Bano, G Attolini, F Rossi, V Stambouli, A silicon carbide nanowire field effect transistor for DNA detection, Nanotechnology 27 (2016) 235501. https://doi.org/10.1088/0957-4484/27/23/235501

[135] K. Rogdakis, S. Y. Lee, M. Bescond, S. K. Lee, E. Bano and K. Zekentes, 3C-Silicon Carbide nanowire FET: An experimental and theoretical Approach, IEEE Trans. on Elec. Dev. 55 (2008) 1970. https://doi.org/10.1109/TED.2008.926667

[136] K. Rogdakis, M. Bescond, E. Bano and K. Zekentes, Theoretical comparison of 3C-SiC and Si nanowire FETs in ballistic and diffusive regimes, Nanotechnology 18 (2007) 475715 and in K. Rogdakis, M. Bescond, K. Zekentes and E. Bano, Mat. Sci. Forum. 600 (2009) 135138. https://doi.org/10.1088/0957-4484/18/47/475715

[137] K. Rogdakis, S. Poli, E. Bano, K. Zekentes and M. Pala, Phonon- and surface-roughness-limited mobility of gate-all-around 3C-SiC and Si nanowire FETs, Nanotechnology 20 (2009) 295202. https://doi.org/10.1088/0957-4484/20/29/295202

[138] H. K. Seong, H. J. Choi, S. K. Lee, J. I. Lee, D. J. Choi, Optical and electrical transport properties in silicon carbide nanowires, Appl. Phys. Lett. 85 (2004) 1256 and in H. K. Seong, H. J. Choi, S. K. Lee, J. I. Lee, D. J. Choi, Mater. Sci. Forum 527 (2006) 771-75. https://doi.org/10.1063/1.1781749

[139] W. M. Zhou, F. Fang, Z. Y. Hou, L. J. Yan, and Y. F. Zhang, Field-Effect Transistor Based on β-SiC Nanowire, IEEE Electron. Dev. Lett. 27 (2006) 463-65. https://doi.org/10.1109/LED.2006.874219

[140] K. Rogdakis, E. Bano, L. Montes, M. Bechelany, D, Cornu and K. Zekentes, Rectifying source and drain contacts for effective carrier transport modulation of extremely doped SiC nanowire FETs, IEEE Trans. on Nanotechnology, 10 (2011) 980-984. and in K. Rogdakis, E. Bano, L. Montes, M. Bechelany, D. Cornu and K. Zekentes, Mater. Sci. Forum 679 (2011) 613-616. https://doi.org/10.1109/TNANO.2010.2091147

[141] Y. Chen, X. Zhang, Q. Zhao, L. He, C. Huang, Z. Xie, P-type 3C-SiC nanowires and their optical and electrical transport properties, Chem. Comm., 47 (2011) 6398-6400. https://doi.org/10.1039/c1cc10863h

[142] Z. Dai, L. Zhang, C. Chen, B. Qian, D. Xu, H. Chen, L. Wei, Y. Zhang, Fabrication of SiC nanowire thin-film transistors using dielectrophoresis, J. Semicond. 33 (2012) 114001. https://doi.org/10.1088/1674-4926/33/11/114001

[143] J. Choi, E. Bano, A. Henry, G. Attolini and K. Zekentes, Comparison of bottom-up and top-down 3C-SiC NWFETs, Mat. Sci. Forum, 858, (2016) 1001-1005. https://doi.org/10.4028/www.scientific.net/MSF.858.1001

[144] E. Spanakis, J. Dialektos, E. Stratakis, V. Zorba, P. Tzanetakis and C. Fotakis, Phys. Status Solidi C 5 (2008) 3309–13. https://doi.org/10.1002/pssc.200779503

[145] R. B. Wu, K. Zhou, J. Wei, Y. Z. Huang, F. Su, J.J. Chen, L. Wang, Growth of tapered SiC nanowires on flexible carbon fabric: toward field emission applications, J. Phys. Chem. C 116 (2012) 12940–5. https://doi.org/10.1021/jp3028935

[146] T. H. Yang, C. H.Chen, A. Chatterjee, H. Y. Li, J. T. Lo, C. T. Wu, K. H. Chen and L. C. Chen, Chem. Phys. Lett. 379 (2003)155–61. https://doi.org/10.1016/j.cplett.2003.08.001

[147] Y. Ryu, Y. Tak, K. Yong, Direct growth of core–shell SiC–SiO2 nanowires and field emission characteristics, Nanotechnology 16 (2005) S370–4. https://doi.org/10.1088/0957-4484/16/7/009

[148] H. Cui, L. Gong, Y. Sun, G. Z. Yang, C. L. Liang, J. Chen, and C. X. Wang, Direct synthesis of novel SiC@Al$_2$O$_3$ core–shell epitaxial nanowires and field emission characteristics, Cryst. Eng. Comm. 13 (2011) 1416–21. https://doi.org/10.1039/C0CE00435A

[149] X.N. Zhang, Y. Q. Chen, Z. P. Xie and W. Y. Yang, Shape and doping enhanced field emission properties of quasialigned 3C-SiC nanowires, J. Phys. Chem. C 114 (2010) 8251–5. https://doi.org/10.1021/jp101067f

[150] G. Y. Li, X. D. Li, Z. D. Chen, J. Wang, H. Wang and R. C. Che,, J. Phys. Chem. C 113, 2009 17655–60. https://doi.org/10.1021/jp904277f

[151] W. Zhou, L. Yan, Y. Wang, Y. Zhang, SiC nanowires: a photocatalytic nanomaterial, Appl. Phys. Lett. 89 (2006) 013105. https://doi.org/10.1063/1.2219139

[152] J. P. Alper, M. S. Kim, M. Vincent, B. Hsia, V. Radmilovic, C. Carraro, R. Maboudian, Silicon carbide nanowires as highly robust electrodes for microsupercapacitors, J. Power. Sources 230 (2013) 298–302. https://doi.org/10.1016/j.jpowsour.2012.12.085

[153] S-C Chiu, H-C. Yu, Y-Y. Li, High electromagnetic wave absorption performance of silicon carbide nanowires in the Gigahertz range, J. Phys. Chem. C 114 (2010) 1947–1952. https://doi.org/10.1021/jp905127t

Keyword Index

	Vol. Page
Acheson Process	2. 1
Aspect Ratio Dependent Etching	2. 175
Barrier Height	1. 127
Bipolar	1. 191
BJT	2. 1
Bottom Up Process	2. 233
Carborundum	2. 1
Catalyst	2. 233
Channeling	2. 107
Chemical Cleaning	1. 1
Chemical Etching	2. 175
Contact Resistivity	1. 27
Core/Shell Heterostructure	2. 233
Cree	2. 1
Crystal Growth	2. 1
Diffusion Barrier	1. 27
DIMOSFET	2. 1
Diode	1. 127
Dopants Activation	2. 107
Dry Etching	2. 175
Electrochemical Etching	1. 1
Electroluminescence	2. 1
Epitaxy	2. 1
Etch Rate	2. 175
Field Effect Mobility	2. 63
Gas Phase Chemistry	2. 175
Gate Oxide	2. 63
GTO	2. 1
Hexagonality	2. 1
High Temperature Electronics	1. 27
High-κ Dielectrics	2. 63

	Vol. Page
History	2. 1
HTCVD	2. 1
IGBT	2. 1
Implantation	2. 107
Implantation Modeling	2. 107
Inverter	2. 1
Ion Etching	2. 175
LED	2. 1
Lely Platelets	2. 1
LETI Method	2. 1
Lifetime Enhancement Thermal Oxidation	2. 1
Luminescence	2. 233
Mask Selectivity	2. 175
Material Properties	2. 1
Metal Carbides	1. 27
Metal Silicides	1. 27
Microloading	2. 175
Micromasking	2. 175
Micropipe	2. 1
Microtrenching	2. 175
Modified Lely Method	2. 1
Moissanite	2. 1
MOSFET	2. 1, 63
Nanocrystal	2. 233
Nanoelectromechanical Systems (NEMS)	2. 233
NanoFET	2. 233
Nanopillar	2. 233
Nanotube	2. 233
Nanowire	2. 233
NWFET	2. 233
Ohmic Contact	1. 27
Photoluminescence	2. 233
Plasma Chemistry	2. 175
Plasma Etching	2. 175

	Vol. Page
Polytypism	2. 1
Porous SiC	1. 1
Post Oxidation Annealing	2. 63
Post-Implantation Annealing	2. 107
Power Devices	1. 191
Radio Detector	2. 1
RAF Growth Process	2. 1
Reliability	1. 191
Residue Free Etching	2. 175
Schottky Barrier	1. 27
Schottky Contact	1. 127
Schottky Diode	2. 1
SiC Devices	2. 1
Sidewall Slope	2. 175
Silicon Carbide (SiC)	2. 1, 107
Silicon Nitride	2. 63
Silicon Oxide	2. 63
Sputtering	2. 175
Stacking Fault	2. 1
Status	1. 191
Step-Controlled Epitaxy	2. 1
Step-Flow Growth	2. 1
Sublimation Sandwich	
Method	2. 1
Surface Passivation	2. 63
System Benefits	1. 191
Technology	2. 1
Transfer Length Method	1. 27
Transition Line Model	1. 27
Trends	1. 191
Unipolar	1. 191
Varistor	2. 1
Wet Etching	1. 1
Wide Band Gap Power	
Electronics	1. 127

Author Index

		Vol. Ch.
Arith, F.	School of Engineering, Newcastle University, Newcastle upon Tyne, United Kingdom	2. 2
Bakowski, M.	RISE Acreo, Kista, Sweden	1. 4
Bosi, M.	IMEM-CNR, Parma, Italy	2. 5
Brezeanu, G.	University POLITEHNICA of Bucharest, Faculty of Electronics, Telecommunications and Information Technology, Bucharest, Romania	1. 3
Gammon, P. M.	School of Engineering, University of Warwick, Coventry, United Kingdom	1. 3 2. 2
Giannazzo, F.	Consiglio Nazionale delle Ricerche – Istituto per la Microelettronica e Microsistemi (CNRIMM), Catania, Italy	1. 3
Godignon, P.	Centre Nacional de Microelectrònica (CNM), Barcelona, Spain	2. 3
Jokubavicius, V.	Department of Physics, Chemistry and Biology (IFM), Linköping University, Linköping, Sweden	1. 1
O'Neill, A.	School of Engineering, Newcastle University, Newcastle upon Tyne, United Kingdom	2. 2
Pezoldt, J.	FG Nanotechnologie, Institut für Mikro- und Nanoelektronik und Institut für Mikro- und Nanotechnologien MacroNano®, Ilmenau, Germany	2. 4
Rascunà. S.	STMicroelectronics, Catania, Italy	1. 3
Roccaforte, F.	Consiglio Nazionale delle Ricerche – Istituto per la Microelettronica e Microsistemi (CNRIMM), Catania, Italy	1. 3

Rogdakis, K. Department of Electrical & Computer Engineering,   2. 5
             Hellenic Mediterranean University, Heraklion, Greece

Russell, S. School of Engineering, University of Warwick, Coventry,   2. 2
             United Kingdom

Saggio, M. STMicroelectronics, Catania, Italy   1. 3

Syväjärvi, M. Department of Physics, Chemistry and Biology (IFM),   1. 1
             Linköping University, Linköping, Sweden

Torregrosa, F. IBS (Ion Beam Services), Peynier, France   2. 3

Urresti, J. School of Engineering, Newcastle University, Newcastle   2. 2
             upon Tyne, United Kingdom

Vasilevskiy, K. School of Engineering, Newcastle University, Newcastle   1. 2
             upon Tyne, United Kingdom   2. 1

Vavasour, O. School of Engineering, University of Warwick, Coventry,   2. 2
             United Kingdom

Veliadis, V. PowerAmerica & North Carolina State University,   2. 4
             Electrical & Computer Engineering Dept., Raleigh NC,
             USA

Wright, N. School of Engineering, Newcastle University, Newcastle   1. 2
             upon Tyne, United Kingdom   2. 1

Yakimova, R. Department of Physics, Chemistry and Biology (IFM),   1. 1
             Linköping University, Linköping, Sweden

Zekentes, K. MRG-IESL/FORTH, Heraklion, Greece;   1. 2
             Grenoble INP, IMEP-LAHC, Grenoble, France   2. 3
                                                                 2. 4
                                                                 2. 5

About the Editors

Konstantin Vasilevskiy received his MSc degree in solid-state physics from Moscow Engineering Physics Institute, USSR, in 1981, and his PhD in physics of semiconductors from the Ioffe Institute, St. Petersburg, Russia, in 2002. After graduation in 1981, he spent two years working in material characterisation by Auger spectroscopy at Scientific-Research Technological Institute, Ryazan, USSR. From 1984 to 1988, he was with Research Institute "Orion", Kyiv, USSR, where he participated in development and small scale production of silicon IMPATT and microwave *p-i-n* diodes. In 1989, he joined the Ioffe Institute, where he was heavily involved in gallium nitride growth and characterization as well as in design and fabrication of discrete semiconductor devices based on silicon carbide and III-V nitrides. From 1999 to 2000, he was working in the Foundation for Research and Technology-Hellas (FORTH) in Heraklion, Greece as a Visiting Researcher. This work aimed at demonstration of microwave oscillations generated by SiC IMPATT diodes. In 2001, he joined School of Engineering at Newcastle University, United Kingdom, where he is currently a Senior Research Associate. His research at Newcastle University has included in-depth investigation of ohmic and Schottky contacts in SiC devices; development of SiC device processing; design, fabrication and characterization of various SiC devices including trenched-and-implanted JFETs, Schottky diodes, SITs with buried gate formed by deep ion implantation, low voltage Zener diodes; SiC MOSFETs with high-k dielectric gate stacks. Besides his activity in wide band gap semiconductors, Dr Vasilevskiy also has conducted research in graphene growth and characterization. He developed local graphene growth on silicon carbide from nickel silicide supersaturated with carbon; fabricated top-gated FETs using bilayer epitaxial graphene and demonstrated their operation at elevated temperatures. Dr Vasilevskiy wrote 3 book chapters and 114 papers in refereed journals and conference proceedings. He is a coeditor of four books and co-inventor of 16 patents granted in the field of wide band gap semiconductor technology.

Konstantinos Zekentes received his undergraduate degree in Physics, from the University of Crete, Greece, and his Ph.D., in Physics of Semiconductors, from the University of Montpellier, France. He is currently a Senior Researcher with the Microelectronics Research Group (MRG) of the Foundation for Research and Technology-Hellas (FORTH) in Heraklion, Crete, Greece and visiting researcher in the Institut de Microélectronique Electromagnétisme et Photonique et le Laboratoire d'Hyperfréquences et de Caractérisation (**IMEP-LaHC**) of CNRS/Grenoble INP/UJF/Université de Savoie. The objective of his current work is the development of SiC-related technology for elaborating high power/high frequency devices as well as SiC-based 1D devices. Dr. Zekentes has more than hundred seventy journal and conference publications and one US patent.

www.ingramcontent.com/pod-product-compliance
Lightning Source LLC
Chambersburg PA
CBHW071332210326
41597CB00015B/1425